THERMAL
ANALYSIS OF
PHARMACEUTICALS

THERMAL ANALYSIS OF PHARMACEUTICALS

Edited by
Duncan Q. M. Craig
Mike Reading

CRC Press
Taylor & Francis Group
Boca Raton London New York

CRC Press is an imprint of the
Taylor & Francis Group, an **informa** business

CRC Press
Taylor & Francis Group
6000 Broken Sound Parkway NW, Suite 300
Boca Raton, FL 33487-2742

First issued in paperback 2020

ISBN-13: 978-0-367-57774-2 (pbk)
ISBN-13: 978-0-8247-5814-1 (hbk)

Visit the Taylor & Francis Web site at
http://www.taylorandfrancis.com

and the CRC Press Web site at
http://www.crcpress.com

Preface

Thermal analysis is one of the mainstay families of techniques for the physical, and on occasions chemical, characterization of pharmaceutical materials. The long history of using thermal methods within the field, and the astonishing variety of applications to which these methods have been put, reflect the necessity of developing reliable and versatile characterization tools for the successful development of pharmaceutical products. Indeed, it is interesting to reflect that one of the key focus areas within the industry, next to drug discovery, is analytical development, an area made even more pertinent by the recent Process Analytical Technologies initiative launched by the United States Food and Drug Administration (FDA) whereby the product-to-market process may be accelerated if appropriate validation of each stage of manufacture is provided.

Given the context of the clear importance of analytical methodologies, it is perhaps useful to consider what thermal methods may and, just as importantly may not, achieve within the pharmaceutical arena. Indeed, it is possible to argue that the strengths and limitations of these methods are intrinsically linked. The first issue relates to the type of phenomenon that thermal methods may usefully study. Although there are examples of these methods yielding chemical information such as extent of cross-linking, purity, or degradation, the vast majority of applications are associated with the characterization of the physical structure and properties of materials. By "physical structure" we mean the study of the arrangement of molecules of (usually) known composition and the events associated with changes in those physical arrangements. Classically, this would include the study of melting or crystallization, although other processes such as glass transitions and associated plasticization phenomena are also widely studied. Usually, these processes are studied not because the operator is necessarily interested in the process *per se* but because that process reflects the structure of the material under ambient or user conditions, a classic example being the detection of polymorphism by studying variances in melting point. Alternatively, the thermal process may lead to the prediction of long-term behavior such as physical stability, examples including the study of glass transitional behavior in order to predict the storage conditions for which the risk of crystallization is minimized.

The second consideration is that thermal methods are extremely versatile, allowing useful characterization of almost any sample. Indeed, the methods may be used to study single or multiple component systems such as finished dosage forms, although care is required in the interpretation of data from the latter. Finally, these methods tend to be simple to use, inexpensive, and operator friendly. This is a highly pertinent practical consideration, as methods whereby absolute specialization is required may be extremely useful but are less likely to have widespread application.

As stated above, the limitations of thermal methods are effectively mirror images of these advantages. These methods involve the application or measurement of changes in heat, hence the resulting data tends to relate to temperatures of transitions or heat flow as the samples undergo such a transition. Consequently, the information gained can often only indirectly be related to the feature of interest such as crystal structure or composition. This therefore requires the operator to extrapolate this information, inevitably leading to assumptions in interpretation. Indeed, it is almost invariably the case that direct chemical or spectroscopic information is not forthcoming unless the thermal method is interfaced with a second measuring technique. This also mirrors the versatility of the approach; such versatility is possible due to the nonspecificity of the measurement. Finally, we would argue that the simplicity of measurement has in some respects proved to be a disadvantage in that thermal methods may be used as "workhorse" methods whereby they are used for simple screening, with little consideration given to the use of the methods in more sophisticated ways. A reasonable analogy comparing thermal methods with other techniques such as solid-state NMR would be learning to play the piano in contrast to a trumpet; it is possible to get a reasonable tune out of a piano with a minimum of effort or training, whereas the playing of the trumpet requires some training to avoid merely producing high-pitched shrieks. However, to play either at a high level of competence requires equivalent skill and patience for both instruments. The point being made is that the sheer accessibility of thermal measurements has arguably caused an underestimation of the level of information that may be obtained if one is willing to spend time exploring these methods in depth.

On that basis it is appropriate to outline the purpose of this book and the intended contribution that it may make to the pharmaceutical field. There are already a large number of excellent texts available in the thermal analysis and polymer fields outlining the principles and uses of these methods. There is a more limited number of (again, excellent) texts that outline the use of these methods for pharmaceutical systems, with particular emphasis on the applications of the methods. In this text we have tried to place more emphasis on the techniques themselves in terms of their use and the interpretation of the corresponding data, while at the same time writing the chapters with a strong leaning toward pharmaceutical scientists.

For example, the most commonly used thermal method is and will almost certainly continue to be differential scanning calorimetry (DSC), hence here we have dedicated three chapters to the technique covering the principles, the optimal use of the method, and pharmaceutical applications. We then include chapters on thermogravimetric analysis, modulated temperature DSC, microcalorimetry, high sensitivity DSC, thermal microscopy, thermally stimulated current, and dynamic mechanical analysis, all of which have attracted great interest within the pharmaceutical field. In all cases, these chapters combine elements of theoretical background, measurement optimization, and pharmaceutical applications. It is our profound hope that in this way we will achieve a suitable mixture of depth, relevance, and accessibility.

It is also necessary for us to define the limits of what this text is intended to cover. In the first instance, we have not attempted to provide a comprehensive review of all pharmaceutical thermal analysis studies but have instead focused on including examples that illustrate a particular point. Secondly, we have not gone into coupled methods to any significant extent; there has been interest in using spectroscopic methods interfaced with thermal approaches and a wide range of such techniques have been explored or even commercialized. Again, we made the decision not to go into these approaches due to their highly diverse nature and our wish to focus on methods that are of most direct relevance to pharmaceutical scientists. Finally, we have not gone into the emerging raft of technologies associated with site-specific thermal analysis, such as microthermal analysis (which, perhaps ironically, the editors are themselves highly involved with at present), for the same reasons of wishing to focus on established techniques within the space available.

It only remains for us to express profound thanks to the various authors associated with the text and to the publishers for their support. Because of a range of personal circumstances this book has been longer in the making than we originally intended; unfortunately, this is far from unusual when editing books, but at the same time we do wish to thank all concerned for their patience, and we hope they and the pharmaceutical community will consider our joint efforts to have been worthwhile.

Duncan Q.M. Craig and Mike Reading

The Editors

Duncan Q.M. Craig, Ph.D., is head of pharmacy at the School of Chemical Sciences and Pharmacy at the University of East Anglia, Norwich, U.K. Previously, he worked at the School of Pharmacy, The Queen's University of Belfast, joining in 1999. There he set up the Pharmaceutical Materials Science Programme within the Drug Delivery and Biomaterials Group. He was previously a reader in pharmaceutical materials science at the School of Pharmacy, University of London.

Dr. Craig's work has involved the development of novel analytical techniques for the study of the physical properties of pharmaceuticals, with particular emphasis on thermal, dielectric, rheological, and microscopic methods. In particular, his interests lie in developing rational approaches to drug delivery system design. This has involved introducing novel techniques into the pharmaceutical arena, with his group being at the forefront of developing pharmaceutical applications for techniques such as modulated temperature DSC and microthermal analysis. He is a former editor of the *Journal of Pharmacy and Pharmacology* and the thermal analysis journal *Thermochimica Acta*. In 2003 Dr. Craig was awarded the Young Investigator Award of the Controlled Release Society, an international organization dedicated to the development of novel drug delivery strategies. He was the science chair for the British Pharmaceutical Conference 2005.

Mike Reading, Ph.D., is internationally recognized for his work on the development of novel thermoanalytical techniques. After receiving a B.Sc. and Ph.D. at Salford University and doing postdoctoral work in France (the CNRS center for calorimetry and thermodynamics, Marseilles), he worked with ICI until 1997. He left to join the IPTME at Loughborough University where he was director of the Advanced Thermal Methods Group. In 2004 he moved to the University of East Anglia to take up a chair in pharmaceutical characterization science.

As a senior research scientist with ICI Paints, Dr. Reading was involved in a wide range of materials science and analysis projects, mainly involving polymers. One outcome from his work was modulated temperature differential scanning calorimetry, which has now become a common, commercially available technique. While still at ICI, then at Loughborough University, he invented, with co-workers Azzedine Hammiche and Hubert Pollock of Lancaster University, a scanning probe microscopy technique known as microthermal analysis. This has also become a commercially available instrument that has won a number of awards for innovation. His current research interests center on using scanning probe microscopy to characterize the structure and chemical composition of samples on a small, especially nano, scale. He is developing and applying to pharmaceutical materials a variety of approaches to nano analysis including, for example, photothermal microspectroscopy.

Contributors

M. Bauer
Sanofi Recherche
Laboratoire de Physique des Polymères
Université Paul Sabatier
Toulouse, France

N. Boutonnet-Fagegaltier
Sanofi Recherche–Montpellier
Laboratoire de Physique des Polymères
Université Paul Sabatier
Toulouse, France

Graham Buckton
School of Pharmacy
University of London
London, U.K.

A. Caron
Sanofi Recherche
Laboratoire de Physique des Polymères
Université Paul Sabatier
Toulouse, France

Duncan Q.M. Craig
School of Chemical Sciences and
  Pharmacy
University of East Anglia
Norwich, Norfolk, U.K.

H. Duplaa
Sanofi-Synthélabo Recherche
Toulouse, France

Simon Gaisford
School of Pharmacy
University of London
London, U.K.

Andrew K. Galwey (Retired)
The Queen's University of Belfast
Medical Biology Centre
Belfast, Northern Ireland, U.K.

David S. Jones
School of Pharmacy
The Queen's University of Belfast
Belfast, Northern Ireland, U.K.

Vicky L. Kett
The School of Pharmacy
The Queen's University of Belfast
Medical Biology Centre
Belfast, Northern Ireland, U.K.

C. Lacabanne
Sanofi Recherche–Montpellier
Laboratoire de Physique des Polymères
Université Paul Sabatier
Toulouse, France

A. Lamure
Sanofi Recherche–Montpellier
Laboratoire de Physique des Polymères
Université Paul Sabatier
Toulouse, France

Trevor Lever
Trevor Lever Consulting
South Horrington Village
Wells, Somerset, U.K.

J. Menegotto
Sanofi-Synthélabo Recherche
Toulouse, France

John R. Murphy
School of Chemical Sciences and
  Pharmacy
University of East Anglia
Norwich, Norfolk, U.K.

Ann W. Newman
SSCI, Inc.
West Lafayette, Indiana, U.S.A.

Mike Reading
School of Chemical Sciences and
  Pharmacy
University of East Anglia
Norwich, Norfolk, U.K.

Imre M. Vitez
Formulations R&D
Cardinal Health
Somerset, New Jersey, U.S.A.

Contents

1 Principles of Differential Scanning Calorimetry

Mike Reading and Duncan Q.M. Craig

CONTENTS

1.1 INTRODUCTION

Differential scanning calorimetry (DSC) is the most widely used method of thermal analysis within the pharmaceutical field. The approach usually involves the application of a linear heating or cooling signal to a sample and the subsequent measurement of the temperature and energy associated with a range of thermal events including melting, crystallization, glass transitions, and decomposition reactions. There are numerous benefits associated with the method, such as the small sample size required, the wide temperature range available with most commercial instruments (typically –120 to 600°C), and the simplicity and rapidity of measurement.

To use the technique optimally, it is essential to have a sound grasp of both the principles underlying the methods and the most appropriate measurement parameters to be used for each sample. This chapter deals with the former consideration in that the basic instrument operation and the principles underlying the measurement process will be described, whereas Chapter 2 outlines the basic issues associated with the choice of experimental parameters and calibration. Chapter 3 then goes on to describe some of the principal applications of DSC within the pharmaceutical field.

1.2 COMMON FORMS OF DSC

1.2.1 Development of DSC

The basic principle underpinning DSC is that a sample is subjected to a heat signal and the response measured in terms of the energy and temperature of the thermal events that take place over the temperature range or time interval under study. The temperature profile may be in the form of heating, cooling, or an isothermal program, with heating being the most widely used approach. Consequently, although the most common use of the technique, certainly within the pharmaceutical sphere, has been to study melting responses by heating the sample at a controlled rate, many studies have also been performed on crystallization, glass transitions, and kinetic reactions such as curing. It should be borne in mind that the most common use of DSC in a global sense is for the characterization of polymers; hence, much of the theory, and indeed the hardware, has been developed with this application in mind. However, the principles involved are equally applicable to pharmaceuticals, inorganic materials, ceramics, and biological systems.

The approach involves placing the sample (usually in quantities of approximately 5–10 mg) in a metal crucible (pan) along with a reference pan (usually empty in the case of DSC but containing an inert reference material in the case of differential thermal analysis [DTA]) in a furnace and heating or cooling at a controlled rate, usually in the region of 5 to 10°C/min. When the sample undergoes a thermal event such as melting or crystallization, the temperature and energy associated with that event is assessed. As this information may be directly related to the solid-state structure, the approach is of great use for the study of pharmaceutical systems in terms of, for example, differentiation between polymorphs and measurement of glass transitions, as will be discussed in more detail in Chapter 3.

The technique is a development of the earlier method of DTA, which has now been largely (but by no means entirely) superseded by DSC. It is, however, helpful in the first instance to consider the underlying principles of DTA to gain a better understanding of heat flux and power compensation DSC.

The basic components of a DTA apparatus are a temperature-controlled furnace containing sample and reference cells and a pair of matched temperature sensors connected to recording apparatus, as indicated in Figure 1.1. The temperature sensors (usually thermocouples) are in contact with the sample and reference or their containers, and the output is amplified and recorded. DTA data may be plotted as a function of sample temperature, reference temperature (as is usually the case), or time. In both DTA and DSC, the measurement relies on the occurrence of a temperature difference between a sample and reference (ΔT) as a result of the thermal event in question.

In the case of DTA, an inert material is placed in the reference pan such that the heat capacity of the sample is approximately matched. Under these conditions, when the sample and reference are being heated at a constant rate, the DTA signal will be small when no transition is occurring. Ideally, the arrangement of the pans in the furnace is exactly symmetrical; so when the heat capacity of sample and reference are exactly the same, the temperature difference is zero.

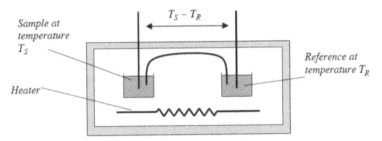

FIGURE 1.1 Typical arrangement for DTA apparatus. (Adapted from Wendlandt, W., *Thermal Analysis*, 3rd ed., Wiley-Interscience, New York, 1986.)

When the sample melts, the rate of temperature increase for the sample will be lower than that of the reference, as the energy imparted by the heating signal will be contributing to the breaking of solid–solid bonds rather than simply the raising of the sample temperature. This is illustrated in Figure 1.2.

In essence, the heat input into the sample contributes to the melting process rather than the increase in temperature, whereas the heat input into the reference continues to lead to a temperature increase. When melting is complete, the sample returns to the programmed temperature; hence, by examining the difference in temperature between the two pans, a peak is seen, as shown in Figure 1.2b.

It is clear from Figure 1.2 that the technique may easily and rapidly detect the temperature at which a particular thermal event is occurring. Indeed, current uses of the technique are based largely on the ability of the method to detect the initial temperatures of thermal processes and to qualitatively characterize them as endothermic or exothermic, reversible or irreversible, first order or higher order, etc. (2). The ability to run experiments in a range of atmospheres has also rendered the approach particularly useful for the construction of phase diagrams.

Clearly, it would be desirable if the area under the peak was a measure of the enthalpy associated with the transition. However, in the case of DTA, the heat path to the sample thermocouple includes the sample itself. The thermal properties of each sample will be different and uncontrolled. In order for the DTA signal to be a measure of heat flow, the thermal resistances between the furnace and both thermocouples must be carefully controlled and predictable so that it can be calibrated and then can remain the same in subsequent experiments. This is impossible in the case of DTA, so it cannot be a quantitative calorimetric technique. Note that the return to baseline of the peak takes a certain amount of time, and during this time the temperature increases; thus the peak appears to have a certain width. In reality this width is a function of the calorimeter and not of the sample (the melting of a pure material occurs at a single temperature, not over a temperature interval). This distortion of peak shape is usually not a problem when interpreting DTA and DSC curves but should be borne in mind when studying sharp transitions.

Differential scanning calorimetry was introduced in the 1960s as a means of overcoming the difficulties associated with DTA. Fundamentally, there are two different types of DSC instruments: heat flux and power compensation. In common with DTA, DSC involves the measurement of the temperature difference between a

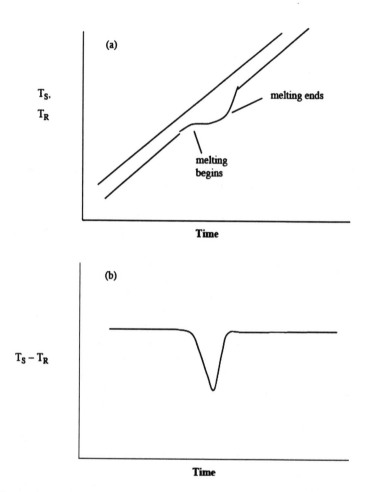

FIGURE 1.2 Typical DTA response showing (a) sample and reference temperature (TS and TR) and (b) temperature differential (TSTR) as a function of time. (Adapted from Wendlandt, W., *Thermal Analysis*, 3rd ed., Wiley-Interscience, New York, 1986.)

sample and reference. The basic innovation associated with the earliest DSC, as originally introduced by the Perkin-Elmer Co. (3,4), was that the temperature difference is kept to near-zero by placing the temperature sensors in a bridge circuit. The electrical power supplied to the sample furnace is varied so that the temperature difference between sample and reference is zero (or as close to zero as possible); hence the term *power compensation*. The reference pan is made to go through a strictly linear temperature program. The control system (ideally) forces the sample to experience the same linear temperature program. Consequently, any differences in power between this and the sample pan plus sample must be related to the sample heat capacity and the enthalpies associated with transitions. Heat flux DSC is very similar to DTA except that the sample and reference thermocouples are not placed inside the pans. Instead they are located as close as possible to the sample and reference pans as part of well-defined heat paths between the furnace and the sample

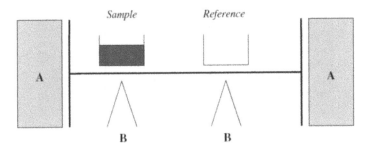

FIGURE 1.3 Schematic of a heat flux DSC. A = furnace, B = thermocouple.

and reference pans. Because of this, calibration is possible, and the heat paths are the same during subsequent experiments.

A large number of texts are available that give more details of both approaches (1,2,5–10), and only a brief outline will be given in the following text.

1.2.2 HEAT FLUX DSC

Heat flux DSC is the conceptually simpler of the two approaches and is represented schematically in Figure 1.3. Typically, two crucibles, one empty and one containing the sample, are placed symmetrically within a furnace with a thermocouple placed in close contact with each. The thermocouples are connected in a back-to-back arrangement such that the voltage developed from the pair is a direct measure of the temperature difference between the sample and reference. As stated previously, the cell should be designed such that the heat paths from the furnace to the sample and reference are identical, and both are also stable and well defined (usually through a metal membrane or armature that also supports the crucible). The equation for heat flow from the furnace to each crucible is then given by

$$dQ/dt = \Delta T/R \tag{1.1}$$

where Q = heat, t = time, ΔT = temperature difference between the furnace and the crucible, and R = thermal resistance of the heat path between the furnace and the crucible.

From Equation 1.1, it follows that the temperature difference between sample and reference is a measure of the difference in heat flow due to the presence of the sample in one of the crucibles, provided that the furnace and heat paths are truly symmetrical. Consequently, this differential heat flow is a measure of the properties of the sample, with all other influences (heat adsorption by the crucible, heat losses through convection, etc.) having been eliminated by use of the comparison with the reference. The ΔT signal requires calibration to provide a heat flow as a function of temperature, and this is usually carried out by use of standards that are usually pure metals with known enthalpies of melting and materials with known heat capacities (see Section 2.4 in Chapter 2).

A common improvement to the simple system described above is to use multiple thermocouple pairs connected in series. This increases the signal roughly in propor-

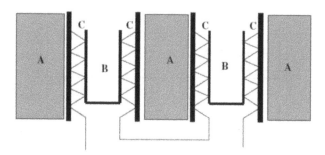

FIGURE 1.4 Schematic of a heat flux DSC with thermopiles. A = furnace, B = sample and reference crucibles, C = sample and reference thermopiles.

tion to the number of sensors used. In one configuration, illustrated in Figure 1.4, the multiple thermocouples are arranged between the source of heat and the sample with the same arrangement for the reference. The difference between the sets of thermocouples (thermopiles) at the sample and reference positions is then measured.

In heat flux instruments, it is generally the furnace that is subjected to the temperature program. Transitions in the sample are detected as a consequence of deviations from following this program. This means that the temperature profile experienced by the sample is not known in advance, although it is measured and can, therefore, be determined after the experiment. In practice these deviations from the preset program are usually small and present no practical problems of interpretation.

1.2.3 POWER COMPENSATION DSC

This type of instrument is represented schematically in Figure 1.5. The first clear difference, compared to the heat flux instrument, is that this approach uses two separate furnaces, one for the sample and a second for the reference. Both are programmed to go through the same temperature profile and the difference in electrical power supplied to the two furnaces is measured. The only difference between them should be that one contains the sample. Consequently, it is again the case that the differential signal is a measure of the sample properties, all other factors having been eliminated by the use of the reference measurement. In principle, because the quantity being measured directly is electrical power, the output requires no calibration. However, in practice, it has been found necessary to calibrate these instruments in the same way as heat flux calorimeters, using standards in order to obtain the most accurate results.

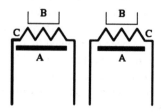

FIGURE 1.5 Schematic of a power compensation DSC. A = furnaces, B = sample and reference crucibles, C = sample and reference platinum resistance thermometers.

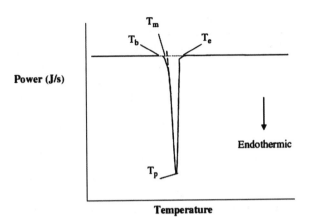

FIGURE 1.6 Typical melting endotherm for a low-molecular-weight, high-purity material.

In contrast to the heat flux approach, in these instruments it is the sample temperature that is programmed, and this should, ideally, conform exactly to the selected temperature profile. In principle, therefore, power compensation instruments provide for a better-defined experiment. In practice, however, the control system requires a finite temperature difference in order to operate and has a finite response time, so perfect control is never achieved.

Both designs of calorimeter have been commercially available for nearly 30 years without either assuming supremacy. This is because the performance of the two different types of instrument is very similar in terms of sensitivity and accuracy. The smaller furnaces of the power compensation instruments mean that they can achieve significantly higher heating and cooling rates, which is an advantage for some experiments. The heat flux instruments tend to have better baseline stability as the single furnace tends to be more robust than the two nominally matched furnaces of the power compensation design. Both instruments therefore have their advantages and disadvantages.

1.2.4 Basic Interpretation of DSC Data

DSC data is typically expressed in terms of power as a function of temperature, as indicated in the schematic diagram shown in Figure 1.6.

Examination of Figure 1.6 indicates the typical features of a simple melting response. At temperatures below the melting event, the trace shows a baseline that indicates the heat flow (power) required to raise the temperature of the sample in order to keep up with the heating program. This is therefore a function of the heat capacity of the sample Cp as this parameter represents the energy required to raise the temperature of the sample by 1K. In terms of the DSC signal, this may be expressed via

$$dQ/dt = Cp \cdot dT/dt \qquad (1.2)$$

where dQ/dt is the heat flow and dT/dt is the heating rate. In principle, therefore, the heat flow value corresponding to the baseline should yield the heat capacity of the sample. In practice it is necessary to carefully calibrate the instrument under the conditions used for the measurement, as there are several factors over and above the heat capacity of the sample that may contribute to the baseline heat flow, including the nonuniformity of the cell and differences in mass between the sample and reference pans. The issue of accurate heat capacity measurement has been the subject of an extensive number of studies (11–15), and only a brief description will be given here. It is helpful to mention that such measurements are not common practice within the pharmaceutical sciences as yet. Given the recent interest in amorphous and partially amorphous systems that has arisen in recent years (e.g.,16–18), it is quite possible that there will be a growing interest in such measurements in the future, given the well-established wealth of information that may be obtained from heat capacity measurements of amorphous systems (9).

Traditional heat capacity measurements using DSC rely on the relationship

$$Cp = K\Delta Y/b \tag{1.3}$$

where K is the calorimetric sensitivity, ΔY is the difference in baseline values obtained in the presence and absence of sample material, and b is the heating rate. K is found by calibration with a substance of known heat capacity, with pure, industrially manufactured sapphire (Al_2O_3) being the most commonly used material. It should be noted that heat capacity will vary with temperature; hence, tables of Cp against T are available for sapphire. Heat capacity determination essentially involves three DSC scans. First, an empty-pan baseline is obtained by simply running two empty pans under the conditions to be used for the determination. Second, the experiment is repeated under identical conditions using the calibrant to obtain K, and third, the unknown is run against the reference to obtain ΔY. For more accurate measurements the experiment is run at a range of scanning speeds. As mentioned previously, this represents only the very basic means of measurement, and the texts listed previously give more detail regarding both calibration and measurement. In addition, the technique of modulated temperature DSC appears to offer several advantages with respect to heat capacity measurement, and these will be discussed in Chapter 4.

On further heating, the melting process begins. Again, however, care is required in terms of data handling. Figure 1.4 shows the response of a high-purity, low-molecular-weight material that by definition will give a narrow melting response. In practice, many materials show relatively broad melting response due either to the nature of the sample or the presence of thermal gradients within the sample pan. Similarly, many materials show multiple peaks. For example, Gelucires, which are pegylated glyceride derivatives used in controlled-release preparations, may exhibit a range of peaks due to the multicomponent nature of the material and the tendency of these components to cocrystallize into distinct structures. An example of a typical Gelucire melting peak is given in Figure 1.7.

The question therefore arises as to where the melting point should be taken from. There are a number of options: the temperature at which deviation from the baseline begins (T_b), the extrapolated onset temperature (T_m), the peak temperature (T_p), and the temperature at which the trace returns to the baseline (T_e; notation used taken

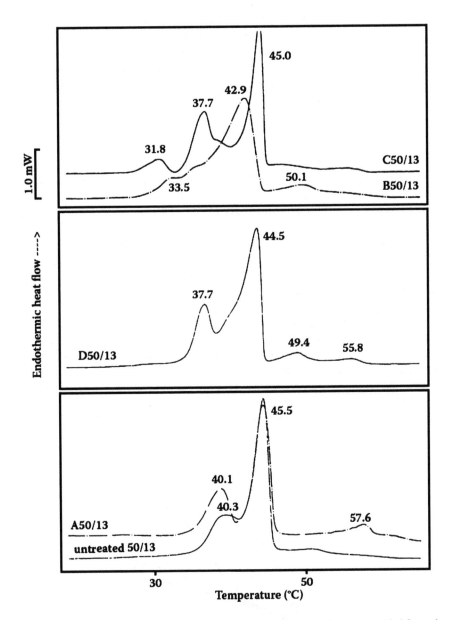

FIGURE 1.7 Melting of Gelucire 50/13. A = solvent crystallized, B = slow-cooled from the melt, C = fast cooled, D = cooled under ambient conditions. (Reproduced from Sutananta, W. et al., *Int. J. Pharm.* 110, 75–91, 1994.)

from (8)). In the context of discussing DTA, Wunderlich (7) recommends that for samples whereby temperature gradients are small and the melting process is sharp, the extrapolated onset should be used, whereas the peak temperature should be used for materials whereby extensive peak broadening is apparent (and by implication for multiple peaks). The key consideration is that the instrument should be calibrated using the same measuring method as is to be used during the experimental run.

The area under the peak A yields the enthalpy change associated with the process according to

$$A = k'm(-\Delta H) \tag{1.4}$$

where k' is the calorimetric sensitivity (an electrical conversion factor), m is the mass of the sample, and ΔH is the enthalpy. Consequently, it is necessary to calibrate the instrument under the same conditions used as those of the experimental run and to measure the area of the enthalpic peak for the calibrant. For many experiments it is acceptable to use indium, which has a heat of fusion of 28.71 J/g. However, if events that take place at temperatures very different from the melting point of indium (156.61°C) are to be studied, it is essential to use an alternative calibrant (see Chapter 2). A further consideration is the relationship between the absolute peak area and the scanning speed. At higher scanning rates, the peaks will appear considerably larger. This may be understood with reference to Figure 1.8.

During a melting process, the heat input into the sample will be utilized to melt that material; hence a temperature differential will develop between the sample and reference as the former remains effectively isothermal during the melt. At low heating rates (T_S', T_R', A'), that lag will be relatively small at any given temperature because the reference temperature T_R' will not have increased greatly in the time period of melting owing to the slow heating rate; hence, A' will be small. For fast heating rates (T_S'', T_R'', A'') the difference between T_S'' and T_R'' will be greater

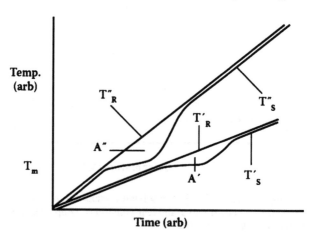

FIGURE 1.8 The effect of heating rate on signal sensitivity (see text for explanation of symbols). (Reproduced from Coleman, N.J. and Craig, D.Q.M., *Int. J. Pharm.* 135, 13, 1996.)

during the melt; hence, A″ will also be large. Overall, therefore, lower heating rates favor greater resolution. Sensitivity is, however, improved by higher heating rates; thus any experiment must use the heating rate that gives the right balance of resolution and sensitivity.

1.3 TYPICAL RESULTS FOR COMMON TRANSFORMATIONS STUDIED BY DSC

It is now appropriate to consider the most commonly encountered types of processes that are studied by DSC and the types of results they yield.

1.3.1 KINETICALLY CONTROLLED RESPONSES

A number of processes of pharmaceutical interest are kinetically controlled, examples including crystallization, curing, and many degradation reactions. Kinetic behavior is described classically by the Arrhenius equation, which may be given in the general form

$$d\alpha/dt = f(\alpha)Ae^{-Ea/RT} \tag{1.5}$$

where α = the extent of the reaction, t = time, $f(\alpha)$ = some function of extent of reaction, A = the preexponential constant, Ea = the activation energy, R = the gas constant, and T = absolute temperature. This type of behavior is associated with the well-known energy barrier model for thermally activated processes. In this model, a material changes from one form to another more thermodynamically stable form but must first overcome an energy barrier that requires an increase in Gibbs free energy. Only a certain fraction of the population of reactant molecules have sufficient energy to do this, and the size of this fraction (and hence the total number of reactant molecules) determines the speed at which the transformation occurs. The fraction of molecules with sufficient energy is dependent upon the temperature in a way given by the form of the Arrhenius equation; thus, this must also be true for the transformation rate. The types of process that can be modeled using this type of expression include chemical reactions, diffusion-controlled processes such as the desorption of a vapor from a solid, and some phase changes such as crystallization. There will be some constant of proportionality, H', such that the rate of heat flow can be directly related to the rate of the process, viz.

$$dQ/dt = H' \cdot d\alpha/dt = H' \cdot f(\alpha)Ae^{-Ea/RT} \tag{1.6}$$

$H' \cdot d\alpha/dt$ is not the total heat flow because it neglects the contribution from the heat capacity of the material. This heat capacity can be considered to be the energy contained in the various vibrational, translational, etc., modes available to the sample. The energy contained in these molecular motions is stored reversibly, which simply means that the amount of energy taken up by them when increasing the temperature by 1°C can be entirely recovered by reducing the temperature by 1°C. This can be

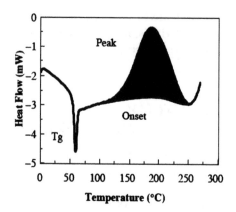

FIGURE 1.9 Example of a chemical reaction, in this case an epoxy cure, as detected by DSC, while the interpolated baseline corrects for Cp. The shaded peak is the exotherm from the chemical reaction. In this case it is proceeded by a glass transition (see chapter text).

contrasted with the enthalpy associated with, for example, a typical chemical reaction that will be either gained by the sample (endothermic) or lost (exothermic) irreversibly. Adding in the heat capacity contribution to the heat flow gives

$$dQ/dt = \beta C_p + H'f(\alpha)Ae^{-Ea/RT} \tag{1.7}$$

where β = heating rate and C_p = heat capacity.

Figure 1.9 gives a typical trace for a chemical reaction showing Arrhenius kinetics. As the reaction proceeds, the heat capacity of the sample usually changes in some way. This is normally approximated with an interpolated baseline as illustrated, which then represents the βC_p contribution to the heat flow. The shaded area gives the enthalpy of the chemical reaction.

The use of DSC for studying such reactions allows the operator the possibility of obtaining not only empirical information regarding the temperature and time relationship for processes of interest but also more detailed kinetic parameters as outlined above. Clearly, the kinetic analysis may be considerably more complex when one considers non-Arrhenius behavior, and also takes into account issues such as the size and shape of the reacting species as may be the case for, for example, excipient compatibility reactions (21). Some more advanced kinetic functions are outlined in Chapter 5.

The scanning method has advantages when studying multiple stage processes that may overlap. For example, studying a higher temperature process is difficult isothermally when a lower temperature process is also present. The early part of any higher temperature isotherm will be dominated by the lower temperature process, and it may well be impossible to delineate even approximately where one process ends and the other starts. Attempts to use a lower-temperature isotherm until the first reaction is finished before proceeding to a higher temperature isotherm inevitably lead to arbitrary decisions on what exact thermal treatment to use. Scanning at multiple heating rates can give information on the kinetics of even complex behavior.

In general, simple one-stage reactions are probably best studied isothermally, whereas more complex systems are best investigated using temperature scanning.

A highly topical area within the pharmaceutical sciences for which DSC measurements may be of use for kinetic studies is in the study of recrystallization, either on cooling from the melt or from the amorphous state. Classically, such reactions are described by Avrami kinetics (22–24), given by

$$\alpha = 1 - \exp[-(Kt)^m] \tag{1.8}$$

where α is the fraction crystallized as a function of time t. K and m are constants, with the latter dependent on crystal growth morphology. The corresponding rate equation may be obtained by simple differentiation with respect to time to yield

$$\left(\frac{d\alpha}{dt}\right) = Km(1-\alpha)[-\ln(1-\alpha)]^{1-1/m} \tag{1.9}$$

This is generally referred to as the Johnson–Mehl–Avrami (JMA) equation and is frequently used for the formalization of thermal analysis crystallization data. This analysis does assume isothermal crystallization conditions and homogeneous nucleation or randomly distributed nuclei for heterogeneous growth. In addition, the growth rate of the new phase should be controlled by temperature and be independent of time. Finally, low anisotropy of the growing crystals is also assumed (25). However, Henderson (26,27) has indicated that the expression can be extended in nonisothermal conditions, whereas Ozawa (28) has proposed expressions for nonisothermal crystal growth from preexisting nuclei. A simple approach that may be easily used to yield comparative (rather than absolute) data is the Kissinger equation (29)

$$d[\ln(\phi/Tc^2)]/d[1/Tc] = -Ea/R \tag{1.10}$$

whereby the recrystallization peak maximum temperature (T_c) is recorded as a function of scanning rate (ϕ) to yield the activation energy E_a. Clearly, there are a number of possible approaches to analyzing kinetic data using DSC. However, caution is required by the operator in that each approach carries concomitant assumptions that may or may not be compatible with the nature of the response involved.

1.3.2 Melting and Other First-Order Phase Transitions

A first-order phase transition is defined as a process whereby the derivative of the change in free energy ΔG with respect to temperature is not equal to zero (30), i.e.,

$$(\delta\Delta G/\delta T) = (-\Delta S) \neq 0 \tag{1.11}$$

where $-\Delta S$ is the entropy change. In essence, the change in state is accompanied by a change in the temperature dependence of the free energy as the system changes from the solid to the liquid state. In practical terms the description of first order

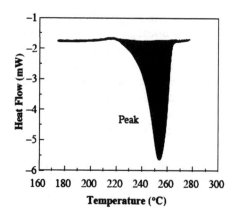

FIGURE 1.10 A typical polymer melt shown as the shaded endotherm.

indicates that the transition in question will occur at a specific temperature and will be independent of heating or cooling rate, although, as will be demonstrated, instrumental factors almost invariably result in an apparent rate dependence. The most common example studied by DSC is melting, which is characterized as a change in specific volume accompanied by a latent heat of fusion. However, other transitions such as liquid crystalline transformations may also be considered to be first order. As indicated earlier, at the melt temperature the sample will remain isothermal until the whole sample has melted. This is because a finite time is required for heat to penetrate the sample; hence, during this process the heat input contributes to the latent heat of fusion rather than increasing the temperature of the sample. The factor that determines the speed of the transition is the rate at which heat can be supplied by the calorimeter. Normally, this is fast, so the transition is very sharp with a small post-transition "tail," the length of which is determined by the speed with which the calorimeter can reestablish the heating program within the sample. Pure materials generally produce very narrow melting peaks, whereas the introduction of soluble impurities usually broadens the peak considerably by lowering the onset temperature; this phenomenon can be used to quantify the amount of impurity present in a sample.

The issue of peak broadening is also extremely important when dealing with polymeric samples. Many polymers produce a range of crystalline forms with different melting temperatures without necessarily having different internal unit cell arrangements, i.e., melting variations may be seen over and above those caused by polymorphism (described in Chapter 4). Typically, a semicrystalline polymer will comprise a distribution of crystallites with differing degrees of perfection and thus different melting temperatures. The melting transition in these materials is broad, as a succession of crystallite populations melts one after the other as the sample temperature reaches their melting temperatures. This kind of broad melting transition is illustrated in Figure 1.10. We can express this as

$$dQt/dt = \beta(C_p + g(t,T)) \qquad (1.12)$$

where g(t,T) = some function of time and temperature that models the contribution to the heat flow from the melting process.

When the melting is rapid with respect to the heating rate, g(t,T) will simply be a function of temperature. This means that, in the case of the distribution of crystallites, the melting contribution to heat flow will be proportional to heating rate in an analogous manner to the heat capacity if no other process occurs. In reality this simple case is rarely encountered. This point will be discussed in more detail in Chapter 4 in terms of the use of temperature modulation. As with Arrhenius processes, a baseline can be interpolated under the peak to approximate the contribution from βC_p (Figure 1.10); hence the area under the peak is a measure of the total heat of fusion.

1.3.3 Second-Order Phase Transitions: The Glass Transition

A second-order transition is defined as a process whereby the derivative of the change in free energy is zero but the second-order derivative is nonzero, i.e.,

$$(\delta^2 \Delta G / \delta T^2) = (-\Delta Cp/T) \neq 0 \qquad (1.13)$$

In this analysis the transition is defined as a step change in the heat capacity of the sample as a function of temperature. By far the most important transition that is generally considered to be second order is the glass transition, T_g. However, for completeness, other examples of second-order transitions include Curie point transitions where a ferromagnetic material becomes paramagnetic, the transition from an electrical superconductor to a normal conductor, and the transition in helium from being a normal liquid to being a superfluid at 2.2 K.

There is still considerable debate with regard to the precise meaning of the term glass transition. At first sight, the step change in C_p that occurs at the glass transition might be interpreted as a discontinuity that in turn would mean that it is a second-order transition. In fact, the transition is gradual as it occurs over about 10 degrees or more. Its position also varies with the heating and cooling rate (and with frequency in MTDSC), which reveals that it is also a kinetic phenomenon; hence, again, the process does not fit the strict definition of a second-order response.

From a phenomenological viewpoint the glass transition is seen as a change in sample flexibility over a narrow range of temperatures, with the material classically appearing brittle and glassy below the glass transition temperature (designated T_g) and more pliable at temperatures >T_g. The classical method of forming glasses is via cooling from the melt. For crystalline systems the cooling process results in nucleation and crystal growth at a temperature corresponding to the melting point T_m (assuming no supercooling takes place). The exothermic crystallization process results in a dramatic decrease in the free volume of the system (defined as the difference between the total volume and that occupied by the constituent molecules) as the molecules become arranged in an ordered lattice. On further cooling, less marked changes in free volume are observed as a result of heat capacity and thermal contraction effects. This is illustrated in Figure 1.11. For a glass-forming system,

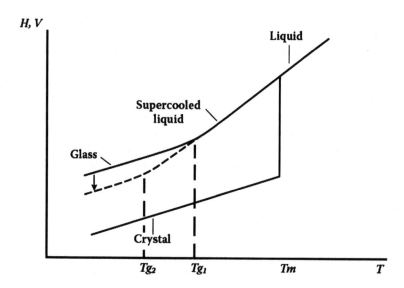

FIGURE 1.11 Schematic representation of the glass transition, comparing the behavior of systems forming glass and a crystalline solid on cooling.

however, the cooling rate may be too rapid in relation to the rate of crystallization for nucleation and growth to occur. Consequently, the system supercools and remains liquid below *Tm*. However, on further cooling, the system reaches a point whereby the molecular mobility decreases in a discontinuous manner, resulting in the system becoming "frozen" as the material assumes the material characteristics of a solid. In fact, the gross molecular conformation remains essentially the same as that of a liquid, the difference lying in a dramatic reduction in the translational and rotational motions of those molecules. The temperature at which this process occurs is cooling rate dependent, with slower rates leading to lower values for T_g, as illustrated in Figure 1.11.

There are a number of theories associated with the fundamental nature of the glass transition, but the issue remains the subject of debate within the field. One approach is to consider the transition in terms of relaxation processes. As the material is cooled in a linear fashion, at temperatures above T_g, the relaxation processes are rapid in relation to the cooling process; hence, molecular mobility remains high (relatively speaking). As the temperature lowers, the relaxation time increases up to a point whereby the rate of relaxation matches the rate of cooling. At this point and below, the large scale motions of the constituent molecules are slow in relation to the cooling program; hence, the molecular mobility relative to the time scale of the measurement is dramatically reduced and the material assumes macroscopic solid-like characteristics. Calorimetrically, this manifests itself as a reduction in the measured heat capacity.

A related explanation involves consideration of the free volume of the system, whereby this parameter is reduced to a critical value, below which free molecular motion may not take place (31–33). The glass transition may also be considered in terms of configurational entropy (34), whereby above T_g the system has additional degrees of freedom which are lost on going through the transition. The issue of the temperature dependence of the configurational entropy has been addressed by Kauzmann (35), who considered whether the glass transition could be infinitely low if the cooling rate was similarly slow. He suggested that a lower limit does indeed exist due to the thermodynamic necessity (via the third law) of the entropy for the glass being higher than the corresponding perfect crystal. This contradiction, known as the Kauzmann paradox, is resolved by there being a lower limit to the T_g, known as the Kauzmann temperature (T_k), which is typically 20 to 50 K below the experimentally derived glass transition. However, the paradox may also be resolved by considering that the extrapolation of the liquid line used by Kauzmann to predict a point at which the entropy of the glass would cross that for the crystal, is simply erroneous. The shape of the liquid line below T_g is not accessible experimentally.

In terms of the practicalities of measurement, the glass transition is essentially a change in the heat capacity of the material; hence (at least theoretically), the transition is seen as a step change in the baseline of a DSC trace (see Equation 1.2). A very simple relaxation model for the increase in heat capacity at the glass transition is given by

$$d\eta/dt = \exp[(\Delta h^*/(RT_g^2)(T - T_g)](T\Delta Cp - \eta)/\tau_g \qquad (1.14)$$

where

$$\eta = \delta + T\Delta Cp, \qquad (1.15)$$

δ = the excess enthalpy relative to the equilibrium value, ΔC_p = the heat capacity change at the glass transition, Δh^* = the apparent activation energy, T_g = the glass temperature, and τ_g = the relaxation time at equilibrium at T_g.

This response is demonstrated above for Vitamin E USP (36), which is a liquid at room temperature but may form a glass on rapid cooling. In this case the step change in baseline corresponding to the glass transition is clearly visible (Figure 1.12).

However, in practice the glass transition is often superimposed by an endothermic response, as indicated in Figure 1.13 for a polylactide microsphere system (37); the change in the baseline before and after the event is clearly visible but, similarly, it is also clear that the potential for confusion with a melting endotherm is considerable. Furthermore, it is difficult to establish the exact position of the glass transition under such a peak with confidence, although methods are available whereby the T_g may be ascertained (20).

When a sample is cooled from above T_g at a given rate, it follows a given enthalpy line as shown in Figure 1.14. This then gives rise to a glass with a certain enthalpy. If a glass with a lower enthalpy than this is heated at the same rate, then an overshoot will result that gives rise to the relaxation peak seen at T_g. This is because this enthalpy must be recovered so the material can return to equilibrium above T_g. There are two common ways of making a glass with a lower enthalpy than that produced by cooling at rate $-b$. The first is to cool it at some rate slower

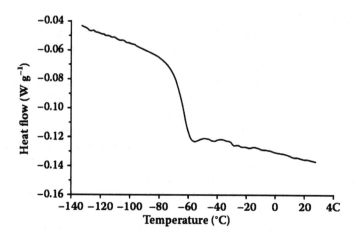

FIGURE 1.12 DSC trace for Vitamin E USP. (Reproduced from Barker, S.A. et al., *J. Pharm. Pharmcol.* **52**, 941, 2000.)

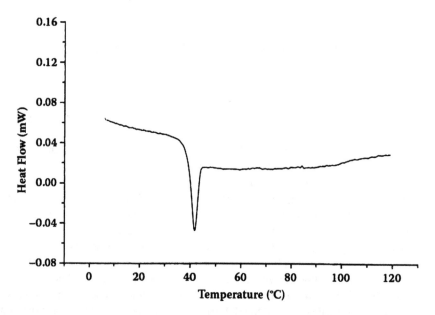

FIGURE 1.13 DSC trace of glassy polylactide microspheres. (Reproduced from Passerini, N. and Craig, D.Q.M., *J. Controlled Release* 2001, **73**, 111–115, 2001.)

than −*b*. On reheating at *b* the enthalpy peak will then be seen. The other is to cool at −*b* then anneal the sample, which will then lose enthalpy as it tries to move closer to the equilibrium line. Again, on reheating, the relaxation peak will be present. If a sample is cooled at −*b*, and then reheated at a slower rate, an exotherm will be observed before T_g as the sample loses enthalpy (anneals) as it is being heated.

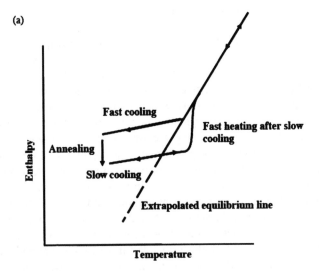

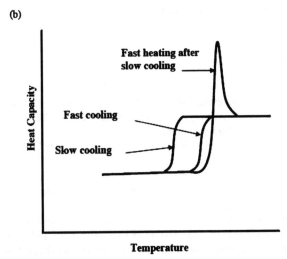

FIGURE 1.14 Schematic representation of (a) the enthalpy curve and (b) the heat capacity through T_g, illustrating the effect of cooling and heating rate.

Over and above such experimental issues, a further reason for the appearance of the endothermic peak is annealing effects. As illustrated in Figure 1.14, the glassy system is above the extrapolated equilibrium enthalpy for the liquid state, and hence there will be a tendency for enthalpic reduction as a function of time toward this equilibrium. Again, heating through T_g for such aged systems results in the appearance of an endothermic peak due to the necessity for heat input for reestablishment with the $> T_g$ enthalpy.

This time-dependent increase in the value of the enthalpic relaxation peak may in itself be used to assess the relaxation time of the system in question; this is discussed in more detail in Chapter 3.

1.4 CONCLUDING COMMENTS

This chapter has attempted to illustrate both the basic principles of DSC and, in general terms, the type of information that the method may provide. The information given here should be considered in conjunction with that imparted in the following three chapters that deal with experimental practice, pharmaceutical applications, and the new development to DSC, modulated temperature DSC. However, we believe it is essential for the principles underpinning the use of the technique to be appreciated to promote best practice when using the instrument. In particular, and without wishing to end the chapter on a negative note, it is apparent that DSC is often used as something of a black box within the pharmaceutical sciences, as illustrated by, for example, the common practice of not labeling the y (power) axis when presenting DSC data and the frequent omission of calibration information. However, the important point to be emphasized in the present context is that DSC represents a highly versatile and reliable approach for the study of a wide range of systems and problems. The following chapters will take this principle further by illustrating both the issues associated with making good measurements and the ways in which information may be extracted for the study of pharmaceutical materials in particular.

REFERENCES

1. Wendlandt, W., *Thermal Analysis*, 3rd ed., Wiley-Interscience, New York, 1986.
2. Gallagher, P.K., Thermoanalytical instrumentation, techniques and methodology, in *Thermal Characterisation of Polymeric Materials*, 2nd ed., Turi, E.A., Ed., Academic Press, San Diego, 1997, pp. 2–205.
3. Watson, E.S., O'Neill, M.J., Justin, J., and Brenner, N., *Anal. Chem.* **36**, 1233–1238, 1964.
4. O'Neill, M.J., *Anal. Chem.* **36**, 1238–1244, 1964.
5. Ford, J.L. and Timmins, P., *Pharmaceutical Thermal Analysis*, Ellis Horwood Ltd., Chichester, U.K., 1989.
6. Dollimore, D., Thermoanalytical instrumentation, in *Analytical Instrumentation Handbook*, Ewing, G.W., Ed., Marcel Dekker, New York, 1990, pp. 905–960.
7. Wunderlich, B., *Thermal Analysis*, Academic Press, San Diego, CA, 1992.
8. Mathot, V.B.F., Ed., *Calorimetry and Thermal Analysis of Polymers*, Hanser Publishers, Munich, 1994.
9. Brown, M., Ed., *Handbook of Thermal Analysis*, Elsevier, Amsterdam, 1998.
10. Vyazovkin, S., *Anal. Chem.* **76**, 3299, 2004.
11. Richardson, M.J., in *Developments in Polymer Characterisation*, Dawkins, J.V., Ed., Vol. 1, Applied Science Publishers, London, 1978, p. 205.
12. Mraw, S.C., *Rev. Sci. Inst.* **53**, 228, 1982.
13. Mathot, V.B.F., *Polymer* **25**, 579, 1984.
14. Jin, Y. and Wunderlich, B., *Thermochim. Acta* **226**, 155–161, 1993.
15. Wunderlich, B., in *Thermal Characterisation of Polymeric Materials*, 2nd ed., Turi, E.A., Ed., 1997, pp. 206–483.
16. Hancock, B.C., Shamblin, S.L., and Zografi, G., *Pharm. Res.* **12**, 799, 1995.
17. Hill, J.J., Shalaev, E.Y., and Zografi, G., *J. Pharm. Sci.* **94**, 1636, 2005.

18. Craig, D.Q.M., Royall, P.G., Kett, V.L., and Hopton, M.L., *Int. J. Pharm.* **179**, 179, 1999.
19. Sutananta, W., Craig, D.Q.M., and Newton, J.M., *Int. J. Pharm.* **110**, 75–91, 1994.
20. Coleman, N.J. and Craig, D.Q.M., *Int. J. Pharm.* **135**, 13, 1996.
21. Richardson, M., DSC on polymers: experimental conditions, in *Calorimetry and Thermal Analysis of Polymers*, Mathot, V.B.F., Ed., Hanser Publishers, Munich, 1994.
22. Avrami, M., *J. Chem. Phys.* **7**, 1103, 1939.
23. Avrami, M., *J. Chem. Phys.* **8**, 212, 1940.
24. Avrami, M., *J. Chem. Phys.* **9**, 177, 1931.
25. Malek, J., *Thermochim. Acta* **355**, 239, 2000.
26. Henderson, D.W., *J. Therm. Anal.* **15**, 325, 1979.
27. Henderson, D.W., *J. Non-Cryst. Solids* **30**, 301, 1979.
28. Ozawa, T., *Bull. Chem. Soc. Jpn.* **57**, 639, 1984.
29. Kissinger, H., *Anal. Chem.* **29**, 1702, 1957.
30. Ehrenfest, P., *Proc. Acad. Sci.*, Amsterdam **36**, 153, 1933.
31. Fox, T.G. and Flory, P.J., *J. Appl. Phys.* **21**, 581, 1951.
32. Fox, T.G. and Flory, P.J., *J. Phys. Chem.* **55**, 221, 1951.
33. Fox, T.G. and Flory, P.J., *J. Polym. Sci.* **14**, 315, 1954.
34. Elliott, S.R., *Physics of Amorphous Materials*, Longman, London, 1983.
35. Kauzmann, W., *Chem. Rev.* **43**, 219, 1948.
36. Barker, S.A., Yuen, K.H., and Craig, D.Q.M., *J. Pharm. Pharmcol.* **52**, 941, 2000.
37. Passerini, N. and Craig, D.Q.M., *J. Controlled Release* **73**, 11–115, 2001.

2 Optimizing DSC Experiments

Trevor Lever

CONTENTS

2.1 INTRODUCTION

A differential scanning calorimeter (DSC) measures the heat flow to or from a sample, usually as the sample is heated. Any change that can affect the heat flow characteristics within the DSC or the sample will therefore affect the data produced by the instrument. This chapter explains the variables that can affect the heat flow signal measured by the DSC and why it is important for the operator to understand them. This chapter is intended to be complementary to both the operator's manual (for whatever instrument is used) as well as the many texts on DSC theory, with the focus being on the practical aspects of making good DSC measurements.

To understand and appreciate the effect that each of the experimental factors have upon the final DSC trace, it is useful to have a basic understanding of the heat flow and heat transfer processes occurring within the DSC during an experiment. The specific heat flow path will vary from one type of DSC design to another, but some general rules will apply. As this topic has been covered in more detail in Chapter 1, it is considered helpful to revisit some of the very basic concepts to provide a context for explaining the influence of experimental parameters.

In a typical experiment the DSC records the heat flow into, or out of, the sample as a function of temperature or time. In the absence of any transition (e.g., melting), the heat flow recorded can be represented as

$$dQ/dt = dQ/dT \cdot dT/dt \qquad (2.1)$$

where dQ/dt is the heat flow, dQ/dT is the heat capacity, and dT/dt is the heating rate. In other words,

$$\text{Heat Flow} = \text{Heat Capacity} \cdot \text{Heating Rate}$$

When a sample undergoes a thermal transition, such as melting, this may be considered as either a gross change in heat capacity of the sample or as an extra term $f(T,t)$ that can be added to Equation 2.1 to cover these kinetic events:

$$dQ/dt = dQ/dT \cdot dT/dt + f(T,t) \qquad (2.2)$$

or

$$\text{Heat Flow} = \text{Heat Capacity} \cdot \text{Heating Rate} + \text{Kinetic Events}$$

As no universal "heat flow meter" exists, it is necessary to make differential measurements by measuring the heat flow between the sample and an inert reference. The magnitude of the heat flow will be governed by the thermal resistance of the system, and the difference in temperature between the sample and reference

$$dQ/dt = (T_R - T_S)/R \qquad (2.3)$$

where T_R and T_S are the reference and sample temperatures, respectively, and R is the thermal resistance between reference and sample.

From these equations it can be seen that key quantities that can affect the measured heat flow signal in a DSC are:

- The heating rate
- The heat capacity
- Kinetic events (e.g., melting or degradation)
- The thermal resistance between the sample and reference
- The temperature difference between the sample and reference

These variables must be understood and controlled to obtain quality DSC data, that is, data that are both reproducible and representative of the sample's properties.

Some of these variables can be controlled, or at least influenced, by the operator. The heating rate is clearly a variable over which the operator has control. The operator can only influence others, such as the thermal resistance of the DSC. For example, by changing the purge gas to one that has higher thermal conductivity, heat transfer is improved.

In a typical DSC experiment, a weighed sample is encapsulated in an aluminum pan. This is then placed into the DSC together with a reference pan. The sample and reference are then heated in a controlled environment, and the heat flow difference between them is recorded. Similar experiments are made with standard materials in order to obtain calibration constants so that quantitative temperature and heat flow data may be obtained in subsequent experiments. The variables and issues arising from this simple description of a typical DSC experiment can be split into three categories: equipment, sample, and experiment. These areas are described in more detail later.

2.2 EQUIPMENT

Besides the DSC, there are other pieces of equipment and ancillaries that need to be maintained and calibrated to ensure that accurate DSC data are produced.

2.2.1 ANALYTICAL BALANCE

The analytical balance that is used to weigh out the sample is often the limiting factor in regard to the accuracy of the heat flow signal produced by the DSC. The balance should be checked and calibrated daily against a known traceable reference weight. Also, the balance should be accurate enough for the amount of sample to be weighed and to provide the quantitative enthalpy data required.

A 5-digit balance should be considered as a minimum requirement when weighing samples in the 10 to 20 mg range, a 6-digit balance for 1 to 10 mg samples, and a 7-digit balance for samples less than 1 mg. Balances should be kept clean. Sample pans should also be cleaned with a brush prior to loading on to the balance to ensure that any sample that may have adhered to the outside of the pan is removed.

2.2.2 PURGE GAS

Most DSC experiments on pharmaceutical materials are carried out in an inert atmosphere, usually nitrogen. Helium, which has a higher thermal conductivity, can be used if the thermal resistance of the DSC needs to be reduced. This has the effect of increasing the resolution of transitions that are close to each other in temperature.

Helium should also be used as a purge gas in DSCs that use a liquid nitrogen cold-finger system for subambient operation. This is because the liquid nitrogen coolant temperature is below the dew point of the nitrogen purge gas. A drawback to using helium over nitrogen as a purge gas is that it takes considerably longer for the DSC system to reach equilibrium and stabilize after the cell has been opened to the air. This is because the difference in thermal conductivity between helium and air is much greater than the difference in thermal conductivity between nitrogen and air.

2.2.3 PURGE FLOW

A constant purge of gas through the DSC ensures that any volatile products evolved during the DSC experiment are swept away from the measuring sensor. It also ensures a nonoxidative and constant environment around the sample area, which helps in maintaining day-to-day baseline reproducibility.

A change in the flow rate of the gas used to purge the DSC can have several effects. First, it is possible that it will change the temperature and enthalpy calibration. The magnitude of this variation will vary from one type of DSC design to another as some instruments preheat the purge gas prior to its entering the DSC cell. Second, for experiments where a volatile substance is evolved from the sample when it is heated, the DSC peak shape will be affected by the speed at which the volatile substance is removed.

A slow decrease in the flow rate into the DSC will be observed if cylinders are used that are fitted with single-stage regulators, as the output pressure will decrease as the cylinder slowly empties. This effect may be minimized by the use of two-stage cylinder regulators or, ideally, the use of calibrated mass flow controllers, which will remove this variable from the experiment altogether.

2.2.4 DSC

2.2.4.1 Temperature Calibration

In any DSC experiment where the sample is heated, the sample and its surroundings are not in thermal equilibrium. The sample temperature will be slightly lower than that of the furnace temperature. The transfer of heat between the furnace, the sample, and the reference is not instantaneous and depends upon the heat transfer characteristics of the particular DSC design. Consequently, a correction is required so that accurate transition temperatures are measured from the DSC heat flow data. This introduces the concept of "thermal lag" within the system. In fact, there are a number of thermal lags to be considered: the lag between the furnace and the bottom of the sample pan, the lag between the bottom of the sample pan and the sample, and then the thermal lag throughout the sample. Various methodologies exist to compensate for one or all of these thermal lags. Temperature gradients within the sample can be minimized, for example, by using small samples and pans with good heat transfer characteristics and slow heating rates. However, the thermal lag can never be removed and so must always be calibrated for.

Temperature calibration requires that traceable standards, with known transition temperatures, be run under exactly the same conditions as those to be used when running samples. The observed DSC heat flow transition temperature for the standard is then compared to the known value and a correction is applied, usually through software. The most common temperature calibration standards used in DSC are the melting points of pure metals such as indium, lead, and tin. These metals are available in high purity and with well-known melting points. Few organizations provide these materials with any melting point certification or traceability, one exception being the U.K. Laboratory of the Government Chemist (LGC).*

This temperature calibration does not take into account thermal lag (and therefore temperature gradients) within the sample, one of the arguments put forward to support the use of organic materials to temperature-calibrate a DSC for pharmaceutical measurements. However, the poor availability of pure and certified organic temperature standards of sufficient accuracy is a powerful argument against using such materials.

Temperature calibration should be performed over the temperature range of interest. Surprisingly, many laboratories (and published papers) still settle for a single-point temperature calibration (invariably, indium [156.6°C]) and assume that the temperature calibration of the instrument is linear over the entire temperature range. Probably, this is not the case. At least two temperature calibration points should be made, and these should embrace the temperature range of interest. It is also common to see an operator calibrate with two materials (indium and some other higher melting point metal such as lead [327.5°C] or tin [231.93°C]) and then analyze samples with transitions in the ambient to 100°C temperature range. It is recommended that one temperature calibration point should be close to room temperature and, additionally, one below room temperature, if subambient measurements are made.

The effect of heating rate on the DSC heat flow signal will be considered in more detail in Subsection 2.4.2. However, it is useful to consider this parameter in terms of calibration protocols. Figure 2.1 shows the onset temperature of the melting peak of indium at seven different heating rates. As the heating rate increases, the thermal lag between the furnace and the sample increases. From Figure 2.1 it can

---

* Laboratory of the Government Chemist, Queen's Road, Teddington, Middlesex TW11 0LY, U.K.; Phone: (+44) 028 943 7393.

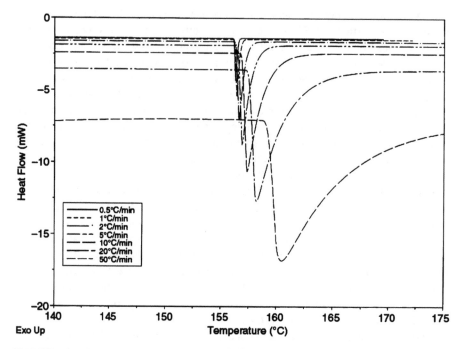

FIGURE 2.1 The effect of heating rate on the onset of melting of indium.

be seen that if this DSC is calibrated at 5°C/min and then measurements made at 20°C/min, the temperature calibration will be substantially in error.

If measurements need to be made at a variety of heating rates during the day, recalibration each time can be a significant inconvenience. One approach to overcome this is to calibrate at 0°C/min. At first this may seem odd, but it does have some advantages. First, it means that experiments can be made at any heating rate, and then a correction may be applied to the data to take into account the heating rate used. This calibration methodology also benefits isothermal experiments (for example, in kinetic studies) as the isothermal temperatures are exactly calibrated as well.

To calibrate at 0°C/min, construct a graph of the onset of melting temperature against heating rate and extrapolate back to the onset temperature for 0°C/min. This value is then entered into the temperature calibration software of the DSC.

2.2.4.2 Enthalpy Calibration

Few materials exist that have accurately known enthalpies and that are also available with traceable certification. Many textbooks will quote heat of fusion values for materials, and it is often possible to obtain high-purity quantities of these materials. Yet it is very difficult to obtain high-purity materials that have traceable and certified enthalpy values. Again, LGCs provide indium that is certified with respect to both enthalpy and temperature. This material can be used, in a single experiment, to calibrate the DSC for both the abscissa and ordinate. The assumption here is that a single enthalpy calibration is acceptable for the entire temperature scale. This can only be tested with other traceable standards, and hopefully in time these will become available.

2.2.5 COOLING ACCESSORIES

Whatever cooling accessory is used, it is important that both it and the DSC have reached thermal equilibrium before any experiments are made. There are a couple of ways to find out how long it takes for a particular cooling system to reach equilibrium:

- Continually cycle indium from 120°C to 200°C at 10°C/min after the cooling system has been turned on. Measure the melting temperature and enthalpy on each cycle until the results are repeatable.
- Monitor the heat flow signal and wait for it to stabilize.

The amount of time required to stabilize will vary considerably between DSCs and cooling systems and may take as long as an hour or two. For busy laboratories that use electrical (mechanical) cooling accessories, the purchase of a power outlet timer can be used to turn the cooling system on an hour or two before the staff start work for the day. A complete calibration procedure should be performed once the DSC and cooling system has stabilized.

2.2.6 AUTOSAMPLERS

It is wise to check the calibration and alignment of any mechanical parts of the autosampler on a regular basis. This should minimize any errors when using this accessory for extended periods of unattended operation. Besides the obvious increase in sample throughput, autosamplers are also of considerable help in running routine calibration and validation samples. One approach is to have weighed calibration and validation samples continually left in the autosampler with several stored run sequences on the data handling system. It then becomes an easy matter to run and calibrate the system overnight for whatever temperature range and pan configuration will be used the following day.

2.3 SAMPLE PREPARATION

The ideal DSC sample would be a thin circular disk, of high thermal conductivity, that did not interact with the DSC cell so that it could be placed directly inside without the need to encapsulate the material in a sample pan. These criteria meet the need for minimizing thermal gradients within the sample and reducing the thermal resistance of the DSC. Unfortunately, few pharmaceutical materials conform to such restrictions. Care must therefore be taken during sample preparation in balancing the need to minimize thermal lag and thermal gradients, meanwhile ensuring that the sample is still representative of the bulk material.

2.3.1 SAMPLE FORM

2.3.1.1 Powders

Ideally, these should be free flowing and, if necessary, of a consistent particle size distribution. If the material contains "lumps," it may be necessary to pass a portion of the sample through two sieves (say, 50 and 100 μM) and run the portion collected in between. The presence of agglomerates or other lumps within a sample can lead to "shoulders" appearing on the DSC heat flow curve during transitions. A pestle and mortar should not be used to grind a sample down into smaller particles until it is certain that this will not affect the stability or polymorphic nature of the material.

2.3.1.2 Liquids

Low-viscosity liquids can be transferred to the hermetic DSC pan using a Pasteur-pipette or similar apparatus. Dipping one end of an opened paper clip into viscous liquids and then placing it onto the base of the sample pan may transfer higher viscosity samples. Care should be taken not to contaminate the lip of the pan with any liquid as this will result in a poor seal.

2.3.1.3 Creams and Emulsions

Poor reproducibility can result if the sample is not homogenous. If possible, stir or shake the container prior to sampling. Use similar techniques to those described for handling liquid samples.

2.3.2 THERMAL HISTORY AND CRYSTALLINITY

The effect of previous heat treatments (thermal history) can significantly affect the shape of the DSC curve for semicrystalline materials. This can make comparisons difficult, unless a similar thermal history has been applied to both samples. This effect is well known for polymers, fats, and waxes and so can become an issue when looking at suspensions or polymer-modified drug delivery systems.

For long-chain hydrocarbons, the amount of order or crystallinity will depend on how slowly the sample has been cooled from the melt back down to room temperature. For samples that are cooled quickly, there is less time available for the sample to assemble into ordered structures. The result will be a material with a smaller fusion endotherm when it is subsequently reheated. If the sample is cooled slowly, there is more time for it to crystallize. Here, the result will be a material with a larger fusion endotherm on reheating.

When studying this type of material, it is clearly important to establish whether a standardized thermal history is appropriate. If this is the case, the simplest method for achieving this is a heat–cool–reheat DSC method. The first heating ramp destroys any previous thermal history (assuming the maximum temperature is sufficient to remove any remaining nuclei without causing sample degradation), the cooling ramp imposes a known thermal history on the sample, and the second heating ramp allows the sample to be measured with a known thermal history.

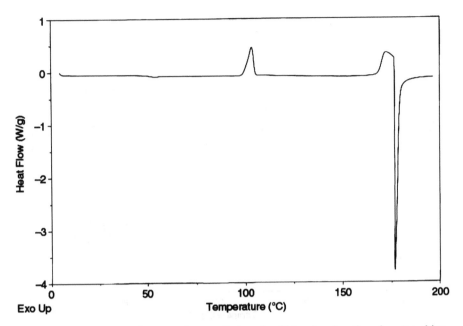

FIGURE 2.2 DSC trace for amorphous salbutamol sulfate, showing the glass transition, recrystallization peak, and melting/degradation peak.

Organic materials can also exist in crystalline and amorphous forms. An amorphous material can give rise to several transitions as it is heated, as indicated in Figure 2.2. The first transition can be observed as an endothermic shift in the baseline. This is the glass transition temperature. As the temperature increases an exothermic peak may be observed if the sample can orientate into a more crystalline form. Finally, any crystalline material present may melt or (as in this case) melt and decompose at higher temperatures.

2.3.3 Sample Size

As the sample mass is increased so does the time increase that is required for that sample to melt, once the material has reached its melting point. When a material is melting, its temperature stays constant. This can be observed as a discontinuity in the temperature signal during the melting transition. Figure 2.3 illustrates these effects by comparing the fusion endotherms for three samples of indium of increasing mass. As the sample mass increases there is an increase in the thermal gradient across the sample. This leads to a decrease in the slope on the leading edge of the heat flow data obtained during melting. Figure 2.4 shows the same data as in Figure 2.3, but this time the data are plotted against temperature rather than time. All three indium samples melt at exactly the same onset temperature, but the peak maximum moves to higher temperatures as the sample mass is increased. The effects of sample size on the transitions measured by DSC are summarized in Table 2.1.

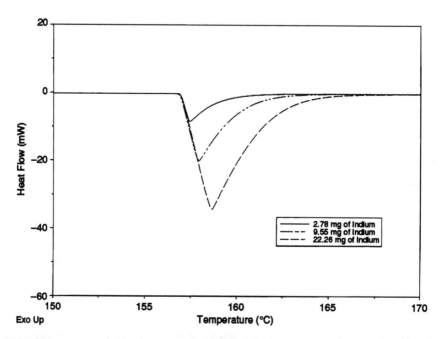

FIGURE 2.3 Comparison of three samples of indium, of varying mass, heated at 10°C/min. Hermetic pan with inverted lid. Nitrogen purge gas at 25 ml/min. Heat flow plotted against temperature.

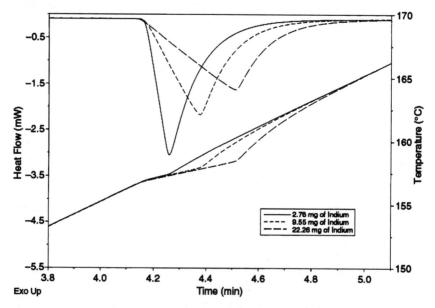

FIGURE 2.4 Comparison of three samples of indium, of varying mass, heated at 10°C/min. Hermetic pan with inverted lid. Nitrogen purge gas at 25 ml/min. Heat flow (uppermost curves) and sample temperature (lower curves) plotted against time.

TABLE 2.1
Summary of the Effect of Changing Sample Mass on the
Transitions Measured by DSC

Sample Size	Sensitivity	Resolution	Onset Temperature	Peak Maximum Temperature
Increase	Increases	Decreases	Constant	Increases
Decrease	Decreases	Increases	Constant	Decreases

2.3.4 SAMPLE PAN

There is a wide range of sample pans available, designed to meet general and specific sample and application needs. It is important to determine which type of DSC pan to use for a given sample or application. It should go without saying that, regardless of which type of pan is used to encapsulate the sample, the DSC should be calibrated by running standards in the same type of pan and under the same experimental conditions. Once the sample and reference pan have been introduced into the DSC, it is important to wait for thermal equilibrium to be reestablished (stable heat flow signal) before starting the experiment. It will take longer for the DSC to reach thermal equilibrium and for the heat flow signal to stabilize as the mass of the sample and pan increases.

The types of pans available can be divided into three general types: general, hermetic, and high-pressure.

2.3.4.1 General Pans

These are usually used for most powder and solid samples, and come in two parts (pan and lid). The pan is sealed by pressing the lid into the pan and crimping or folding the side wall of the pan over the lid, using a dedicated sealing press. Available in various materials, aluminum is typically used for pharmaceutical materials. If the sample reacts with aluminum then either anodized aluminum, platinum, or gold pans may be used. Once sealed, these pans are not airtight. During the sealing process the lid is pushed down on the top of the sample. If the sample is changed (e.g., a polymorphic transformation) by the effect of pressure, then the pan should not be sealed. In this instance the lid may simply be rested on top of the sample. It is important to use a lid as this ensures good heat flow to the top of the sample.

After the pan is removed from the sealing device it may be rubbed along the bottom of a plate to ensure that the pan has a flat bottom and that there will be good heat transfer between the DSC and the pan. For samples that can withstand the pressure, a small metal rod with a diameter slightly smaller than that of the pan may also be used to push down on the lid as the pan is gently rubbed along the metal plate. For samples that will undergo changes with such treatment, or if the sample needs to be hermetically sealed, the use of hermetic pans is recommended.

Most general and hermetic DSC pans are punched out of thin aluminum sheet that is contaminated with small amounts of machine oil. If these pans are used without cleaning, the oil can lead to spurious events appearing on the DSC trace. Consequently, all pans should be cleaned with solvent (nonchlorinated) and stored in a desiccator prior to use.

2.3.4.2 Hermetic Pans

Hermetic pans can be sealed without the need to place any stress (pressure) upon the sample. Again, they come in two parts (pan and lid), and a special sealing press is used to cold-weld the two pieces together. Once sealed, the pan can withstand an internal pressure of around 300 kPa (~3 atm) before rupturing. Slightly thicker aluminum is used in the manufacture of hermetic pans to ensure that a good gas seal is obtained. This leads to an increase in the time constant of the system and a slight loss of resolution. Another point to consider here is that these pans, although airtight, are usually sealed in an air atmosphere. If an inert atmosphere is needed around the sample during the DSC experiment, then the pans should be sealed in a glove box that is purged with nitrogen.

Hermetic pans are also ideal for running liquids, emulsions, and creams. When running liquid samples, care should be taken in ensuring that no liquid gets onto the lid of the pan as this will result in a poor seal. The quality of the seal may be checked by leaving the sealed pan in an analytical balance for 5 to 10 min to ensure that there is no mass lost, prior to placing the sample in the DSC. The quality of the seal can be improved by ensuring that both the hermetic pans and lids are cleaned in solvent and dried prior to use. Additionally, the pans may be kept in a warm oven (~100°C) to temper the aluminum prior to sealing.

Two additional problems arise when handling volatile liquids. The first is loss of sample through volatilization prior to the pan's being sealed; the other is the tendency for some liquids to "wick" up the side of the pan and affect the quality of the hermetic seal produced. Both problems can be minimized by cooling the pans (and sample) prior to sealing and weighing. A small bag of crushed solid carbon dioxide (Cardice) is ideal for achieving this. If necessary, place the pan on top of a small piece of solid Cardice, add the sample (from a cooled pipette or syringe) then add the cooled lid and seal. Check the quality of the hermetic seal by placing the sealed pan on an analytical balance for a few minutes following temperature reequilibration to ensure there is no weight loss.

With some types of hermetic pans, the lid may be inverted and placed into the sample pan. This is ideal for running smaller samples as it produces an hermetic pan with a smaller "dead" volume above the sample, and improves heat transfer to the top of the sample. For very small samples two pans may be used, one inverted and pushed inside the other. This type of configuration is also good for sealing 2 to 3 mg of calibration sample and keeping as a retained standard.

Liquid samples, if heated to too high a temperature, will burst the pan open when the pressure exceeds around ~300 kPa for aluminum hermetic pans and ~600 kPa for gold pans. If higher temperatures or pressures are required, the use of a large volume stainless steel capsule to hold the sample is recommended. Another solution to this problem is to pierce the lid of the pan to enable volatile components

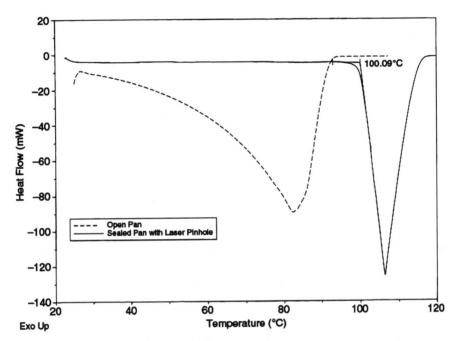

FIGURE 2.5 Comparison of DSC heat flow data for water obtained at 5°C/min in an open pan (dashed line) and a sealed hermetic pan with a 50-μm hole in the lid.

to escape when the pressure inside the pan equals the external pressure. For pure liquids this temperature will be equal to their boiling points. This gives a much sharper peak in the heat flow curve, when compared to crimped pans, with the onset being equal to the boiling point of the evolved gas. It is preferable (if possible) to obtain lids that have laser-drilled holes as this ensures good reproducibility for onset temperatures. Figure 2.5 shows the DSC heat flow data for water in an open pan compared to an hermetic pan containing a laser-drilled pin hole in the lid. If the operators themselves need to make holes in the lids, this is best performed prior to sealing the pan, as it is much easier and reduces the possibility of cross contamination between samples.

The use of the hermetic pan with laser pinhole can be extended by using this configuration to measure boiling points of materials within a pressure DSC cell. If the boiling point is measured at a range of known pressures, the boiling point shifts can be used in the Clausius–Clapeyron equation to obtain quantitative vapor pressure data.

2.3.4.3 Pressure Pans

For pressures higher than 300 kPa (~3 atm), materials other than aluminum, and with different sealing systems, are used. A variety of pan types exist with specific temperature, pressure, and volume restrictions.

2.3.4.3.1 O-Ring Sealed Pans

By using stainless steel as the pan material and an o-ring as the sealing mechanism, these pans can withstand pressures up to 3 MPa. The thermal lag of these pans is much higher than hermetic pans due to the use of thick stainless steel as the pan material. Consequently, sensitivity and resolution are both reduced. Slow to moderate heating rates are also recommended (1–5°C/min) to minimize thermal gradients within the sample. Placing an o-ring around the lid and pushing this to the pan using a special sealing tool seals the pan. Once sealed, these pans cannot be reused. If the sample reacts with stainless steel, it may be placed in an aluminum hermetic pan (with no lid) and this placed into the stainless steel pan prior to sealing.

2.3.4.3.2 Washer Sealed Pans

By increasing the wall thickness and using a metal washer as a soft metal seal, these screw-threaded pans can with hold pressures up to 10 MPa. By using a soft metal seal (usually copper or gold) these pans are reusable but will eventually distort over time leading to poor sealing. As with the o-ring sealed pans, slow to moderate heating rates are required in order to minimize thermal lag and thermal gradients within the sample. The thermal lag in these pans is higher than the o-ring sealed pans, leading to a further reduction in sensitivity and resolution.

2.3.4.3.3 Glass Ampoule Pans

Glass ampoules are used for samples that might react with metallic crucibles, or for assessing stability and reactivity. Often, this is the only reliable way to obtain accurate DSC curves for reactive materials. Here, the sample is placed inside a glass ampoule that has a narrow neck, which is then sealed off with a small Bunsen flame. It is necessary to cool the sample in the bottom of the ampoule during the sealing process to ensure that it is not heated. The sealed glass ampoule is then either placed directly in the DSC or placed inside a metal carrier to ensure good heat transfer to the sample.

2.3.4.4 Reference Pan

The reference pan should always be of the same type as the sample pan. Ideally, the mass of the reference pan should be close to the mass of the sample pan and sample. This will minimize the start-up "hook" observed at the beginning of the DSC experiment (see Subsection 2.4.1). To compensate for the sample mass, one or two extra aluminum sample pan lids may be added to the reference pan.

Once a pan has been sealed, it is difficult to know if it contains a sample or not. It is recommended that a small "mark" may be scratched on the lid of any empty reference pan in order that it may be easily identified. It can be very costly to discard perfectly good gold or platinum reference pans just because one cannot see inside them.

The exact type of pan available for each DSC will be specific to the supplier. It is prudent to consult with the manufacturer as to the range of pans available and follow their specific guideline as to which pan to use for a given application. Figure 2.6 and Table 2.2 summarize the various types of pan configurations typically used for pharmaceutical measurements.

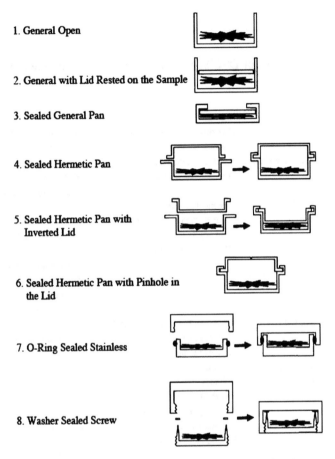

1. General Open

2. General with Lid Rested on the Sample

3. Sealed General Pan

4. Sealed Hermetic Pan

5. Sealed Hermetic Pan with Inverted Lid

6. Sealed Hermetic Pan with Pinhole in the Lid

7. O-Ring Sealed Stainless

8. Washer Sealed Screw

FIGURE 2.6 Summary of typical DSC pan configurations used in pharmaceutical DSC experiments.

2.4 EXPERIMENTAL PROCEDURE

2.4.1 START-UP HOOK

Figure 2.7 shows the initial heat flow data for three separate DSC experiments started at 120°C. Heating rates of 5, 10, and 20°C/min were used for individual experiments. It can be seen that as the heating rate is increased the initial transient (start-up hook) increases in magnitude and the heat flow signal also increases in magnitude. This arises because at the start of each experiment, power is increased to the furnace so that it heats up. The heating rate increases at the beginning of the experiment until the required experimental heating rate is obtained. The observed transient can be predicted from Equation 2.1, i.e., Heat Flow = Heat Capacity · Heating Rate.

The magnitude of start-up hook can be minimized by matching the mass of sample and reference pans. Figure 2.8 shows the heat flow data for a series of experiments on a pharmaceutical material where the reference mass has been

TABLE 2.2
Summary of the Types of DSC Sample Pans That Are Available and Their Typical Uses

Pan Type	Application	Comments
General open pan	Free diffusion of volatiles (if the sample depth is small)	Avoid this configuration if possible
General pan with lid "rested" on sample	Sample is sensitive to the pressure of sealing	Use hermetic pans unless highest resolution and sensitivity is required
Sealed general pan with lid	General DSC experiments	
Sealed hermetic pan	Liquids, creams, suspensions and solids that are sensitive to the pressure of sealing	Maximum pressure for aluminum pans is ~300 kPa (3 atm)
		Maximum pressure for gold pans is ~600 kPa (6 atm)
		Maximum temperature is 600°C
		Typical volume is ~ 40 µl
Sealed hermetic pan with inverted lid	As above, but a smaller dead volume is achieved and heat transfer is better	Use the inverted lid unless a larger sample size is required
Sealed hermetic pan with pinhole in the lid	Liquids Solvates Free/bound water Boiling point measurements	
O-ring sealed stainless steel pan	Liquids above 120°C Protein denaturation	Maximum pressure is ~3 MPa
		Maximum temperature is ~250°C
		Typical volume is ~100 µl
		If the sample reacts with stainless steel use an aluminum hermetic pan to hold sample; alternatively, use a glass ampoule
Washer sealed screw threaded pan	Thermal stability and reactivity Hazards evaluation	Maximum pressure is ~10 MPa
		Maximum temperature is ~300°C
		Typical volume is ~70 µl
		If sample reacts with stainless steel use glass ampoule
Sealed glass ampoule	Very volatile liquids Thermal stability and reactivity Hazards evaluation Materials that react with metals	Maximum pressure is ~20 MPa
		Typical maximum temperature is ~600°C
		Typical volume ~50 µl

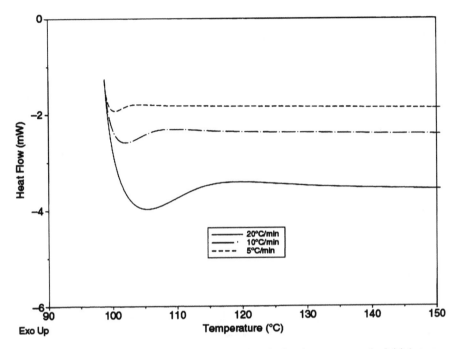

FIGURE 2.7 Comparison of the effect of increasing the heating rate upon the initial start up transient of a DSC. Empty general pans were placed in both the sample and reference positions. Nitrogen purge gas at 25 ml/min.

changed between experiments. When the sample is run with no reference pan (i.e., the sample mass is larger than the reference mass), a large start-up hook is observed in the endothermic direction. When a reference pan is used that is of a larger mass than the sample, a start-up hook is observed in the exothermic direction. The magnitude of the start-up hook is minimized when the mass of the reference and sample are similar. Note that the quantitative data are not affected by the mismatch of sample and reference masses; the calculated enthalpy of fusion for this pharmaceutical material is constant.

Figure 2.9 shows an expansion of the early part of the data from a DSC experiment made at 10°C/min. The actual heating rate has been calculated from the derivative of the sample thermocouple at 30-sec intervals. For this system (specific DSC instrument, sample pan, purge gas, etc.), the experimental heating rate and heat flow signal have stabilized after around 2.5 min. This time is typical for experiments using general or hermetic pans containing a few milligrams of sample. The 2.5-min stabilization time is a good rule of thumb, and applies to most general heating rates. As DSC experiments should be started well below the expected onset temperature of the first transition, the "2.5 rule" can be used to determine the maximum start temperature for any experiment. For example, if the first transition in a sample is 55°C, then at a heating rate of 10°C/min the maximum starting temperature would be 30°C (55 – [10 × 2.5]). If a heating rate of 20°C/min is used, the maximum start

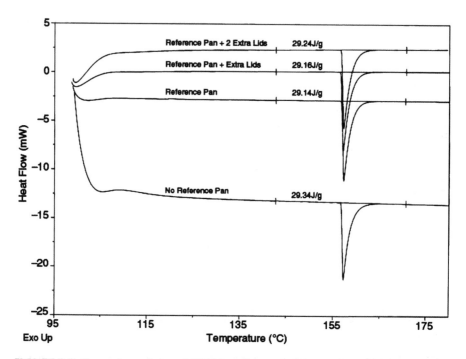

FIGURE 2.8 Comparison of several DSC heat flow curves for a potential drug candidate. No reference pan; reference pan matched to sample pan weight; matched reference pan mass plus an additional aluminum lid; matched reference pan plus two additional aluminum lids. All runs are at 10°C/min. Nitrogen purge gas at 10 ml/min.

temperature would be 5°C. This starting temperature would then require the use of a cooling accessory.

2.4.2 HEATING RATE

Figure 2.10 shows the DSC data for indium as a function of heating rate (not temperature calibrated for each scanning speed). Several points can be observed as the heating rate is increased:

1. The baseline curves are increasingly offset.
2. The height, magnitude and width of the melting peak increases.
3. The melting transition is observed at higher temperatures.

The calculated values from the curves in Figure 2.10 are given in Table 2.3.

Several important considerations and trade-offs in designing DSC experiments result from these observations. As the heating rate is increased, the width of the melting transition increases, and so resolution (the ability to measure transitions that are close together) is decreased. Further, the peak height increases, and so the detection limit (the ability to differentiate a transition above instrumental noise) is

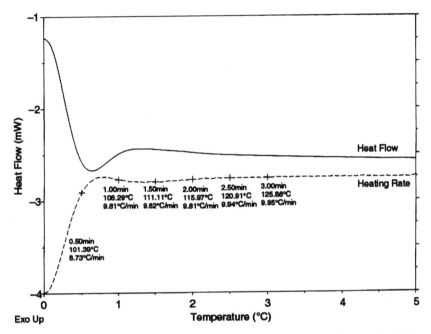

FIGURE 2.9 Heat flow and derivative temperature data for the first 5 min of a DSC experiment on 2.780 mg of indium heated at 10°C/min in an aluminum general pan. Nitrogen purge gas at 25 ml/min.

increased. Finally, the experimental time is decreased. The converse of the above is true if the heating rate is decreased. These conclusions are summarized in Table 2.4.

A compromise heating rate of 10°C/min is a reasonable starting point for most DSC investigations. In the pharmaceutical sciences, where samples often have transitions temperatures that are separated by only a few degrees (e.g., polymorphs), resolution usually wins out over increased experimental time and decreased sensitivity. In these cases, slower heating rates are employed.

2.4.3 DATA HANDLING

With the computerization of DSC instrumentation, additional experimental parameters need to be considered. Data that are stored on the computer consists of an array of time, temperature, and heat flow values passed to it from the DSC microprocessor. The values may be treated or assessed mathematically, and it is important to understand the effect these values will have on the stored heat flow data. It is not possible to cover the specifics of every commercial system here but some general points may be made.

2.4.3.1 Data Sampling Rates

In order to reduce the size of data files, there is usually a software function that permits the selection of either the number of data points to be collected or a specific

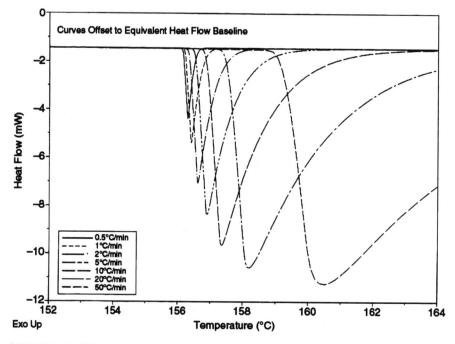

FIGURE 2.10 Effect of heating rate on the DSC data for 2.78 mg of indium in an aluminum general pan. Nitrogen purge gas at 25 ml/min.

TABLE 2.3
Effect of Changing the Heating Rate on the Transition Values for Indium Measured by DSC

Heating Rate	Onset Temperature (°C)	Enthalpy (J/gm)	Peak Maximum Temperature (°C)	Peak Height (mw)	Peak Width at Half Height (°C)	Experimental Time (min)
0.5	156.15	28.76	156.29	−2.948	0.19	200
1.0	156.21	28.70	156.40	−3.939	0.28	100
2.5	156.34	28.77	156.61	−5.609	0.49	40
5.0	156.54	28.75	156.88	−6.935	0.77	20
10.0	156.87	28.74	157.35	−8.204	1.27	10
20.0	157.53	28.64	15.21	−9.114	2.25	5
50.0	159.24	28.78	160.51	−9.756	5.10	2

TABLE 2.4
Summary of the Effect That Changing the Heating Rate Has upon Transitions Measured by DSC

Heating Rate	Resolution	Sensitivity	Experimental Time
Increase	Decreases	Increases	Decreases
Decrease	Increases	Decreases	Increases

TABLE 2.5
The Effect That Changing the Data Sampling Rate Has upon the Measured Values for the Melting of Indium

Data Sampling Rate (points/sec)	Onset Temperature (°C)	Enthalpy (J/gm)	Peak Maximum (°C)	Peak Height (µW)	Peak Half Width at Half Height	Number of Data Points Used in Integration
5.0	156.64	28.79	157.16	−7.915	1.32	420
2.0	156.66	28.77	157.16	−7.822	1.34	140
1.0	156.63	28.77	157.25	−7.827	1.34	84
0.5	156.57	28.22	157.45	−7.414	1.43	42
0.2	156.61	27.99	157.27	−5.981	1.90	17

data-sampling rate. As the number of data points collected is reduced for an experiment, each thermal transition is described by fewer data points. This results in a decrease in the measured enthalpy of DSC peaks, as part of the peak area is missing. The sharper the peak is, the greater the error in the measurement. This is illustrated in Table 2.5, which provides data on the effect that changing the data sampling rate has on the calculated values for the melting transition of a sample of indium.

The effect of reducing the sampling rate from five points per second to one point per second appears to be minimal on the enthalpy calculation. Further reduction in the sampling rate, below one point per second, results in a decrease in the value obtained for the heat of fusion for indium. Other calculated values, such as peak maximum temperature, peak width at half height, and peak height are also affected.

2.4.3.2 Threshold Values

To reduce the amount of data that are stored in the experimental data file, some data handling systems offer the possibility of using "thresholds" during the DSC experiment. In this situation, the previous data point is compared to the current sampled data point. If the temperature or heat flow difference is not significant (when compared to the threshold value) the data are not stored. In essence the system attempts to store only significant data, i.e., data that are different from the previous value.

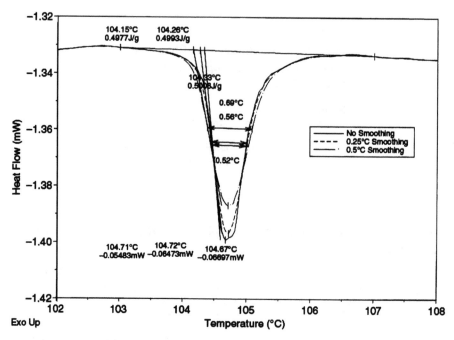

FIGURE 2.11 Comparison of the calculated values obtained as increased smoothing is applied to stored DSC data.

2.4.3.3 Smoothing

The frequency at which the data are stored on the PC may be quite different from the frequency at which the DSC microprocessor samples the data. In order to reduce noise and data storage requirements, these data are likely to be averaged in some way inside the DSC microprocessor. This averaging may be carried out a second time by the instrument control software. The operator usually has control over the amount of smoothing (or filtering) that takes place in the computer prior to storage. This is quite a separate process from smoothing recalled data using a data analysis program. The effect of smoothing the raw data can be seen in Figure 2.11. As the level of smoothing (averaging) is increased, the peak height is reduced, and the peak width increased, leading to a loss of resolution. The onset temperature for the transition is also moved to lower temperatures. Very high smoothing levels also lead to a reduction in the calculated enthalpy of the transition. Of course, the level of noise in the DSC heat flow signal is also reduced.

2.4.3.4 Purge Gas and Flow Rate

The majority of DSC experiments are performed under an inert atmosphere, usually nitrogen. This creates an inert and easily reproducible atmosphere around both the sample and reference during the experiment. The purge gas should pass through a drier to remove any moisture prior to entering the DSC. This is especially important if working at subambient temperatures, as the moisture will condense and lead to corrosion of the DSC with time. It should also be noted that if the purge gas conditions are changed the system should be recalibrated.

2.4.4 TEMPERATURE MODULATION

The development of modulated temperature differential scanning calorimetry (MTDSC) has introduced additional practical considerations into the DSC experiment and is discussed in more detail in Chapter 4. In a typical MTDSC experiment, a periodic perturbation is added to the temperature programmer, typically a sine wave. This has the effect of producing a modulated heating rate, which leads to a modulated heat flow signal being produced as the output from the DSC. Rearranging Equation 2.1 gives:

$$\text{Heat Capacity} = \text{Heat Flow/Heating Rate}$$

or for MTDSC experiments

$$\text{Heat Capacity} = \text{Modulated Heat Flow/Modulated Heating Rate}$$

The heat capacity component can then be removed from the averaged modulated heat flow to produce a kinetic heat flow signal. The heat capacity and the kinetic heat flow signals are often described (respectively) as the "reversing" and "nonreversing" contributions to the traditional heat flow signal, as measured in traditional DSC experiments.

An additional calibration constant is required for accurate MTDSC experiments; this is the heat capacity calibration. The heat capacity constant is calculated as the ratio of the theoretical heat capacity of a known standard to the measured heat capacity of the material. The heat capacity constant is sensitive to changes in the modulation conditions, especially the frequency of modulation.

Better heat transfer is achieved if helium is used as a purge gas. The heat capacity constant is very sensitive to small changes in the flow rate of helium and so helium should only be used as a purge gas if mass flow controllers are used to control the flow rate.

Finally, the temperature modulation conditions should be selected so that there are at least six cycles across the width of the transition. Otherwise artifacts will be introduced into the deconvolution of the modulated heat flow into the reversing and nonreversing signals. This invariably leads to the operator having to use slow heating rate (1–3°C/min) when analyzing pharmaceutical materials. A more detailed discussion on the selection of appropriate experimental conditions can be found elsewhere (1–3).

2.5 SCREAM

Haines (4) has suggested a useful acronym to remember the key parameters that are important in making good general thermal analysis experiments. The author has taken the liberty of modifying his acronym SCRAM (expanded to follows) to include some minor additions specific to making good DSC experiments:

Sample — form, mass, particle size, pan, etc.
Calibration — baseline, temperature, heat flow, heat capacity
Reference — matched to the sample mass and pan
Experimental Method — heating rate, start temperature, data handling considerations
Atmosphere — purge gas and flow rate
Modulation — amplitude, period, underlying heating rate, number of cycles under a transition

2.6   ESTABLISHING VALIDATION TESTS

It is important to have some validation tests when things start to (or appear to) go wrong. These validation tests can be used to check the calibration and performance of the DSC system. They should be run on a periodic basis, and the results kept as an historical log of the DSC system performance. There are three validation tests that are advisable to run on a regular basis: baseline, calibration, and performance. The data from the validation tests are invaluable for assessing system performance when a strange or erroneous result presents itself.

For example, it may be that the operator is running the same QC test on a regular batch of product and one day obtains an unexpected result. Maybe one sees an extra peak, or the material melts at a few degrees higher. Before it can be concluded that the sample is out of specification, the experiment should be repeated. Then, it is necessary to ascertain whether the results from validation experiments are consistent with stored historical data. If the sample data is still reproducible after this, the sample can be considered to have failed specifications.

2.6.1   BASELINE VALIDATION

Obtaining a simple baseline experiment over the temperature range that is used for sample work is a quick way of ensuring that the DSC is not contaminated. Simply record the no-sample baseline with the purge gas flowing and any cooling accessories running that would normally be used. With a new DSC (or after replacing the DSC sensor/cell/head) there will often be some "bedding in" of the system. However, as the DSC ages there will only be some slight changes in baseline performance. Additional peaks or bumps on the baseline are indicative that the DSC needs cleaning. If these persist after cleaning (follow the specific instructions from the manufacturer), look at the gas lines and any cooling accessory as a possible source of the problem. If validation records for the system are maintained, look for items that changed since the last good baseline was obtained. Maybe the DSC system has been moved or the purge lines have been changed, or a new operator has recently used the DSC.

Typical DSC baseline data from room temperature to 300°C at 5°C/min are shown in Figure 2.12. Here, the heat flow data has been plotted against temperature,

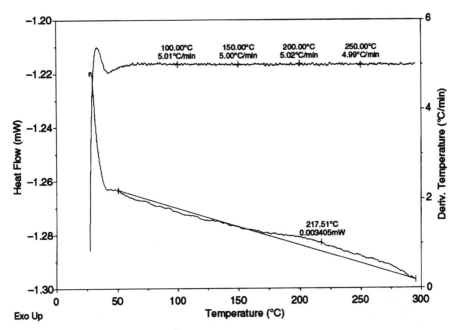

FIGURE 2.12 Typical validation DSC baseline data obtained from ambient to 300°C at 5°C/min. Empty general pans were placed in both the sample and reference positions. Nitrogen purge gas at 25 ml/min. Heat flow and derivative sample temperature (heating rate) plotted against temperature.

together with the time derivative of the temperature signal. Although there is some baseline slope, the curvature is less than 3 μW, and there are no spurious peaks that would indicate contamination. The value of the heating rate at 50°C intervals has also been calculated from the derivative of the temperature signal to validate the accuracy of the temperature programmer. An area of baseline region may also be expanded so that the peak-to-peak noise may be calculated. The exact values here are not critical. It is more important to observe any changes in the curves that occur over time.

2.6.2 CALIBRATION VALIDATION

Having established that the baseline performance of the DSC is acceptable, the next step is to run a known calibration material (invariably indium) to check the temperature and enthalpy calibration. The author is aware of at least one laboratory that has used the same piece of indium for over a year and obtained excellent validation data (5). The indium sample is run at the beginning of every week with the DSC uncalibrated (i.e., measuring signals that are not processed by the temperature calibration algorithm in software), and graphs are plotted of the onset temperature and enthalpy against time. A reduction in the calculated enthalpy value (in the absence of any calibration software) is indicative of a change in the sensitivity of DSC.

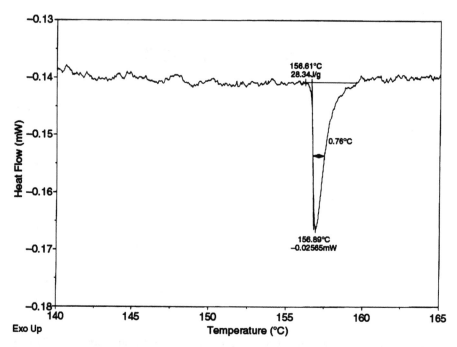

FIGURE 2.13 DSC data for 5 μg of indium heated at 10°C/min. Crimped general DSC pans and nitrogen purge gas at 20 ml/min.

Some laboratories also choose to run a very small mass of indium, as this permits the assessment of signal to noise under normal operational conditions. Figure 2.13 shows such indium validation data for a 5-μg sample.

2.6.3 PERFORMANCE VALIDATION

Pharmaceutical scientists often raise concerns that indium is not representative of the type of sample that they would typically run on a day-to-day basis, hence the need for a performance standard. This is a sample where the thermal behavior is known, the sample can be obtained in high purity, and ideally has two or more peaks close together so resolution can be measured. If such a material exists as one of a company's own product, then this may be an acceptable performance validation material for that organization or laboratory.

One performance validation material that is often used is 4,4′-azoxydianisol, which undergoes a solid-to-liquid crystal (anisotropic) transformation at around 115°C, followed by an anisotropic to isotropic conversion at around 134°C. 4,4′-Azoxydianisol can be thermally cycled and used repeatedly. This sample enables resolution, sensitivity, and signal and noise to be established quickly on any DSC system. Figure 2.14 shows the data from a single 4,4-azoxydianisol sample, but displayed at two different sensitivities, so that small transitions at 135°C can easily be seen. This material has been suggested as a useful sample to compare the performance of various DSC systems (6).

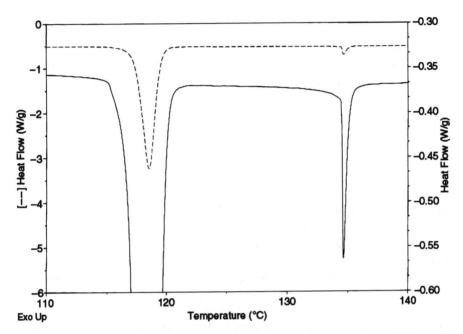

FIGURE 2.14 DSC data for 2.64 mg of the liquid crystal azodianisol heated at 10°C/min in an hermetic pan with lid inverted. Nitrogen purge gas at 25 ml/min. Data are shown full scale (left-hand axis) and 20 times expansion (right-hand axis).

It is recommended to run these validation tests on a regular basis, perhaps weekly, and the results archived for future comparison and reference. These experiments do not take too much time (about an hour of instrument time) and the process can be run overnight if the DSC has an autosampler. When a new operator is assigned to running the DSC, ensure that they are competent in running these validation tests prior to analyzing other laboratory samples.

2.7 REPORTING RESULTS

Detailed suggested guidelines for reporting results can be found in *For Better Thermal Analysis* produced by the International Confederation for Thermal Analysis and Calorimetry (7). A summary of the key parameters for DSC experiments is given here for convenience:

All DSC results must be accompanied by the following information:

- Identification of all substances by definitive name, formula, or equivalent compositional data
- A statement of the source of all substances, their history, and purity if known
- A statement regarding how the equipment has been calibrated
- A statement of the exact temperature program used
- Identification of sample atmosphere, flow rate, and purity

- A statement of the dimensions of the sample pan
- Identification of the apparatus used by type, manufacturer, and model number
- The sample weight (preferably before and after the experiment) and any sample pretreatment

2.8 REGULAR TASKS

Instead of "Regular Tasks," this section could equally as well have been titled "Regulatory Tasks," as there is an increasing demand for improving and documenting "best practice" within the pharmaceutical laboratory. Each laboratory needs to define and implement its own approach and, hopefully, this section may serve as an *aide memoir* for the academic and industrial worker.

Two log books should be maintained for the DSC system. The first should contain calibration and validation data, and the second should contain a record of all service and maintenance visits, together with a log of any changes made to the system. Such changes include software or firmware upgrades and relocation of the equipment to another laboratory, as well as any parts replacement.

After any changes, always perform a complete set of calibration and validation experiments before starting on any sample work. Detailed below are the typical regular tasks for an active laboratory:

Daily
    Indium calibration (temperature and enthalpy)
    Balance calibration check
Weekly
    Full temperature range calibration
    Enthalpy calibration
    Baseline calibration
    Check flow meters and gas bottles
    Back-up data and methods
Monthly
    Clean DSC cell
    Review any calibration "Drift"
    Check gas line drying agents
Annually
    Preventative maintenance visit

REFERENCES

1. Reading, M. and Hourston, D.J., *Modulated-Temperature Differential Scanning Calorimetry: Theoretical and Practical Applications in Polymer Characterisation*, Springer, Dordrecht, The Netherlands, 2006.
2. Hill, V.L., Craig, D.Q.M., and Feely, L., *Int. J. Pharm.* 190, 21, 1999.

3. TA Instruments Inc., Modulated DSC® Compendium: Basic Theory and Experimental Conditions, Literature Piece TA-210, New Castle, DE, 1994.
4. Haines, P.J., *Thermal Methods of Analysis: Principles, Applications and Problems*, Blackie Academic and Professional, Glasgow, 1995, p. 19.
5. Willcocks, P. and Luscombe, I.D., ICI Wilton, Middlesborough, U.K., personal communication, 2001.
6. van Ekeren, P.J., Hol, C.M., and Witteveen, A.J., *J. Therm. Anal.* 49, 1105, 1997.
7. Hill, J.O., Ed., *For Better Thermal Analysis and Calorimetry*, 3rd ed., ICTAC, Vancouver, Canada, 1991.

SUGGESTED FURTHER READING

Brown, M.E., Ed., Differential thermal analysis and differential scanning calorimetry, in *Handbook of Thermal Analysis and Calorimetry: Principles and Practice*, Vol. 1, Haines, P.J., Reading, M., and Wilburn, F.W., Elsevier, Amsterdam, 1998, chap. 5.

Brown, M.E., *Introduction to Thermal Analysis: Techniques and Applications*, Chapman and Hall, New York, 1998.

Charsley, E.L. and Warrington, S.B., Eds., Pharmaceutical applications of thermal analysis, in *Thermal Analysis: Techniques and Applications*, Hardy, M.J., Ed., Royal Society of Chemistry, London, 1992, chap. 9.

Dodd, J.W. and Tonge, K.H., *Thermal Methods*, ACOL/Wiley, London, 1987.

Ford, J.L. and Timmins, P., *Pharmaceutical Thermal Analysis*, Ellis Horwood, Chichester, 1989.

Haines, P.J., *Thermal Methods of Analysis: Principles, Applications and Problems*, Blackie Academic and Professional, Glasgow, 1995.

Hatakeyama, T. and Quinn, F.X., *Thermal Methods*, Wiley, Chichester, 1994.

McNaughton, J.L. and Mortimer, C.T., *Differential Scanning Calorimetry*, Vol 10 of IRS Physical Chemistry Series 2, Butterworths, London, 1975.

Wendlandt, W.W., *Thermal Analysis*, 3rd ed., John Wiley & Sons, New York, 1986.

Wunderlich, B., *Thermal Analysis*, Academic Press, New York, 1990.

Widman, G. and Reisen, R., *Thermal Analysis: Terms, Methods, Applications*, Huthig, Heidelberg, 1987.

3 Pharmaceutical Applications of DSC

Duncan Q.M. Craig

CONTENTS

3.1 INTRODUCTION

In this chapter, some of the considerations relating to the use of differential scanning calorimetry (DSC) for the study of pharmaceutical systems will be outlined. This is an extremely large topic in its own right, and it is not logistically feasible to list every study that could be placed in this category. On that basis, the discussion has been confined to the most widespread uses of DSC within the pharmaceutical sciences, namely polymorphism, pharmaceutical hydrates, and glassy systems. For a description of many of the pharmaceutical systems not covered in this chapter, the reader is referred to the following general texts. First, the book by Ford and Timmins (1) provides a thorough account of pharmaceutical systems that have been studied using thermal techniques. Second, the special issues of *Thermochimica Acta* (2,3) specifically dealing with pharmaceutical thermal analysis are recommended as these provide a comprehensive account of a selection of the topics covered here. During the course of the chapter, further texts in which more information on specific subjects is available will also be recommended.

Virtually all pharmaceutical companies of substantial size will own a DSC, whereas the majority of academic pharmaceutical departments will either own one or have access to it. As far as the pharmaceutical industry is concerned, the breadth and extent of use of this instrument may vary considerably; some will only use it as a routine screening for polymorphism whereas others will have researchers dedicated specifically to thermal and related characterization techniques. The level of sophistication with which the technique is used varies accordingly, with some companies using the instrument almost entirely as a screening tool with little or no specific data interpretation, whereas others will use the method for more advanced predictive studies such as characterization of the glass transitions of freeze-dried systems. Similar variation may be found in academic departments, with a limited number of groups exploring the use of DSC as a subject in its own right.

There are some broad points that can usefully be made prior to discussing specific examples. In particular, it must be remembered that DSC is a nonspecific technique, measuring the total energy associated with a thermal event at any particular temperature or time interval. While techniques such as modulated temperature DSC provide a means of deconvoluting the processes associated with these events (discussed in Chapter 4), the fact remains that one is almost invariably attempting to derive specific structural information from heat flow data. Such derivation is therefore dependent on the interpretative skills of the operator as the information from the raw data is indirect; this probably represents the single largest disadvantage of the method. The second point is that the use of heating or cooling scans almost invariably means that the sample properties are being measured at a temperature other than that of immediate interest. For example, a study designed to ascertain the solid state structure of a material at room temperature will involve heating the sample to elevated temperatures to allow the structure to be ascertained via the melting point. In other words, the majority of pharmaceutical DSC studies involve extrapolation of the information obtained to another temperature of interest. This is not necessarily a disadvantage as such but, as will be discussed, may require consideration when dealing with multicomponent systems.

Having mentioned these broad limitations, it is also important to emphasize the remarkable versatility of the technique, again as a result of the fact that the method is essentially measuring heat flow processes that are applicable to virtually all systems. Indeed, it could well be argued that the method is underexploited within the pharmaceutical field, and it is hoped that this and the other texts mentioned previously will raise awareness of some of the possibilities that the technique affords which are not currently routine practice.

3.2 THE DETECTION AND CHARACTERIZATION OF POLYMORPHISM

In this section, the specific subject of thermal analysis of pharmaceutical polymorphs is discussed; readers are referred to the thorough text by Giron (4) for further information on this topic. A knowledge of the polymorphic behavior of a drug is an essential facet of dosage form development in terms of product performance (particularly dissolution and bioavailability), stability, and regulatory considerations. As a result of the differences in lattice structure, polymorphic forms will tend to melt at different temperatures, thereby presenting the possibility of using DSC as a means of detecting such forms. Indeed, many examples of this are available in the literature and the detection of polymorphism remains one of the principal uses of DSC within the pharmaceutical industry. If, for example, two melting peaks are observed for drug on a single heating run, then the formulator may justifiably consider this to be an indication of the possibility of polymorphism, although it is by no means definitive proof, for which techniques such as x-ray diffraction (XRD) may be used. Indeed, one of the major drawbacks of using DSC for this application is that the data obtained are often ambiguous, as other phenomena such as hydrate formation may result in the appearance of two peaks, whereas in practice the two forms may not be seen in a single run. In this case, the only evidence for polymorphism may be a relatively small change in melting point compared to other batches. However, it is also possible to use DSC for the characterization of polymorphism with a far greater degree of sophistication than is currently common practice; the principles underpinning such analysis will be outlined here.

3.2.1 THE MELTING OF ENANTIOTROPIC AND MONOTROPIC POLYMORPHS

If a drug has two enantiotropic polymorphic forms, either may be stable depending on the temperature at a given pressure, whereas for monotropic polymorphism only one form is stable irrespective of temperature. The study of polymorphism is further complicated by the fact that both thermodynamic and kinetic factors must be considered, as some metastable forms are sufficiently kinetically stable as to render them extremely difficult to differentiate from thermodynamically stable configurations. Clearly, however, the more detailed the knowledge of the thermodynamic and kinetic behavior of the different forms, the greater the ability to predict the likely behavior on storage.

The difference between enantiotropic and monotropic forms may be visualized in terms of differences in the temperature-dependent free energy relationship between the respective forms. For enantiotropic polymorphs, there exists a unique temperature (T_0) below the melting point of either form at which the free energy of the two is the same. Consequently, above or below this temperature, either one or the other form will be thermodynamically stable. For monotropic polymorphs,

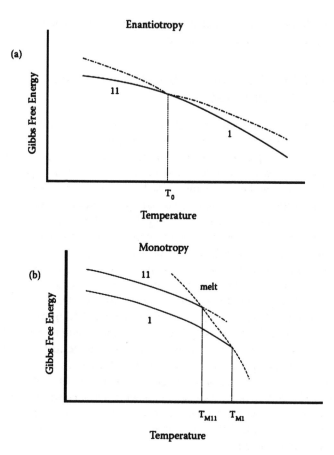

FIGURE 3.1 The relationship between Gibbs free energy (G) and temperature for: (a) enantiotropic and (b) monotropic polymorphic forms (1 and 11). The solid lines indicate thermodynamically stable forms and the dashed line metastable; the long-dashed line represents the free energy of the melt. (Based on Giron, D., *Thermochim. Acta*, 248, 1, 1995.)

however, the free energy of one form is always lower than that of the other, irrespective of temperature, hence, the higher energy polymorph will invariably be metastable with respect to the lower energy system. The free energy relationship with temperature may be summarized with reference to Figure 3.1a and Figure 3.1b, using the nomenclature of Form 1 and Form 11 for convenience where 1 is the higher melting polymorph; for the sake of simplicity, the discussion will focus on transformations between two polymorphs, although it should be appreciated that many materials may exist in several forms.

Figure 3.1a shows the free energy curves for the reversible enantiotropic Forms 1 and 11, with the crossover temperature T_0 representing the temperature at which the free energy values are equal. Consequently, below T_0 the lower-melting Form 11 is stable whereas above it the higher-melting Form 1 is stable. However, owing to kinetic constraints the transformation will not be instantaneous and metastable forms of both 1 and 11 may exist; the dotted lines indicate the free energy of such forms. For the enantiotropic systems, there should theoretically be no melting point for 11 as the transformation temperature T_0 occurs below the intersection with the liquid curve. In practice, however, metastable Form 11 may melt at a temperature above T_0. For irreversible monotropic systems, there is no intersection between the two curves below the melting point (Figure 3.1b); hence, Form 11 will always be metastable to Form 1 form in the solid state. Figure 3.1b also shows the free energy of the liquid state; the melting point is given by the intersection of this curve with the solid-state free energy curves. For the monotropic systems, there is no equivalent crossover in the free energy curves; hence, Form 11 is invariably metastable with respect to Form 1. It is also seen that the metastable Form 11 will melt at a lower temperature than the stable Form 1.

The differences between enantiotropic and monotropic systems also have consequences in terms of understanding and interpreting DSC data. Figure 3.2a and Figure 3.2b show the equivalent enthalpy changes for Form 1 and Form 11; hence, the heat of fusion will be given by the difference in enthalpy between the two solid curves and that of the liquid at the melting point of either form. There are several possible scenarios associated with the transformations that either system may undergo, depending on both thermodynamic and kinetic parameters. Some of these will now be outlined with reference to both the appearance of the DSC peaks (Figure 3.3a and Figure 3.3b) and the associated underlying thermodynamic and kinetic parameters.

For enantiotropic systems (Figure 3.1a, Figure 3.2a, and Figure 3.3), the system may undergo a solid-state transformation from Form 11 to 1 at T_0 (Figure 3.1a) followed by melting at T_{M1} (Figure 3.2a). Examination of Figure 3.2a indicates that this melting transformation will be endothermic and the associated energy will be given by ΔH_{F1}. The appearance of the DSC trace is shown in Figure 3.3, whereby an endothermic transition corresponding to the transformation of Form 11 to 1 is seen at T_0, followed by melting of Form 1 at T_{M1}. If one only considers the thermodynamic stability of the system, this simple behavior should be universal for enantiotropic systems because only one form will be stable above T_0. In practice, however, this may not be the case owing to kinetic considerations resulting in the (now metastable) Form 11 persisting above T_0. In this case, the scenario shown in Figure 3.3b is observed, whereby a single endothermic process is observed corresponding to the melting of Form 11 that has persisted beyond T_0 for kinetic reasons. The enthalpy of this transition will be given by ΔH_{F11} (Figure 3.2a). However, when the metastable Form 11 melts at T_{11}, it is possible that Form 1 will recrystallize from the melt, which will then itself melt at T_1 (Figure 3.3b).

The third scenario (Figure 3.3c) represents the situation when Form 1 is present below T_0, in which case it will be metastable with respect to 11 and may undergo a spontaneous solid state transformation to the latter. Examination of Figure 3.2a indicates that in this case the transformation will be exothermic. Form 11 may then undergo a second solid state transformation to 1 at T_0 that will then melt at T_{M1}. Finally, metastable Form 1 may be present at the outset of the experiment and remain kinetically stable throughout the DSC run; it will then simply melt at T_{M1} (Figure 3.3d).

A similar analysis may be used for monotropic systems (Figure 3.4). In this case, the lower melting point 11 polymorph is metastable to the higher melting 1

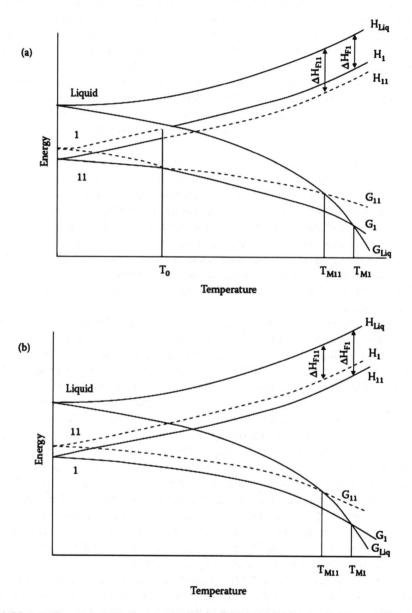

FIGURE 3.2 The relationship between enthalpy (H) and temperature for: (a) enantiotropic and (b) monotropic polymorphic forms (1 and 11). The solid lines indicate thermodynamically stable forms and the dashed line metastable. (Based on Giron, D., *Thermochim. Acta*, 248, 1, 1995; Burger, A., *Acta Pharm. Technol.*, 28, 1, 1982.)

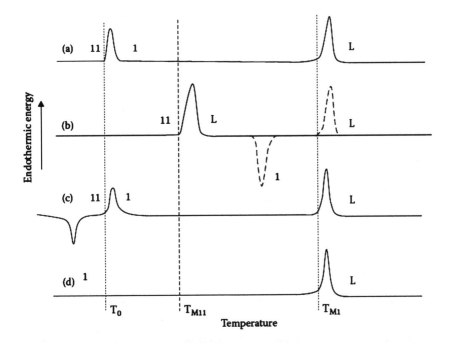

FIGURE 3.3 Schematic DSC traces showing the thermal behavior of enantiotropic systems under four sets of conditions: (a) solid state transformation of 11 to 1 followed by melting of 1, (b) melting of metastable 11 followed by recrystallization to 1, (c) solid-state transformation of 1 to 11 followed by reverse transformation at T_0, with melting at TM1; and (d) persistence of metastable 1 below T_0 followed by melting at TM1. (Based on Giron, D., *Thermochim. Acta*, 248, 1, 1995.)

over the whole temperature range. If the polymorph is in the stable Form 1, only a single melting peak will be observed corresponding to T_{M1} (Figure 3.4a). If, however, the sample is in the metastable Form 11, the scenario outlined in Figure 3.4b may be observed, whereby a kinetically controlled solid state transformation is seen from 11 to 1; examination of Figure 3.2b indicates that this will be exothermic. Form 1 will then melt at T_{M1}. The third possibility is that Form 11 will persist until T_{M11}, at which temperature it will melt and recrystallize into Form 1 that will again melt at T_{M1}. Note that according to Figure 3.2, the heat of fusion of Form 1 is less than that of Form 11 for enantiotropic systems and *vice versa* for monotropic systems.

3.2.2 PRACTICAL CONSIDERATIONS

As outlined in Chapter 2, the conditions used when running DSC experiments may have a profound influence on the signals obtained. The study of polymorphism presents no exception to this, with one particularly important factor being the heating rate. A number of studies have investigated the effects of heating rate on the measured response of systems exhibiting polymorphism. For example, when studying carbamazepine, Krahn and Mielck (5) demonstrated that the use of slow heating rates resulted in the appearance of two peaks, owing to the melt-recrystallization process

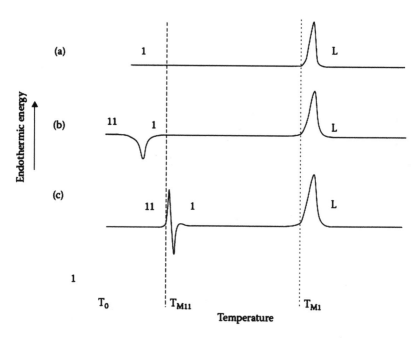

FIGURE 3.4 Schematic DSC traces showing the thermal behavior of monotropic systems under three sets of conditions: (a) melting of 1, (b) solid-state transformation of 11 to 1 followed by melting of 1, and (c) melting of 11 followed by recrystallization and then melting of 1. (Based on Giron, D., *Thermochim. Acta*, 248, 1, 1995.)

occurring within the timescale of the experiment. Similarly, Giron (6), studying the polymorphic transformation of butylhydroxyanisole, showed two endotherms on heating, the appearance and proportion of which depended on the heating rate. At higher rates a single peak was seen at approximately 63°C, whereas at lower rates an additional smaller second peak was seen at approximately 66°C. This was ascribed to the lower heating rates resulting in there being sufficient time for transformation to the higher melting form to take place.

Behme et al. (7) clearly demonstrated the importance of the scanning speed in their study of gepirone hydrochloride, whereby marked differences in the DSC traces can be seen depending on the heating rate (Figure 3.5). In particular, at higher rates the melt-recrystallization of the lower melting into the higher melting form may be clearly visualized, this interpretation being confirmed using hot-stage microscopy. At slower rates, a solid–solid transformation is seen (again confirmed using hot-stage microscopy); hence, it was concluded that at room temperature the metastable form is present that undergoes melt/recrystallization at fast scan rates. At slower rates, however, the system has time to undergo a solid-state transformation into the more stable form, which then melts at 212°C.

A further issue is the purity of the samples. Olives et al. (8) have demonstrated that the addition 5–500 ppm of metals, which are commonly eluted from glass containers to ethanol-water solutions of tolbutamide, can prevent the appearance of polymorph B. Furthermore, the addition of Ca^{2+} to the solution prior to crystallization

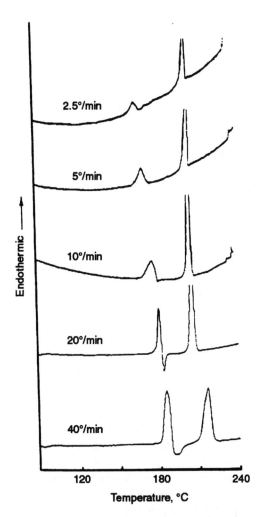

FIGURE 3.5 DSC traces for gepirone hydrochloride run at a range of scanning speeds. (Reproduced from Behme, R.J. et al., *J. Pharm. Sci.*, 74, 1041, 1985. With permission.)

led to what appeared to be a new polymorphic form of the drug. Over and above the clear implications of this study for drug manufacture, the study implies that trace metals may alter the interconversion of polymorphic forms.

The issues of sample encapsulation and residual moisture are also of considerable importance. Both issues are illustrated by the study of Schinzer et al. (9) who examined the interconversions of solvates and polymorphs of the quinolone antibiotic premafloxacin. Form 1 showed a series of transformations on heating (Figure 3.6). More specifically, two melt/recrystallizations were observed, ascribed to conversions to Form 11 and Form 111, respectively, followed by melting of Form 111. However, examination of equivalent samples equilibrated under conditions of elevated humidity using different pan types yielded markedly different results (Figure 3.7). More specifically, the use of sealed pans resulted in the entrapment of moisture,

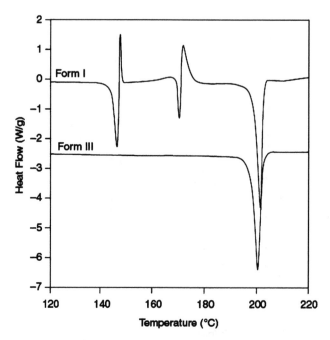

FIGURE 3.6 DSC traces of dry premafloxacin samples encased in perforated aluminum pans. (Reproduced from Schinzer, W.C. et al., *J. Pharm. Sci.*, 86, 1426, 1997. With permission.)

which in turn altered the recrystallization behavior. The authors also emphasized the usefulness of thermal microscopy as a complementary technique for these studies as it was possible to directly visualize recrystallization of the different forms from the melts, thereby preventing confusion with solid-state transformation reactions. In this thorough investigation, the authors describe the use of the additional techniques such as solution calorimetry and microcalorimetry to examine the same systems, thereby demonstrating the value of using a combined approach to better understand the behavior of a complex polymorphic system.

3.2.3 Examples of Drug Polymorphism

A number of texts are available that describe the basic concepts and regulatory aspects associated with drug polymorphism (e.g., 10,11). Similarly, there have been numerous studies describing the identification of both drug and excipient polymorphic forms using DSC, which have been comprehensively classified by Giron (4). The following examples have been chosen on the basis of their exemplifying practical considerations and are therefore only illustrative; there is no intention to provide a comprehensive description of all the studies available within the present text. For further information on more specific aspects of polymorphism, one can refer to texts by Buttar et al. (12) who discuss conformational energy differences between polymorphs, and try Botoshansky et al. (13) who outline a combination of structural (x-ray diffraction) and thermodynamic (DSC) studies of enantiotropic polymorphism of *N*-anilinophthalimide and *N*-(*N*=-methylanilino)phthalimide. Blagden et al. (14)

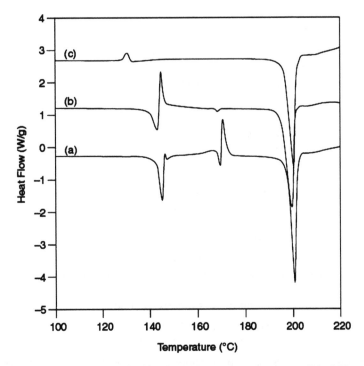

FIGURE 3.7 DSC traces of form 1 premafloxacin: (a) equilibrated at 53% RH and encapsulated in a perforated pan, (b) equilibrated at 53% RH and encapsulated in a sealed pan, and (c) equilibrated at 75% RH and encapsulated in a sealed pan. (Reproduced from Schinzer, W.C. et al., *J. Pharm. Sci.*, 86, 1426, 1997. With permission.)

have used a graph set approach to classify the polymorphs of sulfathiazole in order to examine the solvent-dependence of the polymorphic form generated.

The key issues in the study of drug polymorphism are the identification of different polymorphs (from a regulatory and patent viewpoint as much as from a formulation one), the determination of polymorph stability, and the assessment of processing conditions on polymorph generation. Whereas DSC is widely used as a means of routine screening for polymorphs, it is by no means universally effective as a means of identification. For example, several studies have indicated that DSC shows very similar traces for the various polymorphs of cimetidine (e.g., 15); hence, the generation of apparently identical peaks is not a guarantee of the absence of different crystal modifications. In other cases, however, the differentiation between polymorphs may be more obvious.

In terms of stability, determination of whether the forms are monotropic or enantiotropic is of considerable benefit, as such knowledge will theoretically aid the formulator in determining the likely stability of the material at any stated temperature. In practice, the differentiation between these forms may be a nontrivial task due to the existence of kinetically stable metastable polymorphs. A number of authors (e.g., 4,16–18) have used the guidelines developed by Burger (19), which are summarized in Table 3.1.

TABLE 3.1
Summary of Means of Differentiating between
Monotropic and Enantiotropic Polymorphism, with
"1" Being the Higher Melting Form

Enantiotropy	Monotropy
1 stable > transition	1 always stable
11 stable < transition	11 not stable at any temperature
Transition reversible	Transition irreversible
Solubility 1 higher < transition	Solubility 1 always lower than 11
Transition 11 to 1 endothermic[a]	Transition 11 to 1 exothermic
$\Delta H_F^1 < \Delta H_F^{11}$	$\Delta H_F^1 > \Delta H_F^{11}$ [b]
IR peak 1 before 11	IR peak 1 after 11
Density 1 < Density 11	Density 1 > Density 11

[a] This does not take into account the possibility of persistent metastable forms. If an enantiotropic Form 1 persists to a temperature below T_0, then it may undergo an exothermic transformation to Form 11.
[b] This can be seen from Figure 3.2a and Figure 3.2b.

Source: Adapted from References 4 and 19.

A number of studies have successfully differentiated between monotropic and enantiotropic forms, an early example being that of Burger (20) who studied chlormaphenicol palmitate, which exists in three forms: A(1), which is stable at room temperature but has a poor clinical efficacy; B(11), which is metastable; and C(111), which is highly unstable and difficult to observe. The author reported an exothermic transition corresponding to the conversion of B to A, concluding that the transformation was monotropic. Mitchell (21) studied the interconversion of two forms of anhydrous metoclopramide. The author reported two endothermic transitions on heating Form A, the first (125°C) corresponding to a solid-solid transition and the second (147°C) representing the melting of Form B. Cooling molten Form B resulted in that form being retained (seen by the disappearance of the lower temperature peak). However, on storage over 3 d at room temperature, the lower temperature peak reappeared, indicating conversion to Form A that was more stable at this temperature. The preceding evidence points toward the two forms being enantiotropic in nature (the notation identifying the polymorphs has been changed to avoid confusion with earlier sections). Similarly, Chang et al. (17) studied the thermal behavior of a range of dehydroepiandrosterone polymorphs and solvates. The authors reported three polymorphic forms of the drug and used Burger's rule "that if the higher melting form has the lower heat of fusion, the two forms are usually enantiotropic, otherwise they are monotropic" (22) to assign all three polymorphs as being monotropic.

3.2.4 THERMODYNAMIC ANALYSIS OF POLYMORPHIC TRANSITIONS

A further refinement to the concept of identifying different polymorphic transitions has been the study of the thermodynamic stability of different forms. Yu (23) has used the melting data of two polymorphic forms to calculate the Gibbs free energy difference between the forms over a range of temperatures, thereby allowing extrapolation to storage temperatures. In essence, the free energy difference between the two polymorphic forms is calculated via knowledge of the melting temperature, enthalpy, and heat capacity change on melting from

$$\Delta G_0 = \Delta H_{MI} \, (T_{MII}/T_{MI} - 1) + (C_{p,L} - C_{p,1})[T_{MI} - T_{MII} - T_{MII} \, \ln \, (T_{MI}/T_{MII})] \quad (3.1)$$

where ΔG_0 is the free energy of the process

$$\text{Form } 11 \, (T_{MII}) \rightarrow \text{Form } 1 \, (T_{MII}) \qquad\qquad (3.2)$$

with Form 11 being the lower melting form. The enthalpies, temperatures, and heat capacities are given by ΔH, T, and C_p with subscripts referring to the melting point (M) and physical form (1, or 11, or liquid, L). The corresponding enthalpies and entropy values for the process are given by

$$\Delta H_0 = \Delta H_{MII} - \Delta H_{MI} + (C_{p,L} - C_{p,1})(T_{MI} - T_{MII}) \qquad (3.3)$$

and

$$\Delta S_0 = \Delta H_{MII}/(T_{MII} - \Delta H_{MI}/T_{MI} + (C_{p,L} - C_{p,1}) \, \ln \, (T_{MI}/T_{MII})] \qquad (3.4)$$

The free energy at any temperature T is then calculated via

$$\Delta G(T) = \Delta G_0 - \Delta S_0(T - T_{MII}) \qquad\qquad (3.5)$$

The author also includes consideration of nonlinearity between ΔG and T. By knowing the free energy at any temperature T, it is possible to predict the relative stabilities of the two forms. Furthermore, it is possible to extrapolate the free energy to a temperature whereby ΔG is zero, this corresponding to the transition temperature between the two forms. If this temperature occurs above the melting point of the system, it is monotropic. In this case, the intersection is referred to as the virtual transition point. Yu (23) went on to apply this method to 96 pairs of polymorphs and found a remarkable level of agreement with experimentally derived values of the transition temperature. This approach has also been adopted by Giordano et al. (18) in their study of piroxicam pivalate, the authors calculating a transition temperature of approximately 32°C. Although still at an early stage, this approach does appear to be of considerable interest and practical relevance. A further innovation is the work of Mao et al. (24) whereby the solubility ratio of two enantiotropic polymorphic forms may be calculated based on derivation of the free energy difference between the forms using the temperature and energy of the solid–solid transition. Recently, Yu et al. (25) have adapted the model to measure free energy differences between polymorphs via measurement of the eutectic melt with a common additive.

3.2.5 MIXTURES OF POLYMORPHS

A further important issue is the identification and quantification of mixes of poly-morphs in a single sample. Clearly, only one polymorph will be thermodynamically stable at any given temperature (although strictly speaking, both will be equally stable at the enantiotropic transition temperature). However, the metastable form may be sufficiently kinetically stable so as to be present as a mix with the more stable polymorph. There have been a limited number of examples of such coexistence of forms in the literature. For example, Csakurda-Harmathy and Thege (26) recom-mend the use of DSC as a simple means of identifying the presence of form A chloramphenicol palmitate in samples of B, although clearly this method is best suited to systems that show clearly defined differences in the melting behavior of the constituent polymorphs. Ikeda et al. (27) studied the thromboxane antagonist seratrodast, showing an exothermic solid state transformation process between the two polymorphic forms. The authors used the magnitude of the enthalpy associated with this exotherm to estimate the proportion of polymorphs present, from which an Arrhenius plot for the rate of transformation between the two forms was obtained.

3.2.6 POLYMORPHIC TRANSITIONS INDUCED BY GRINDING
AND COMPRESSION

A number of studies have examined the effects of grinding and compression on the polymorphic form of drugs. Chan and Doelker (28) subjected a total of 32 drugs that are known to exhibit polymorphism to a trituration test in order to examine the extent to which grinding results in polymorphic change. The authors reported that approximately a third of the drugs examined showed such alterations (although, obviously, this figure could differ depending on the trituration method used). Caf-feine, sulfabenzamide, and maprotiline hydrochloride were then chosen as test mate-rials to examine the effects of compaction on polymorphic form. This study is of particular interest as the authors examined the distribution of polymorphs through different surface regions of the tablet, demonstrating variations depending on, for example, whether the upper or lower tablet surface (with reference to the compaction cycle) were examined.

Pirttimäki et al. (29) studied the extent of polymorphic transformation of anhy-drous caffeine as a function of grinding time and compaction pressure using x-ray diffraction, demonstrating conversion of metastable to stable form as a function of these parameters. Similarly, Kaneniwa and Otsuka (30) examined the effects of grinding on chloramphenicol palmitate. The authors isolated and characterized the three recognized polymorphs (A, B, and C) using x-ray diffraction and DSC. The DSC studies showed melting peaks at 90.3°C and 86.7°C for Form A and Form B, respectively, whereas Form C showed an exothermic peak at 64.5°C followed by melting at 86.3°C, implying a monotropic conversion into Form B. Tamura et al. (31) had formerly reported conversion of Form B to Form A on storage at 82°C, indicating that Form B is itself metastable at this temperature. Kaneniwa and Otsuka (30) went on to study the melting point dependence of the three forms on grinding, showing the conversion of Forms B and C to the stable Form A. Figure 3.8 shows

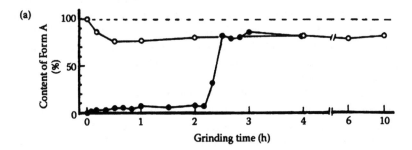

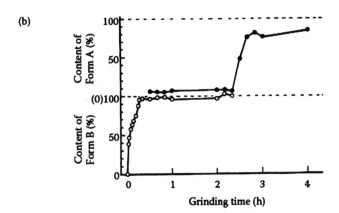

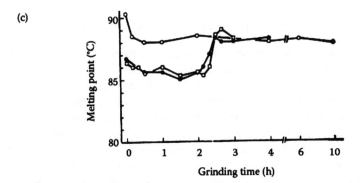

FIGURE 3.8 Interconversion of polymorphic forms of chloramphenicol palmitate induced by grinding: (a) effect of grinding on forms A and B (XRD data) (○ Form A, ● Form B), (b) effect of grinding on Form C (XRD data) (○ content of Form B, ● content of Form A), and (c) change in melting point of forms A (○), B (●) and C (□) on grinding. (Reproduced from Kaneniwa, N. and Otsuka, M., *Chem. Pharm. Bull.*, 33, 1660, 1985.)

the conversion through these forms on grinding, showing that the XRD and DSC studies yield complementary information. Clearly, in this case the XRD studies were able to yield more specific data on the different forms. However, DSC affords the possibility of obtaining thermodynamic data between the various forms; hence, by using the two techniques in conjunction it is possible to obtain a detailed profile of the interconversion process.

In a later study, Otsuka et al. (32) examined the effects of grinding on the polymorphic forms of cephalexin, chloramphenicol palmitate, and indomethacin. Cephalexin was converted into an amorphous form whereas chloramphenicol palmitate tranformed from the metastable B and C forms to the stable A form. Indomethacin was found to transform to an amorphous form on grinding at 4°C but formed a metastable polymorph on grinding at 30°C.

3.2.7 OVERVIEW

The characterization of polymorphism remains a highly important aspect of formulation science, although the extent to which polymorphic behavior may be predicted is still limited. DSC undoubtedly has uses in terms of identification of different forms, although some workers believe that the advantages of using DSC over x-ray diffraction purely for identification lie more in convenience than reliability. It is reasonable to suggest that the real advantage in DSC studies lies in more sophisticated thermodynamic analysis of the data, particularly in terms of characterizing monotropic or enantiotropic behavior, which is not easily achieved using XRD and in identifying the transition temperature for enantiotropic systems. The possibility of being able to predict polymorphic behavior in the manner described by, for example, Yu (23) is also of undoubted interest and may represent a useful future direction for the field.

3.3 PHARMACEUTICAL HYDRATES AND SOLVATES

Solvates (or pseudopolymorphs) are solid adducts containing the parent compound and solvent, whereby the solvent is incorporated into the crystal lattice of the parent substance. If the solvent is water, which is usually the case for pharmaceutical systems, the material is termed a *hydrate*. The properties and biological behavior of such compounds may vary considerably from the anhydrous material; hence, it is essential to develop effective means of characterizing these systems. DSC, along with thermogravimetric analysis and thermomicroscopy (described in Chapter 5 and Chapter 6, respectively) are among the most commonly used thermal techniques in this application, whereas XRD, IR, and other spectroscopic have also been widely used. For more information on pharmaceutical hydrates, the reviews by Haleblian and McCrone (10), Giron (4), and Khankari and Grant (33) are highly recommended. Also, more information is available in the work by Zhu and Grant (34,35), who investigate the role of the activity of water in aqueous cosolvent mixes in determining hydrate formation. It should be emphasized that whereas the following comments are confined to the use of DSC in the study of hydrates, it is highly recommended that several techniques be used in conjunction in order to provide complementary information.

The use of DSC alone raises difficulties in interpretation with respect to differentiating between endotherms due to desolvation and those due to melting. Furthermore, differentiation between a true solvate and sorbed solvent may not always be straightforward. One simple way of distinguishing the process of solvent loss from melting is to simply cool the sample down and reheat; if solvent has been lost from the pan, then there should not be an equivalent exotherm on cooling nor should there be an equivalent endotherm on reheating. This assumes that the sample may be temperature-cycled without extensive decomposition, and that the material will not undergo extensive supercooling; if the substance forms a glass, the crystallization and remelting peaks may also be expected to disappear. Differentiating between surface water and water of crystallization is less simple. One method is to store the sample under different humidities prior to measurement. If there is a reasonable relationship between storage humidity and loss peak size, then this would indicate the water is adsorbed rather than present within the crystal lattice in stoichiometric proportions. However, this carries the assumption that storage of the material under different humidities will not change the hydrate form, which may well not be the case. A variation on this theme is the approach of Brown and Hardy (36) who lowered the temperature of solid samples to subzero temperatures to ascertain whether trace amounts of ice could be detected, which would in turn indicate "clustered" water.

3.3.1 THERMAL PROPERTIES OF SOLVATES AND THEIR MEASUREMENT

The desolvation of a material can result in one of the following three outcomes: the remaining crystal lattice may remain essentially unchanged compared to the solvate, the dehydrated form may be partially crystalline or amorphous and, finally, the dehydrated form may have a lattice that is different from the original material (37). Consequently, there are a number of scenarios that may be observed when heating a solvate in a DSC (4). A typical DSC trace for a hydrate may involve an endothermic peak corresponding to the loss of water and a further endotherm at higher temperatures corresponding to the melting of the anhydrous form, assuming that no decomposition takes place prior to melting. The temperature at which the water may be lost can vary considerably and may occur well above 100°C. Alternatively, the desolvation process may occur at or above the melting temperature, in which case an exotherm may be seen as the anhydrous form recrystallizes from the melt (or possibly from an amorphous form). A further consideration is the possibility of different polymorphic forms of solvates, i.e., the solvates are chemically identical but exist in different unit structures; these are discussed in more detail in a later section.

All the normal considerations associated with DSC measurements are applicable to the study of hydrates, although particular care must be given to the choice of pan used. The use of hermetically sealed pans often leads to suppression of the water loss peak because of the buildup of headspace pressure within the vessel, which in turn alters the equilibrium of the dehydration or evaporation process. In this respect, several authors have recommended the use of open pans to study hydrates as the water loss peak is clearly seen under these circumstances. A further possibility is the use of pin-holed pans, which are now commercially available, whereby water loss through a small hole drilled in the pan lid allows the evaporation to take place in a reproducible manner. The use of pin holes represents a slightly less crude approach than the use of open pans and also avoids many of the problems of baseline drift associated with the latter. Crimped pans may also be used, although our own experience is that the reproducibility may be inferior to pin-holed pans. Furthermore, Giron (4) has emphasized the importance of considering particle size when determining the dehydration onset.

A final experimental condition that has not yet received widespread consideration within the pharmaceutical sciences is the flow rate of the purge gas. A study by Kitaoka et al. (38) discusses marked differences in the melting behavior of dehydrated levofloxacin hemihydrate, depending on the experimental conditions used (described in more depth in the following section). These differences may be ascribed to a number of factors including the efficiency of removal of water of dehydration and differences in the thermal conductivity of the purged cell. Irrespective of the explanation, the choice of purge-gas flow rate (and indeed the gas itself) is an experimental parameter that needs to be considered.

3.3.2 EXAMPLES OF STUDIES INVOLVING PHARMACEUTICAL HYDRATES AND SOLVATES

Pharmaceutical examples of studies involving solvates are often linked with investigations into polymorphism; hence, there is some overlap with the previous section. In addition, the chapter on thermogravimetric analysis in this book also deals with this subject. The examples given here have been chosen specifically on the basis of their illustrating the use of DSC, rather than outlining the properties of pharmaceutical hydrates in general.

3.3.2.1 Drugs as Hydrates

A number of studies have described the use of DSC for the study of drug hydrates. For example, Zhu et al. (39) have investigated the hydrate forms of nedocromil magnesium. In particular, these authors have examined the pentahydrate, heptahydrate, and decahydrate using a wide range of complementary techniques. The pentahydrate showed two distinct dehydration steps corresponding to two binding states of the incorporated water, whereas both the heptahydrate and decahydrate showed a small dehydration step at approximately 50°C prior to the principal loss process (Figure 3.9). This study also serves to illustrate the importance of the choice of pan type, as can be easily visualized by comparing the systems in open and crimped pans.

McMahon et al. (40) have studied carbamazepine, which may be present as two enantiotropic polymorphs and as a dihydrate. The authors compared the properties of dihydrates produced from Form 1 and Form 111 carbamazepine in order to investigate the influence of the choice of the anhydrous polymorphic form used to produce the dihydrate (by suspension in distilled water). The authors reported that holding the sample at 25°C in a TGA resulted in the generation of anhydrous form 1, irrespective of the source of the dihydrate. DSC studies, however, indicated

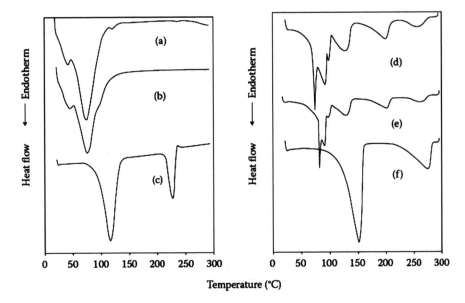

FIGURE 3.9 DSC curves of nedocromil magnesium hydrates: (a) heptahydrate (open pan), (b) decahydrate (open pan), (c) pentahydrate (open pan), (d) heptahydrate (crimped pan), (e) decahydrate (crimped pan), and (f) pentahydrate (crimped pan). (Reproduced from Zhu, H.J. and Grant, D.J.W., *Int. J. Pharm.*, 135, 151, 1996. With permission.)

differences in the postdehydration profile depending on the source of the dihydrate, as shown in Figure 3.10 where dihydrate 111 and 1 refer to dihydrates formed from Form 111 and Form 1, respectively.

The authors suggested that the lower temperature endotherm observed for dihydrate 111 was due to initial formation of Form 111 followed by transformation to Form 1 at ~160°C. However, no discernible differences were detected between the two dihydrates using Raman spectroscopy and XRD. The authors suggested that the differences in structure following dehydration may be a result of persistence of small amounts of anhydrous material in the dihydrate, which may then "seed" the formation of the original polymorphic form, depending on environmental conditions. In environments whereby water is held in relatively close proximity to the solid such as in the (pin-holed) DSC pans, the water drives the reversion of the carbamazepine to the original polymorphic form. When water is lost relatively freely, however, such as in the TGA, the water is not present to facilitate the conversion, and Form 1 is favored in both cases. The proposed relationship between the different forms and dihydrate are shown in Figure 3.11. The study therefore provides a valuable insight into the possible effects of thermal and structural history on the behavior of solvated drugs, with particular emphasis on the possible role of the generation of amorphous material in the determination of subsequent structure. In a recent review, Zhang et al. (41) discuss the effects of processing on transformation between hydrates and polymorphic forms of drugs, including carbamazepine.

Mitchell (21) investigated metoclopramide hydrochloride, which occurs as a monohydrate. The author reported that depending on the conditions used the water

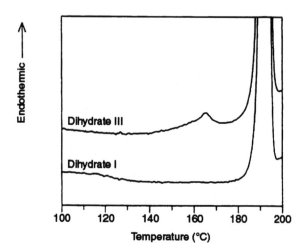

FIGURE 3.10 DSC curves for carbamazepine dihydrate prepared from Form 111 and Form 1. (Reproduced from McMahon, L.E., Timmins, P., Williams, A.C., and York, P., *J. Pharm. Sci.*, 85, 1064, 1996. With permission.)

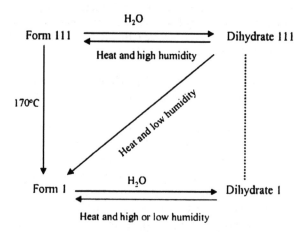

FIGURE 3.11 Proposed relationship between carbamazepine Form 111 and Form 1 and dihydrate. (Reproduced from McMahon, L.E. et al., *J. Pharm. Sci.*, 85, 1064, 1996. With permission.)

of crystallization could be retained above the melting temperature of the solid. This study is also of interest as it explored the effects of experimental parameters (scanning rate, choice of pans, and thermal history) on the thermal response in some detail. Schmidt and Schwarz (42) studied two polymorphic forms of the local anesthetic hydroxyprocaine hydrochloride, linking the thermodynamic stability of the polymorphic forms via free energy diagrams with the tendency to form the hydrate; in particular, the authors noted that the less stable enantiotrope was the more hygroscopic and more likely to form the hydrate. Pudipeddi and Serajuddin (43) have reviewed the literature on the link between polymorphism, hydrate for-

mation, and drug solubility, and in the same paper, more information on this important subject can be found.

3.3.2.2 Hydrates with Fractional Stoichiometries

A number of drugs may exist as hemihydrates or other fractional hydration states.
For example, Kristl et al. (44) studied the pseudopolymorphism of acyclovir, reporting that this drug exists as a 3:2 acyclovir:water fractional hydrate, whereas Mazuel
(45) describes the behavior of norfloxacin that, in the anhydrous form, picks up
water rapidly to form a range of hydrates. These include a sesquihydrate, a hemi-
pentahydrate, a pentahydrate, a hemihydrate, and a dihydrate. Digoxin has been
prepared in the 1/4 H_2O and 1/2 H_2O states by Botha and Flanagan (46) and
characterized using DSC and thermogravimetric analysis (TGA). Similarly, Kitaoka
et al. (38) studied the monohydrate and hemihydrate forms of levofloxacin, the latter
being the pharmaceutically preferred form. Kitaoka et al. (38) reported that dehydration of the hemihydrate under different conditions (gas flow rate and heating
rate) led to marked changes in the melting endotherms (e.g., Figure 3.12), although
the melting of the monohydrate appeared to show little dependence on the experi-

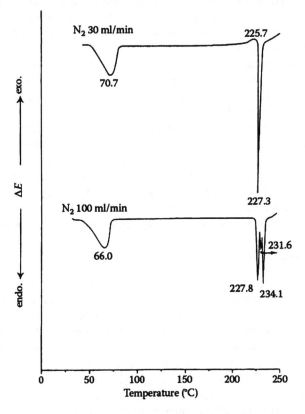

FIGURE 3.12 Effect of N_2 flow rate on the DSC curve of levofloxacin hemihydrate. (Reproduced from Kitaoka, H. et al., *Chem. Pharm. Bull.*, 43, 649, 1995. With permission.)

mental conditions used. The authors attributed these changes to the generation of different polymorphs via partial lattice collapse on dehydration, leading to changes in the nucleation behavior of the solid. Hulsmann et al. (47) studied dispersions of beta-estadiol hemihydrate as a solid dispersion in melt-extruded dosage forms, whereas Dong et al. (48) studied methanolic hemihydrate forms of the artificial sweetener aspartame.

3.2.2.3 Hydrate Polymorphism

In addition to the formation of different stoichiometric hydrate forms, it is also possible that a single hydrate may exhibit polymorphism. This has been described for dioxane solvates of oxazepam (49), hydrates of nitrofurantoin (50,51), and tranilast (52). Jasti et al. (16) studied the polymorphic forms of the cytotoxic agent etoposide. The authors reported that this drug is supplied as a monohydrate (etoposide 1) that converts to the dehydrated form etoposide 1a over a range of temperatures up to 115°C, followed by melting and recrystallization to anhydrous Form 11a at 198 and 206°C, respectively. On cooling from the melt (269°C) under ambient conditions, a second hydrate was formed (Form 11) that remained stable over a 6-month storage period. X-ray diffraction studies indicated that Form 1 and Form 11 exhibited different lattice spacings, whereas Form 1a and Form 11a showed very similar diffraction spectra to the equivalent hydrates. These observations led the authors to conclude that Form 1 and Form 11 represent hydrates of the polymorphs 1a and 11a. The similarity of the lattice spacings for the hydrates and the equivalent anhydrous forms was considered to be compatible with the statement by Garner (37) that the lattice of the dehydrated residue may be nearly identical to the hydrated form, as outlined previously.

Otsuka and Matsuda (50) studied the effects of grinding on the crystalline form of nitrofurantoin, suggesting that the monohydrate Form 1 undergoes a solid-state transition to monohydrate Form 11 on grinding under high-humidity conditions. Similarly, Kawashima et al. (52) describe the formation of three different monohydrates of tranilast, which, on dehydration, yield differing crystal or amorphous forms of the drug. Leung et al. (53,54) describe two forms of aspartame hemihydrate that may be interconverted by ball milling or heating to 160°C in the presence of steam.

3.3.2.4 Nonaqueous Solvates

In addition to hydrates, it is also important to consider the formation of nonaqueous solvates. Clearly, such systems will almost invariably be pharmaceutically undesirable for toxicological reasons. However, their formation and structure must be considered both in terms of the possibility of their generation during the manufacturing process and also desolvation leading to the production of new polymorphs. For example, Caira et al. (55) studied the products of recrystallization of tenoxicam from ethanol and acetonitrile, the former yielding the Form 1 polymorph and the latter, a solvate that on desolvation produced a further polymorphic form. A subsequent study by Caira et al. (56) examined the drug aprazolam, which can exist as a range of solvates depending on the preparation conditions. Comparisons were made

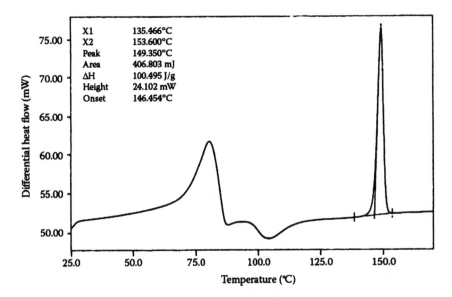

FIGURE 3.13 DSC trace of the methanol solvate of terfenadine. (Reproduced from Hakanen, A. and Laine, E., *Thermochim. Acta*, 248, 217, 1995. With permission.)

between a dihydrate and nonstoichiometric solvates with ethanol and acetonitrile. Interestingly, the DSC scans showed a similar pattern for the three forms, whereas x-ray diffraction indicated that the two nonaqueous solvates were isomorphous. Desolvation, however, led to the formation of the same polymorph from the hydrate and ethanolate but a different polymorph from the acetonitrile solvate.

Another interesting example of nonaqueous solvate formation is given by Hakenen and Laine (57). The authors reported a study into the polymorphic and solvate forms of the H1 receptor antagonist terfenadine. This drug had been reported to exist in at least two polymorphic forms, depending on the solvent of crystallization, with the additional possibility of organic solvate formation. Crystallization from ethanol and acetone gave single peaks at 149 and 153°C, respectively, whereas crystallization from methanol gave a solvate with the profile shown in Figure 3.13. The authors argued that, on the basis of TGA data, the first endothermic peak was due to solvent evaporation, and the ensuing exotherm was due to a structural rearrangement following desolvation. The authors suggested that the evaporation process could be modeled from the DSC data, using the common differential form of an empirical rate equation of order n

$$da/dt = k_T(1 - a)^n \tag{3.6}$$

(where k_T is the rate constant for a reaction with order n). By replacing the da/dt term with $dH/dt \cdot \Delta H^{-1}$, where dH/dt is the differential change in enthalpy and ΔH is the total enthalpy of the transition, it was possible to use linear regression to obtain a value for k_T that, in this case, was determined to be 380 kJ/mole. The authors also used XRD with a hot stage assembly, observing the change in integrated diffraction intensities with time and finding a very similar value for k_T as was noted for the DSC analysis.

3.3.2.5 Pharmaceutical Excipients

In addition to considerations relating to drug hydrates, a number of excipients may exist as hydrates with concomitant effects on their product performance. The two most extensively studied such materials are magnesium stearate and lactose. Given that the latter is discussed in a subsequent subsection, this subsection will be confined to a discussion of magnesium stearate.

Magnesium stearate is extensively used as a pharmaceutical lubricant because of its low cost and high efficacy. However, there are a number of problems associated with its use. In particular, the chemical and physical complexity of commercial batches (58–61) lead to severe interbatch variation problems (62). More specifically, commercial magnesium stearate is a mix of several fatty acid salts that may vary in proportion, the main components being magnesium stearate and palmitate, with magnesium oleate, linoleate, arachidate, and myristate being present in smaller quantities. In terms of the physical structure of commercial magnesium stearate, it is well established that this material may exist in a range of hydrate states, with several authors reporting that commercial batches may also be only partially crystalline (e.g., 63,64). However, specific structural elucidation is difficult when dealing with the chemically inhomogeneous commercial batches. Brittain (64) studied four batches of commercial material using DSC and x-ray diffraction and reported a relationship between the hydrate form and lubricant efficacy. Similarly, Wada and Matsubara (65) studied the DSC responses of 23 batches of commercial material in an attempt to develop the technique as a batch-screening approach.

There has been considerable interest in using DSC (and TGA) to study pure batches of magnesium stearate, with a view to understanding the fundamental properties of the different hydrates and exploring the possibility of using pure materials as commercial products. Ertel and Cartensen (66,67) prepared a dihydrate, anhydrate, and trihydrate form of the material, reporting a single peak for the anhydrate form at 125.1°C and two peaks for the dihyrate and trihydrate, the lower endotherm appearing at 100.1 and 82.9°C, respectively. These two peaks were shown to represent the water loss for the hydrated forms and the melting peak. This seminal study provided the basis for interpreting the thermal behavior of commercial batches, although given the presence of several chemical species in such materials, it remains difficult to define exactly the hydration state of these batches.

Miller and York (68) investigated the thermal and morphological properties of pure batches of magnesium stearate and magnesium palmitate prepared under different pH environments. The authors reported morphological differences for batches prepared in an acid environment from those prepared under alkaline conditions in that the former had a platelike appearance, whereas the latter were more irregular in structure. DSC studies indicated that before drying at 90°C, the DSC traces for alkaline-produced magnesium stearate and palmitate showed four endotherms, the first two being attributed to loss of bound water in two forms (Figure 3.14). This conclusion was supported by TGA studies and by the disappearance of these peaks

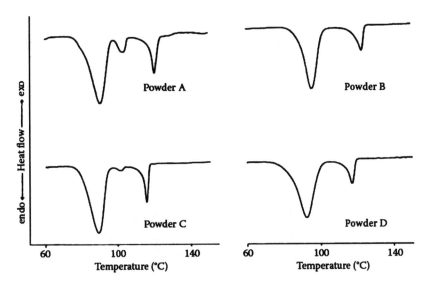

FIGURE 3.14 DSC traces for alkaline produced (A), acid produced (B), magnesium stearate and alkaline produced (C), and acid produced (D) magnesium palmitate. (Reproduced from Miller, T.A. and York, P., *Int. J. Pharm.*, 23, 55, 1985. With permission.)

on drying. Acid-produced magnesium stearate and palmitate, however, showed only one water loss process, indicating that the water is bound in one form only. The authors suggested that, for the alkaline-produced systems, the majority of the water is present in a 2:1 ratio with the fatty acid molecules, with a smaller proportion being present in a 1:1 ratio. Later studies (69) investigated the relationship between the frictional properties and the physicochemical and morphological properties of the batches.

In addition, the authors used the approach of Kissinger (70) to obtain the activation energy of the water loss process. In brief, the peak temperature of the major water loss peak T_m is obtained at a range of scanning rates (N). Log (N/T_m^2) is then plotted against $1/T_m$ to yield a straight line with slope $-E_a/2.303R$, where E_a is the activation energy and R is the gas constant (Figure 3.15).

This study is of interest for several reasons, not the least of which are the demonstration of the complexity of the DSC traces for pure materials (with implications for interpreting DSC traces for commercial batches) and the suggestion that water may exist in more than one bound form in a pure material. This latter point has been further explored by Rajala and Laine (71) who compared two commercial batches of magnesium stearate with a pure grade material, reporting that the two commercial batches showed very different water loss profiles, indicating the presence of a range of binding conformations. The location of the bound water in magnesium stearate samples has been studied by Müller (62) and Sharpe et al. (72). These authors have suggested that the water is present between the intermolecular fatty acid planes, as indicated in Figure 3.16. The differences in the binding of the water is believed to account for the comparative ease of dehydration of the trihydrate compared to the dihydrate.

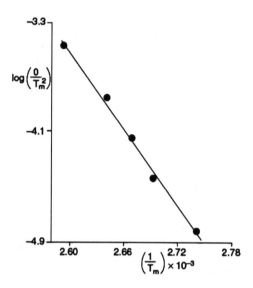

FIGURE 3.15 Plot showing relationship between heating rate N and peak maximum temperature for water loss (T_m) from alkaline-produced magnesium stearate. (Reproduced from Miller, T.A. and York, P., *Int. J. Pharm.*, 23, 55, 1985. With permission.)

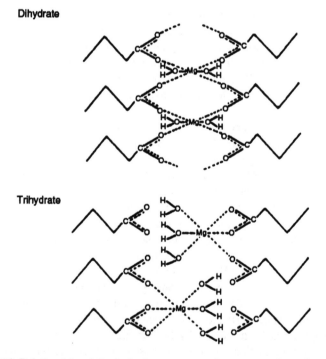

FIGURE 3.16 Relationship of the hydration waters to the stearate planes in the dihydrate and trihydrate phases. (Reproduced from Sharpe, S.A. et al., *Struct. Chem.*, 8, 73, 1997. With permission.)

3.4 PHARMACEUTICAL GLASSES

In recent years, the study of glassy pharmaceuticals has become an issue of increasing importance. This is arguably for several reasons. First, there has been a growing realization that an understanding of the amorphous state is necessary for the optimization of a wide range of dosage forms such as freeze- and spray-dried systems, as well as for many polymeric materials such as polylactic acid microspheres and film coats. Second, amorphous drugs may have improved dissolution characteristics compared to crystalline materials, with the concomitant possibility of improved bioavailability. Third, it is possible to generate amorphous material in otherwise crystalline substances, for example, by grinding or drying. This amorphous fraction may then alter the processing properties and stability of the material. In all these cases, a fundamental understanding of the glassy state is essential to optimize, or indeed prevent, the generation of amorphous material.

As discussed in Chapter 1, this section will place emphasis specifically on DSC studies of amorphous drugs and spray- and freeze-dried systems rather than the theoretical aspects of glass formation and behavior. There will, however, inevitably be a degree of overlap with other chapters owing to the wide range of thermal techniques that are used to characterize such materials.

3.4.1 THE ANALYSIS OF AMORPHOUS DRUGS USING DSC

There have been a number of studies on amorphous drugs and excipients involving DSC. For more information on the nature of glassy pharmaceuticals in general terms, see the reviews by Hancock and Zografi (73) and Craig et al. (74). A number of drugs may be prepared in the amorphous state, and a reference list of many of these has been compiled by Kerc and Srcic (75) (Table 3.2). These range from largely empirical studies to investigations in which detailed analysis of the glassy behavior has been attempted. An example of the latter is the study by Kerc et al. (76) of the thermal properties of glassy felodipine. In particular, these authors placed emphasis on identifying the influence of formation and measuring variables on the DSC traces. For example, the authors examined the effect of increasing the heating rate on the thermal behavior, showing the expected endotherm at higher rates (Figure 3.17). Similarly, these authors examined the effect of heating rate on felodipine recrystallization, using the heat evolution and variable heating rate methods. In brief, the heat evolution method involves assessment of the reaction rate constant at any temperature during the process by consideration of the proportion of the total recrystallization exotherm (ΔH_{tot}) completed at temperature t. The value of k is given by Equation 3.7

$$ k = \frac{dH}{dt} \cdot \frac{1}{\Delta H_{tot} \left(\dfrac{\Delta H_{rest}}{\Delta H_{tot}} \right)^n} \tag{3.7} $$

where (ΔH_{rest}) is nonrecrystallized fraction at time t, and n is the order of the reaction. The authors then attempted to fit the derived data to an Arrhenius plot using values

TABLE 3.2
Glass Transition (T_g) and Melting Point
(T_m) Data for a Range of Pharmaceuticals

Pharmaceutical	T_g/K	T_m/K	T_g/T_m
Glycerol	180	291	0.62
Aspirin	243	408	0.59
Dibucaine	246	336	0.73
Mephenesin	247	340	0.73
Antipyrine	256	380	0.67
Ribose	263	360	0.73
Sorbitol I	270	384	0.70
Methyltestosterone	270	421	0.64
Sorbitol II	271	367	0.74
Phenylbutazone	277	377	0.73
Quinine ethylcarbonate	278	362	0.77
Progesterone	279	399	0.70
Pentobarbital	279	408	0.69
Atropine	281	379	0.74
Ethacrynic acid	282	398	0.71
Citric acid	283	432	0.72
Xylose	283	426	0.66
Tolbutamide	284	403	0.70
Hexobarbital	286	423	0.68
Amobarbital	286	432	0.66
Fructose	286	373	0.77
Tolnaphtate	287	384	0.75
Nimodipine	288	389	0.74
Tartaric acid	289	430	0.67
Flufenamic acid	290	406	0.71
Santonin	290	434	0.67
Ergocalciferol	290	376	0.77
Proxyphylline	295	403	0.73
Acetaminophen	295	447	0.66
Cholecalciferol	296	352	0.84
Paracetamol	297	347	0.86
Eserine	297	378	0.79
Nialamide	297	427	0.70
Chlorotrianisene	298	393	0.76
Chlorimphenicol I	301	349	0.86
Acetaminophen	302	441	0.69
Glucose	303	419	0.72
Nitrendipine	303	429	0.71
Sulfisoxazole	306	460	0.67
Chloramphenicol II	306	414	0.74
Stilbestrol	308	439	0.70
Estradiol-17β-cypionate	309	425	0.73
Dextrose	310	432	0.72

TABLE 3.2 *(Continued)*
Glass Transition (T$_g$) and Melting Point
(T$_m$) Data for a Range of Pharmaceuticals

Pharmaceutical	T$_g$/K	T$_m$/K	T$_g$/T$_m$
Diphylline	315	438	0.72
Phenobarbital	315	452	0.70
Maltose	316	375	0.84
Felodipine	316	407	0.76
Phenobarbital	321	443	0.72
Norethynodrel	324	453	0.72
Quinidine	326	445	0.73
Sucrose	329	453	0.73
Spironolactone	331	478	0.69
Salicin	333	466	0.71
Sulfathiazole	334	471	0.71
Chlormadinone acetate	334	483	0.69
β-Estradiol-3-benzoate	336	472	0.71
Amlodipine besylate	337	467	0.72
Sulfadimethoxine	339	465	0.73
Glibenclamide	344	447	0.77
Dehydrocholic acid	348	502	0.69
Cellobiose	350	498	0.70
Trehalose	350	476	0.74
17β-Estradiol	354	445	0.80
Nicardipin hydrochloride	358	440	0.81
Griseofulvin I	362	422	0.86
Brucine	365	451	0.81
Griseofulvin II	370	497	0.74
Chenodeoxycholic acid	371	436	0.85
Deoxycholic acid	377	447	0.84
Ursodeoxycholic acid	378	477	0.79
Cholic acid	393	473	0.83

Source: Reproduced from Kerc, J. and Srcic, S., *Thermochim. Acta*, 248, 81–95, 1995.

of *n* between 0 and 3, with a view to selecting the value that yielded the most linear plot. The two approaches allowed the authors to estimate the activation energy of the recrystallization process to be in the region of 80 to 140 kJmol^{-1}, depending on the method and experimental conditions used. Interestingly, these authors also studied the influence of glass preparation on the corresponding dissolution behavior, showing that the more rapidly cooled material showed a more rapid release rate. The relationship between the nature of the glassy material and the corresponding drug dissolution rate is an area that to date remains surprisingly unexplored.

A related issue is the effect of additives on the glass transition. This is of importance for two principal reasons. First, it is extremely helpful to have an

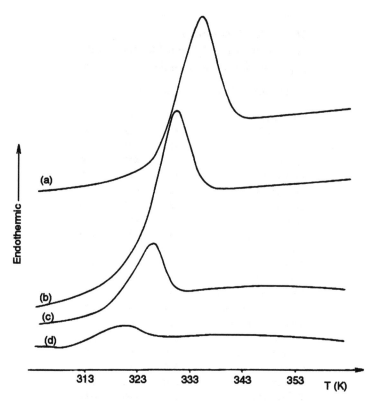

FIGURE 3.17 Heating-rate dependence of the DSC trace for glassy felodipine: (a) 80, (b) 40, (c) 20, and (d) 10 K/min. (Reproduced from Ker, J. et al., *Int. J. Pharm.*, 68, 25–33, 1991. With permission.)

understanding of the effect of sorbed moisture on the glass transition. This issue has been explored in depth by Hancock and Zografi (77), utilizing the Gordon–Taylor and related relationships to examine the T_g-lowering effect of water on a range of excipients. Second, it is important in its contribution to understanding the behavior of binary systems. For example, Timko and Lordi (78) and Fukuoka et al. (79) used the Gordon–Taylor equation to examine the degree of interaction between the components of a range of binary systems including hexobarbital-dextrose, acetaminophen-citric acid, sulfonamide-citric acid, sulfathiazole-citric acid (78), phenobarbital-salicin, and antipyrine-indomethacin (79). Of these systems, only phenobarbital–salicin showed ideal Gordon–Taylor behavior (Figure 3.18); hence, the authors concluded that the components were noninteractive in these mixes.

The stability of glass-forming systems is also a matter of considerable importance. Early studies have explored the association between the glassy behavior and the chemical stability (80), whereas a number of investigations have examined the recrystallization behavior of glassy drugs and excipients. In essence, the increased molecular mobility above the glass transition temperature renders recrystallization

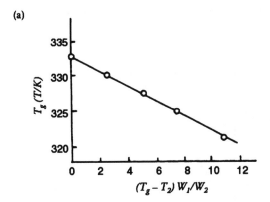

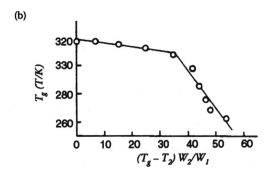

FIGURE 3.18 The variation of the glass transition temperature of binary systems with composition, presented in terms of the Gordon–Taylor relationship (a) phenobarbital-salicin and (b) antipirin-phenobarbital. (Reproduced from Fukuoka, E., Makita, M., and Yamamura, S., *Chem. Pharm. Bull.*, 37, 1047, 1989. With permission.)

considerably more likely, hence, it is essential to have a knowledge of T_g in relation to the storage temperature and to attempt to ensure that the system is maintained below T_g to minimize the risk of physical instability. However, merely knowing the T_g of the dry material is in itself insufficient for two reasons. First, the presence of trace quantities of water may result in a lowering of the T_g to a temperature sufficiently close to (or even below) the storage temperature. Consequently, it is extremely useful to have a knowledge of the water sorption profile of the material and to take appropriate precautions in terms of storage conditions. Second, the system may still exhibit molecular mobility below T_g; hence, storage below this temperature does not guarantee stability. This was demonstrated by Hancock et al. (81), who used measurement of the relaxation endotherm on annealing as a means of determining the relaxation time below T_g (Figure 3.19). These authors recommended that the system be stored at least 50°C below T_g to ensure stability, if possible.

It is possible to make an approximate prediction of the temperature and water content combinations that may or may not lead to instability using the method of Royall et al. (82). These authors describe an adaptation of the Gordon–Taylor

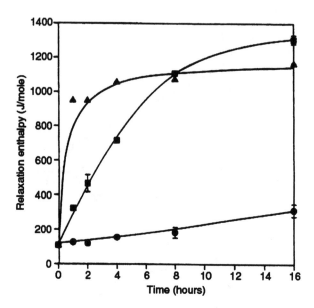

FIGURE 3.19 Variation in the relaxation enthalpy with storage time for indomethacin: ●, T_g = 47, ■, T_g = 32, and ▲, T_g = 16. (Reproduced from Hancock, B.C., Shamblin, S.L., and Zografi, G., *Pharm. Res.*, 12, 799, 1995. With permission.)

equation that allows calculation of the water content w_c that will lead to the T_g of the system being lowered to the storage temperature T_{st} where

$$w_c = \left[1 + \frac{T_{g2}\Delta_2(T_{st} - 135)}{135(T_{g2} - T_{st})}\right]^{-1} \tag{3.8}$$

with Δ_2 and T_{g2} being the density and glass transition of the dry material. A similar expression may be used to calculate the water content w_c' that will result in the T_g falling to a value 50°C above that of the storage temperature, hence, allowing a greater safety margin.

$$w_c' = \left[1 + \frac{T_{g2}\Delta_2(T_{st} - 85)}{135(T_{g2} - T_{st} - 50)}\right]^{-1} \tag{3.9}$$

These expressions may be utilized to devise a phase diagram that shows the combinations of sorbed water and storage temperature resulting in low risk, some risk, and considerable risk of devitrification and hence recrystallization (Figure 3.20).

A further consideration is the study of the generation of amorphous material in otherwise crystalline drugs using DSC. It is now well established that many pharmaceutical materials may contain small quantities of amorphous material that may

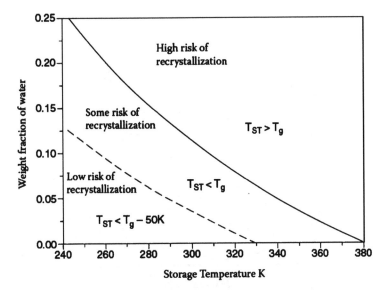

FIGURE 3.20 Diagram showing combination of sorbed water and storage temperature that will lead to varying risks of recrystallization for saquinavir. (Reproduced from Royall, P.G., Craig, D.Q.M., and Doherty, C., *Int. J. Pharm.*, 192, 39, 1999. With permission.)

have a profound effect on the physical properties of that material (83). Depending on the quantity of such material present, DSC may be used to detect and possibly quantify the proportion of glassy material present. Early differential thermal analysis (DTA) studies by Otsuka and Kaneniwa (84) investigated the effects of grinding on the properties of cephalexin, with a broadening of the dehydration peak seen prior to thermal decomposition with increased grinding time (Figure 3.21). X-ray diffraction studies indicated that the grinding process led to an increased level of amorphous material.

More recent studies include those of Mosharraf (85) who used a number of techniques, including DSC, to investigate the generation and properties of partially amorphous griseofulvin. Figure 3.22 shows the DSC response of grisefulvin that has been previously processed by quenching or milling, with the glass transition (with relaxation endotherm) clearly visible prior to the recrystallization endotherm. This work is of particular interest as it not only combines several complementary techniques for the study of a single drug system but also attempts to relate the physical structure of the material to the solubility.

Overall, therefore, there is considerable interest in the study of amorphous drugs and excipients using DSC. In the first instance, knowledge of T_g allows some prediction of the physical stability of the material. However, more sophisticated analysis allows characterization of the effects of plasticizers such as water, the relaxation behavior of the material, and the generation of amorphous material via processing. This last aspect will be considered in more detail in the following two subsections, with particular emphasis on spray drying and freeze drying.

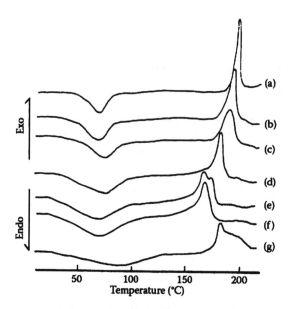

FIGURE 3.21 DTA traces for cephalexin after storage at 20% RH at 35°C for 2 weeks: (a) monohydrate, (b) ground for 3 min, (c) ground for 5 min, (d) ground for 30 min, (e) ground for 60 min, (f) ground for 90 min, and (g) amorphous material. (Reproduced from Otsuka, M. and Kaneniwa, N., *Chem. Pharm. Bull.*, 32, 1071, 1984. With permission.)

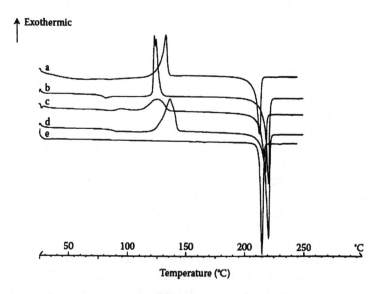

FIGURE 3.22 DSC traces of griseofulvin: (a) quenched, (b) quenched after 9 months storage under ambient conditions, (c) quenched followed by size reduction, (d) quenched (slower solidification), and (e) raw material. (Reproduced from Mosharraf, M., The Effects of Solid State Properties on Solubility and In Vitro Dissolution Behaviour of Suspended Sparingly Soluble Drugs, Ph.D. thesis, University of Uppsala, 1999. With permission.)

3.4.2 THE ANALYSIS OF SPRAY-DRIED SYSTEMS USING DSC

Spray drying is widely used within the pharmaceutical sciences as a means of converting liquids into powder via atomization into a hot air stream. The liquid droplets are dried prior to contact with the walls of the chamber; hence, the solidification process is very rapid and results in the first instance in spherical or near-spherical particles (86,87). Consequently, spray drying may result in profound changes to the physical properties of the material compared to the unprocessed solid form, both in terms of morphology and lattice structure (88). In particular, spray-dried products may be partially or completely amorphous or, alternatively, may result in the generation of a range of polymorphic forms. Examples of the latter include a study by Corrigan et al. (89) on the spray drying of phenobarbitone from ethanolic solution, with the authors reporting the generation of a material that showed characteristics of the Form 111 polymorph after processing in contrast to the commercially available Form 11. However, the majority of DSC studies on spray-dried systems have focused on the generation of amorphous material from the process. Examples of drugs prepared in this way include digitoxin (90), 9.3" diacetylmide-camycin (91), and thiazide diuretics (92).

Yamaguchi et al. (93) studied the effect of spray-drying processing variables on the glassy behavior of 4"-O-(4-methoxyphenyl) acetyltylosin in terms of the stability of the amorphous solid. These authors reported that by changing the inlet temperature of the dryer the glass transition temperature, recrystallization temperature, and recrystallization enthalpy were altered (Figure 3.23), and they suggested that the differences were related to the nucleation profile of the material prepared, using the various protocols. Interestingly, they reported that the physical stability and the dissolution behavior of the drug varied with spray-drying conditions. A further very interesting observation is that the densities of the more stable systems were lower

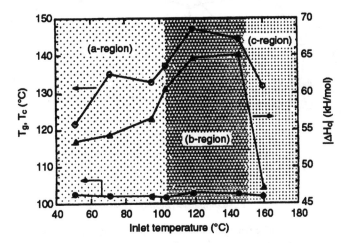

FIGURE 3.23 Comparison of T_g (●), temperature of crystallization (T_c) (○), and enthalpy of crystallization (ΔH_c) (▲) of spray-dried MAT as a function of inlet temperature. (Reproduced from Yamaguchi, T. et al., *Int. J. Pharm.*, 85, 87, 1992. With permission.)

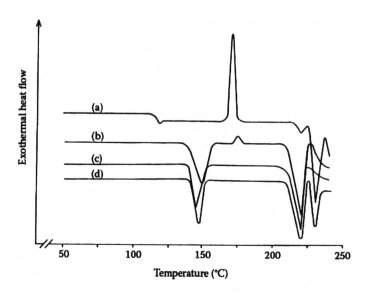

FIGURE 3.24 DSC curves for: (a) 100% amorphous spray-dried lactose, (b) commercial 15% amorphous spray-dried lactose, (c) crystalline α-lactose monohydrate, and (d) crystallized 15% amorphous spray-dried lactose. (Reproduced from Sebhatu, T., Elamin, A.A., and Ahlneck, C., *Pharm. Res.*, 11, 1233, 1994. With permission.)

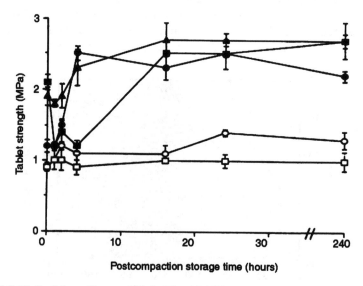

FIGURE 3.25 Radial tensile strength of tablets of 15% amorphous spray-dried lactose, as a function of postcompaction storage at 57% RH. Prior to compaction, the powder was stored for (●) 0 h, (▲) 2 h, (■) 4 h, (○) 6 h, and (□) 336 h at 57% RH. (Reproduced from Sebhatu, T., Elamin, A.A., and Ahlneck, C., *Pharm. Res.*, 11, 1233, 1994. With permission.)

than those that recrystallized rapidly; this may imply that the stable glasses were closer to the equilibrium state than those that exhibited more rapid devitrification.

Sebhatu et al. (94) studied the transformation of amorphous regions of spray-dried lactose on exposure to humidity, suggesting that water uptake occurs preferentially in the amorphous regions. The authors suggest that this uptake may be sufficient to lead to devitrification and recrystallization of the lactose, with concomitant implications for product performance. In this interesting study, the authors compared the thermal behavior of a range of lactose samples, showing the glass transition and relaxation endotherm for the 100% amorphous sample, followed by recrystallization and melting the dehydration of the monohydrate and the melting peak. The commercial 15% amorphous spray-dried lactose showed a dehydration peak and a smaller recrystallization exotherm, reflecting the lower amorphous content. These authors also examined the relationship between the glass transition of the material and the mechanical strength of compacted tablets, an area that has arguably not received the attention it deserves. They found that pre-exposure of 15% amorphous lactose to 57% humidity for 2 to 4 h prior to compaction led to an increase in tablet strength, this being attributed to an increase in plastic flow in the amorphous regions. Postcompaction storage of these samples led to an increase in strength, which was attributed to recrystallization of the amorphous regions, thereby forming solid bridges within the tablet. Longer prior exposure to 57% RH led to weaker tablets, this being attributed to recrystallization prior to tabletting. This hypothesis is supported by the observation that the storage of these tablets did not lead to changes in mechanical strength.

Corrigan (88) has also discussed the use of spray-dried binary systems. For example, polyvinylpyrrolidone (PVP) may be cospray-dried with drugs to alter their physicochemical properties. More specifically, a number of drugs may be prepared in an amorphous form in the presence of PVP, as indicated in Figure 3.26 by the decrease in the heat of fusion with PVP content. Corrigan (88) discussed the applicability of the disruption index (d.i.) suggested by York and Grant (95) and Grant and York (96). This is essentially a means of assessing the influence of incorporating a guest molecule proportionality between the entropy of fusion (or solution) of the material in question and the ideal entropy of mixing. This is given by

$$d.i. = -\delta(\Delta S_f)/\delta(\Delta S_m^{ideal}) \qquad (3.10)$$

where ΔS_f and ΔS_m^{ideal} are the entropies of fusion and ideal mixing. A larger value of $d.i.$ will therefore indicate the degree to which the system has been disrupted. This has been applied to spray-dried systems, as shown in Figure 3.27 for griseofulvin-adipic acid systems (88).

Other studies have included the formation of complexes using spray-drying techniques, including the preparation of cyclodextrin complexes. DSC may be used to determine the extent of drug inclusion within the cyclodextrin ring system, again via monitoring of the decrease in the heat of fusion (97).

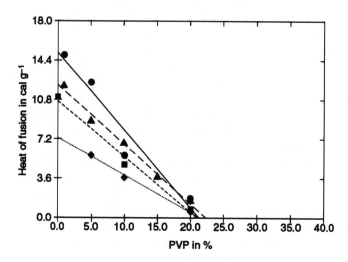

FIGURE 3.26 Relationship between heat of fusion and PVP content for spray dried PVP-drug systems: (●) phenobarbitone, (▲) hydroflumethiazide, (■) cholesterol, and (♦) dipyridamol. (Reproduced from Corrigan, O.I., *Thermochim. Acta*, 248, 245–258, 1995. With permission.)

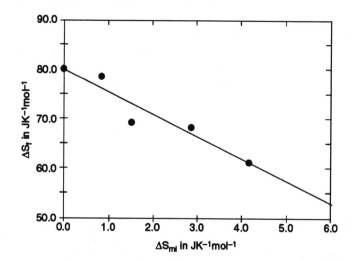

FIGURE 3.27 Correlation between the entropy of fusion (ΔS_f) and the ideal entropy of mixing (ΔS_m) for griseofulvin–adipic acid spray-dried systems. (Reproduced from Corrigan, O.I., *Thermochim. Acta*, 248, 245–258, 1995. With permission.)

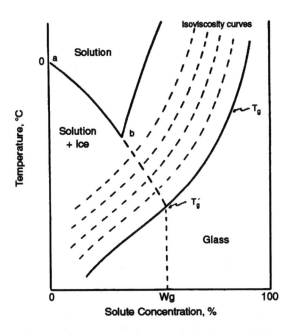

FIGURE 3.28 Schematic representation of the state diagram of a glass-forming binary system. (Reproduced from Nail, S.L. and Gatlin, L.A., in *Pharmaceutical Dosage Forms: Parenteral Medications*, Avis, K.E., Lieberman, H.A., and Lachman, L., Eds., Marcel Dekker, New York, 1993, pp. 163–233. With permission.)

3.4.3 DSC STUDIES OF FREEZE-DRIED SYSTEMS

Freeze drying is a highly important unit operation that allows drying of heat-sensitive drugs and biological material from the frozen state via sublimation of ice. The most common use of the method is for the preparation of injectables in a dry form, which may be easily reconstituted, although other applications such as the preparation of rapidly dissolving tablets have also been described (98). A number of excellent texts are available on the subject of freeze drying (99–105). Nail and Gatlin (99) have summarized the main benefits of the technique, including minimization of chemical decomposition owing to the low temperatures involved, rapid dissolution on reconstitution because of the high surface area of the porous product, and ease of sterilization. The process is widely used for the preparation of proteinaceous injectable systems, whereby the low temperatures employed help to minimize degradation. However, some protein inactivation is usual on freeze drying, hence, it is almost invariably necessary to include cryo- and/or lyoprotectants such as sucrose and trehalose to prevent significant loss of activity.

The process itself involves initial freezing followed by primary and secondary drying. On freezing an aqueous solution, ice forms, which results in an increase in the concentration of solute. The second component may crystallize out as a eutectic with water, or alternatively, if nucleation does not occur within the timescale of the cooling process, then eventually the remaining solution will form a glass with a

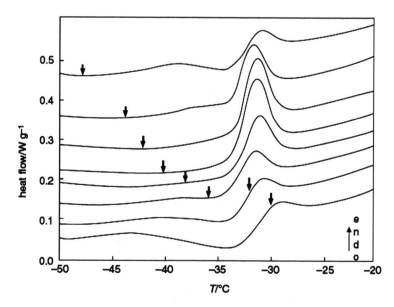

FIGURE 3.29 DSC traces for a series of 40% aqueous sucrose solutions annealed for 16 h at different temperatures. The annealing temperature is indicated by an arrow on each DSC trace. (Reproduced from Ablett, S., Izzard, M.J., and Lillford, P.J., *J. Chem. Soc., Faraday Trans.*, 88, 789, 1992. With permission.)

viscosity sufficiently high (typically ~10^{13} Pa · s) to inhibit further ice formation (Figure 3.28). The characteristics of this glass are of considerable importance in determining the properties of the product.

A number of techniques may be used to study such frozen systems, DSC being among the most widely used. In essence, on ice formation the solution becomes increasingly concentrated until a saturation value is reached (C_g'), whereby the system is said to be in the maximally freeze-concentrated state; the corresponding value for the glass transition is denoted T_g' (Figure 3.28). It is considered essential to maintain the product below this value in order to prevent product collapse. This problem has been discussed in detail (106–110) and is associated with viscous flow of the amorphous material; this phenomenon leads to an inelegant product but may also increase reconstitution times and residual water levels. However, the lower the drying temperature, the higher the cost of the process, whereas, in general, drying below –40°C is not practically feasible (108). More specifically, Pikal (109) has suggested that a temperature increase of 1°C may lead to a reduction in primary drying time of 13%. It is, therefore, essential to dry the material below the collapse temperature (T_c), which, for most practical purposes, is approximately 20°C above T_g' (107).

There has been some discussion regarding the measurement of T_g' using DSC (110). Early studies reviewed in Slade and Levine (111) reported a discontinuity in the heat flow curve of a range of 20% frozen solutions on heating. Later work by Roos and Karel (112,113) and Ablett et al. (114,115), however, suggested that two transitions were apparent in close proximity to each other. They argued that the

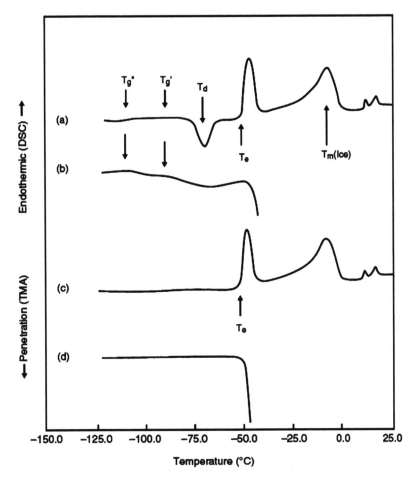

FIGURE 3.30 DSC (a) and TMA (b) traces for 10% calcium chloride solution, showing the glass transitions (Tg', Tg''), the recrystallization process (T_d), the eutectic temperature (T_e), and the melting of ice (T_m) and equivalent traces after annealing at −60°C (c,d). (Reproduced from Chang, B.S. and Randall, C.S., *Cryobiology*, 29, 632, 1992. With permission.)

lower of the two is the true T_g', with the higher-temperature transition representing ice dissolution induced by the increased molecular mobility. This is shown in Figure 3.29, whereby at sucrose concentrations >66% only the glass transition of the maximally concentrated systems is seen (i.e., no ice formation occurs). At lower concentrations, however, the glass transition is followed by a second larger transition that precedes the ice-melting response. However, the nature of higher-temperature transition is still a matter of some controversy, with Blond and Simatos (116) arguing that both transitions are associated with the glass transition, with the higher temperature process overlapping the incipient ice-melting relaxation.

A further consideration with regard to the use of DSC for the measurement of T_g' is that as the temperature is lowered and the viscosity of the system increases, the crystallization process becomes kinetically hindered, leading to lower T_g and

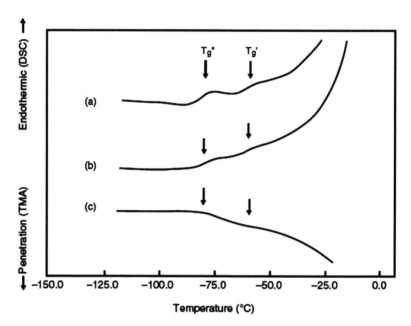

FIGURE 3.31 DSC traces for 10% sodium acetate solution, showing the glass transitions (Tg', Tg'') before (a) and after (b) annealing at –60°C with TMA data (c) for the preannealed systems. (Reproduced from Chang, B.S. and Randall, C.S., *Cryobiology*, 29, 632, 1992. With permission.)

apparent C_g' values. Annealing of the system at temperature immediately at or above T_g' allows crystallization to proceed to completion; hence, the "equilibrium" glassy state is formed; in this context, the term *equilibrium* refers to the limiting behavior of the system, even though by definition the glassy system will be metastable (115). Ablett et al. (114) used this principal to measure T_g' and C_g' via annealing at a range of temperatures. The authors argued that by annealing the sample at T_g', the maximum freeze-concentrated material would be generated as below this temperature recrystallization would not be complete, whereas above this temperature ice dissolution would lower the glass transition. Consequently, the annealing temperature at which the measured T_g is maximal will correspond to the value of T_g'. These authors reported a maximum in T_g at –40°C, which corresponded to a C_g' value of 81.2% sucrose, obtained by inspection of the supplemented phase diagram. More specifically, the equilibrium T_g values are plotted for a range of concentrations of sucrose and C_g'/T_g' obtained from the intersection of this curve with the melting curve.

In association with this approach, Sahagian and Goff (117) used DSC and TMA to study the influence of freezing rate on the formation and stability of the glassy state in freeze-concentrated sucrose solutions. These authors focused on the physical aging of the frozen systems with respect to the enthalpic changes seen on recrystallization. When comparing rapidly and slowly frozen sucrose solutions, the magnitude of the exotherm was found to be considerably greater for the former owing to the less extensive crystallization on cooling. Similarly, the value of the measured T_g was lower for the rapidly cooled systems because of the plasticization effect of the excess unfrozen water. On annealing at a range of temperatures, greater time dependence was seen for the rapidly cooled systems, reflecting the excess enthalpy of these systems as the pseudoequilibrium state is approached. Interestingly, these authors also reported an increase in mechanical resistance past the higher-temperature premelting transition that they attributed to a second-order rearrangement associated with this transition, thus supporting the concept of this thermal event being a result of more than one process.

Chang and Randall (118) studied a range of salts and other excipients in terms of their subambient behavior in aqueous solution, suggesting that the salt solutions may be categorized in three ways. Salts such as disodium phosphate, sodium bromide, and potassium chloride triethanolamine tend to be crystallized rapidly under normal freezing conditions, leading to eutectic formation with water. Such systems therefore form a crystalline matrix in the frozen product that shows no thermal events below the eutectic temperature, using either DSC or thermomechanical analysis, the latter technique being particularly useful in this application as a means of ascertaining the temperature at which collapse may occur. Many salts, however, only partially crystallize during freezing. Aqueous solutions of salts such as calcium chloride, sodium chloride, glycine, and tris base show glass transitions and recrystallization on heating. An example of this is shown in Figure 3.30, whereby thermal events ascribed to glass transitions are seen at −109 and −95°C, followed by recrystallization at −72°C. These materials may therefore be of use if annealing is performed to allow the eutectic to form, as indicated in Figure 3.30c and Figure 3.30d. Finally, salts such as sodium acetate, citric acid, and potassium citrate form amorphous systems on freezing and tend to remain in the glassy state on thawing (Figure 3.31). Collapse will therefore tend to commence at T_g''. Clearly, it is essential to have a knowledge of which type of system is present.

It is also essential to consider the glass transition temperature of the freeze-dried product, particularly with a view to predicting physical and chemical storage stability. The relationship between the physical stability of freeze-dried formulations and T_g must be considered not only in terms of recrystallization but also of product collapse. For example, te Booy et al. (119) showed that storage of sucrose-containing freeze-dried formulations above T_g may result in product shrinkage, collapse, or excipient recrystallization.

In terms of chemical stability, Izutsu et al. (120) and Corveleyn and Remon (121) have discussed the loss of protein stability owing to crystallization of the lyoprotectant. More specifically, it has been suggested that the protective effect of these agents is associated with their being in the amorphous state in the dried product, hence, devitrification and recrytallization leads to loss of activity. Irrespective of recrystallization, storage above or close to the T_g of the product will lead to greater molecular mobility. Duddu and Dal Monte (122) have suggested that the aggregation behavior of a chimeric monoclonal antibody in a sucrose and trehalose formulations may be directly associated with the T_g of these formulations. Interestingly, these authors have suggested that increasing the protein and sugar ratio of the formulation led to an increase in T_g, which, in turn, improved the stability of the product.

A highly interesting development of the use of critical temperatures to predict stability has been presented in a series of papers by the Yoshioka group (123–126).

These authors have discussed the significance of the critical mobility temperature (T_{mc}) at which the lyophilized protein formulation begins to exhibit a Lorenzian relaxation process, this increased molecular mobility being strongly associated with phenomena such as protein aggregation. Indeed, Yoshioka et al. (123) indicated that the Williams–Landel–Ferry (WLF) equation is followed more precisely for such formulations if T_g is replaced by T_{mc}. The T_{mc} values for α-globulin formulations containing a range of excipients were obtained (124), with consideration given to the influence of sorbed water on the T_{mc} and T_g. The authors reported T_{mc} to be approximately 23°C to 34°C lower than the corresponding T_g values, implying that these formulations may exhibit sufficient mobility to lead to degradation well below T_g. These authors therefore argue that the T_{mc} may be more useful than T_g as a means of understanding the relationship between storage temperature and stability of proteinaceous lyophilized formulations.

3.5 CONCLUSIONS

As mentioned at the beginning of the chapter, it is not feasible to cover all aspects of the pharmaceutical uses of DSC in the space available; hence, emphasis has been placed on what are probably the most important applications of the technique. Clearly, DSC may be used as a simple screening method, for example, in the identification of polymorphs or glass transitions. However, there are numerous ways in which the approach can yield a more sophisticated analysis with comparatively little additional effort. Examples outlined here include the predictability of the thermodynamic parameters associated with polymorphic transitions and the assessment of the relaxation behavior of glassy pharmaceuticals. Both these examples have very tangible practical implications for the prediction of stability; hence, such studies are not simply academic exercises but may be of considerable benefit in formulation optimization.

REFERENCES

1. Ford, J.L. and Timmins, P., *Pharmaceutical Thermal Analysis*, Ellis Horwood Ltd., Chichester, 1989.
2. Ford, J.L., Ed., *Thermochim. Acta*, special issue, 264, 1995.
3. Craig, D.Q.M. and Thompson, K., Eds., *Thermochim. Acta*, special issue, 380, 2001.
4. Giron, D., *Thermochim. Acta*, 248, 1, 1995.
5. Krahn, F.U. and Mielck, J.B., *Int. J. Pharm.*, 53, 25, 1989.
6. Giron, D., *S.T.P. Pharma*, 6, 87, 1990.
7. Behme, R.J., Brooke, D., Farney, R.F., and Kensler, T.T., *J. Pharm. Sci.*, 74, 1041, 1985.
8. Olives, A.I., Martin, M.A., del Castillo, B., and Barba, C., *J. Pharm. Biomed. Anal.*, 14, 1069, 1996.
9. Schinzer, W.C., Bergren, M.S., Aldrich, D.S., Chao, R.S., Dunn, M.J., Jeganathan, A., and Madden, L.M., *J. Pharm. Sci.*, 86, 1426, 1997.
10. Haleblian, J. and McCrone, W., *J. Pharm. Sci.*, 58, 911, 1969.
11. Datta, S. and Grant, D.J.W., *Nat. Rev. Drug Discov.*, 3, 42, 2004.

12. Buttar, D., Charlton, M.H., Docherty, R., and Starbuck, J., *J. Chem. Soc., Perkin Trans.* 2(4), 763, 1998.
13. Botoshansky, M., Ellern, A., Gasper, N., Henck, J.O., and Herstein, F.H., *Acta Crystallogr. Sect. B: Struct. Sci.*, 54, 277, 1998.
14. Blagden, N., Davey, R.J., Lieberman, H.F., Williams, L., Payne, R., Roberts, R., Rowe, R., and Docherty, R., *J. Chem. Soc.: Faraday Trans.*, 94, 1035, 1998.
15. Bauer-Brandl, A., *Int. J. Pharm.*, 140, 195, 1996.
16. Jasti, B.R., Du, J., and Vasavada, R.C., *Int. J. Pharm.*, 118, 161, 1995.
17. Chang, L.C., Caira, M.R., and Guillory, J.K., *J. Pharm. Sci.*, 84, 1169, 1995.
18. Giordano, F., Gazzaniga, A., Moyano, J.R., Ventuta, P., Zanoi, M., Peveri, T., and Carima, L., *J. Pharm. Sci.*, 87, 333, 1998.
19. Burger, A., *Acta Pharm. Technol.*, 28, 1, 1982.
20. Burger, A., *Sci. Pharm.*, 45, 269, 1977.
21. Mitchell, A.G., *J. Pharm. Pharmacol.*, 37, 601, 1985.
22. Burger, A. and Ramberger, R., *Mikrochim. Acta*, 2, 273, 1979.
23. Yu, L., *J. Pharm. Sci.*, 84, 966, 1995.
24. Mao, C., Pinal, R., and Morris, K., *Pharm. Res.*, 22, 1149, 2005.
25. Yu, L., Huang, J., and Jones, K.J., *J. Phys. Chem. B*, 109, 19915, 2005.
26. Csakurda-Harmathy, Z. and Thege, K., *J. Therm. Anal.*, 50, 867, 1997.
27. Ikeda, Y., Nishijima, K., Tokai, N., and Nishi, K., *Yakugaku Zasshi: J. Pharm. Soc. Jpn.*, 115, 937, 1995.
28. Chan, H. and Doelker, E., *Drug Dev. Ind. Pharm.*, 11, 315, 1985.
29. Pirttimäki, J., Laine, E., Ketolainen, J., and Paronen, P., *Int. J. Pharm.*, 95, 93, 1993.
30. Kaneniwa, N. and Otsuka, M., *Chem. Pharm. Bull.*, 33, 1660, 1985.
31. Tamura, C. and Kuwano, H., *Yakugaku Zasshi*, 81, 755, 1961.
32. Otsuka, M., Otasuka, K., and Kaneniwa, N., *Drug Dev. Ind. Pharm.*, 20, 1649, 1994.
33. Khankari, R.K. and Grant, D.J.W., *Thermochim. Acta*, 248, 61, 1995.
34. Zhu, H.J. and Grant, D.J.W., *Int. J. Pharm.*, 135, 151, 1996.
35. Zhu, H.J., Ceaninia, Y., and Grant, D.J.W., *Int. J. Pharm.*, 139, 33, 1996.
36. Brown, D.E. and Hardy, M.J., *Thermochim. Acta*, 90, 149, 1985.
37. Garner, G.E., The kinetics of endothermic solid reactions, in *Chemistry of the Solid State*, Academic Press, New York, 1955.
38. Kitaoka, H., Wada, C., Moroi, R., and Hakusui, H., *Chem. Pharm. Bull.*, 43, 649, 1995.
39. Zhu, H.J., Khankari, R.K., Padden, B.E., Munson, E.J., Gleason, W.B., and Grant, D.J.W., *J. Pharm. Sci.*, 85, 1026, 1996.
40. McMahon, L.E., Timmins, P., Williams, A.C., and York, P., *J. Pharm. Sci.*, 85, 1064, 1996.
41. Zhang, G.G.Z., Law, D., Schmitt, E.A., and Qiu, Y.H., *Adv. Drug Del. Rev.*, 56, 371, 2004.
42. Schmidt, A.C. and Schwarz, I., *J. Mol. Struct.*, 748, 153, 2005.
43. Pudipeddi, M. and Serajuddin, A.T.M., *J. Pharm. Sci.*, 94, 929, 2005.
44. Kristl, A., Srcic, S., Vrecer, F., Sustar, B., and Vojnovic, D., *Int. J. Pharm.*, 139, 231, 1996.
45. Mazuel, C., *Anal. Profiles Drug Substances*, 20, 557, 1991.
46. Botha, S.A. and Flanagan, D.R., *Int. J. Pharm.*, 82, 185, 1992.
47. Hulsmann, S., Backensfeld, T., Keitel, S., and Bodmeier, R., *Eur. J. Pharm. Biopharm.*, 49, 237, 2000.
48. Dong, Z.D., Young, V.G., Padden, B.E., Schroeder, S.A., Prakash, I., Munson, E.J., and Grant, D.J.W., *J. Chem. Crystallogr.*, 29, 967, 1999.

49. Hourri, O. and Masse, J., *Thermochim. Acta*, 216, 213, 1993.
50. Otsuka, M. and Matsuda, Y., *J. Pharm. Pharmacol.*, 45, 406, 1993.
51. Pienaar, E.W., Caira, M.R. and Lötter, A.P., *J. Cryst. Spectrosc. Res.*, 23, 739, 1993.
52. Kawashima, Y., Niwa, T., Takeuchi, H., Hino, T., Itoh, Y., and Furuyama, S., *J. Pharm. Sci.*, 80, 472, 1991.
53. Leung, S.S., Padden, B.E., Munson, E.J., and Grant, D.J.W., *J. Pharm. Sci.*, 87, 501, 1998.
54. Leung, S.S., Padden, B.E., Munson, E.J., and Grant, D.J.W., *J. Pharm. Sci.*, 87, 508, 1998.
55. Caira, M.R., Nassimbeni, L.R., and Timme, M., *J. Pharm. Sci.*, 84, 884, 1995.
56. Caira, M.R., Easter, B., Honiball, S., Horne, A., and Nassimbeni, L.R., *J. Pharm. Sci.*, 84, 1379, 1995.
57. Hakanen, A. and Laine, E., *Thermochim. Acta*, 248, 217, 1995.
58. Pilpel, N., *Manuf. Chem. Aerosol News*, 42, 37, 1971.
59. Mroso, P.V., Li Wan Po, A., and Irwin, W.J., *J. Pharm. Sci.*, 71, 1096, 1982.
60. Hölzer, A.W. and Sjögren, J., *Acta Pharm. Suec.*, 18, 139, 1981.
61. Butcher, A.E. and Jones, T.M., *J. Pharm. Pharmacol.*, 24, 1P, 1982.
62. Müller, B.W., *Pharm. Ind.*, 39, 161, 1977.
63. Carli, F. and Colombo, I., *J. Pharm. Pharmacol.*, 35, 404, 1983.
64. Brittain, H.G., *Drug Dev. Ind. Pharm.*, 15, 2083, 1989.
65. Wada, Y. and Matsubara, T., *Powder Technol.*, 78, 109, 1994.
66. Ertel, K.D. and Cartensen, J.T., *Int. J. Pharm.*, 42, 171, 1988.
67. Ertel, K.D. and Cartensen, J.T., *J. Pharm. Sci.*, 77, 625, 1988.
68. Miller, T.A. and York, P., *Int. J. Pharm.*, 23, 55, 1985.
69. Miller, T.A. and York, P., *Powder Technol.*, 44, 55, 1985.
70. Kissinger, H.E., *Anal. Chem.*, 29, 1702, 1957.
71. Rajala, R. and Laine, E., *Thermochim. Acta*, 248, 177, 1995.
72. Sharpe, S.A., Celik, M., Newman, A., and Brittain, H.G., *Struct. Chem.*, 8, 73, 1997.
73. Hancock, B.C. and Zografi, G., *J. Pharm. Sci.* 86, 1, 1997.
74. Craig, D.Q.M., Royall, P.G., Kett, V.L., and Hopton, M.L., *Int. J. Pharm.*, 179, 179, 1999.
75. Kerc, J. and Srcic, S., *Thermochim. Acta*, 248, 81–95, 1995.
76. Kerc, J., Srcic, S., Mohar, M., and Smidkorbar, J., *Int. J. Pharm.*, 68, 25–33, 1991.
77. Hancock, B.C. and Zografi, G., *Pharm. Res.*, 11, 471, 1994.
78. Timko, R.J. and Lordi, N.G., *Drug Dev. Ind. Pharm.*, 10, 425, 1984.
79. Fukuoka, E., Makita, M., and Yamamura, S., *Chem. Pharm. Bull.*, 37, 1047, 1989.
80. Pikal, M.J., Lukes, A.L., and Lang, J.E., *J. Pharm. Sci.*, 66, 1312, 1977.
81. Hancock, B.C., Shamblin, S.L., and Zografi, G., *Pharm. Res.*, 12, 799, 1995.
82. Royall, P.G., Craig, D.Q.M., and Doherty, C., *Int. J. Pharm.*, 192, 39, 1999.
83. Ahlneck, C. and Zografi, G., *Int. J. Pharm.*, 62, 87, 1990.
84. Otsuka, M. and Kaneniwa, N., *Chem. Pharm. Bull.*, 32, 1071, 1984.
85. Mosharraf, M., The Effects of Solid State Properties on Solubility and In Vitro Dissolution Behaviour of Suspended Sparingly Soluble Drugs, Ph.D. thesis, University of Uppsala, 1999.
86. Masters, K., *Spray Drying*, 2nd ed., John Wiley & Sons, New York, 1976.
87. Masters, K., *Spray Drying Handbook*, 5th ed., Longman, John Wiley & Sons, New York, 1991.
88. Corrigan, O.I., *Thermochim. Acta*, 248, 245–258, 1995.
89. Corrigan, O.I., Sabra, K., and Holohan, E.M., *Drug Dev. Ind. Pharm.*, 9, 1, 1983.
90. Nurnburg, E., *Prog. Colloid Polym. Sci.*, 59, 55, 1976.

91. Sato, T., Okodo, A., Sekiguchi, K., and Tsada, Y., *Chem. Pharm. Bull.*, 29, 2675, 1981.
92. Corrigan, O.I., Holohan, E.M., and Sabra, K., *Int. J. Pharm.*, 18, 195, 1984.
93. Yamaguchi, T., Nishimura, M., Okamoto, R., Takeuchi, T., and Yamamoto, K., *Int. J. Pharm.*, 85, 87, 1992.
94. Sebhatu, T., Elamin, A.A., and Ahlneck, C., *Pharm. Res.*, 11, 1233, 1994.
95. York, P. and Grant, D.J.W., *Int. J. Pharm.*, 25, 57, 1985.
96. Grant, D.J.W. and York, P., *Int. J. Pharm.*, 30, 161, 1986.
97. Bootsma, H.P.R., Frijlink, H.W., Eissens, A., Proost, J.H., van Doorne, H., and Lerk, C.F., *Int. J. Pharm.*, 51, 213, 1989.
98. Corveleyn, S. and Remon, J.P., *Int. J. Pharm.*, 152, 215, 1997.
99. Nail, S.L. and Gatlin, L.A., in *Pharmaceutical Dosage Forms: Parenteral Medications*, Avis, K.E., Lieberman, H.A., and Lachman, L., Eds., Marcel Dekker, New York, 1993, pp. 163–233.
100. Levine, H. and Slade, L., *Cryo-Letters*, 9, 21, 1988.
101. Franks, F., *Eur. J. Pharm. Biopharm.*, 45, 221, 1998.
102. Wang, W., *Int. J. Pharm.*, 203, 1, 2000.
103. Tang, X.L. and Pikal, M.J., *Pharm. Res.*, 21, 191, 2004.
104. Schwegman, J.J., Hardwick, L.M., and Akers, M.J., *Pharm. Dev. Technol.*, 10, 151, 2005.
105. Hill, J.J., Shalaev, E.Y., and Zografi, G., *J. Pharm. Sci.*, 94, 1636, 2005.
106. Franks, F., *Cryo-Letters*, 11, 93, 1990.
107. Sun, W.Q., *Cryo-Letters*, 18, 99, 1997.
108. Carpenter, J.F., Pikal, M.J., Chang, B.S., and Randolph, T.W., *Pharm. Res.*, 14, 969, 1997.
109. Pikal, M.J., *J. Parenter. Sci.Technol.*, 39, 115, 1985.
110. Tang, X.L. and Pikal, M.J., *Pharm. Res.*, 22, 1176, 2005.
111. Slade, L. and Levine, H., *Crit. Rev. Food Sci. Nutr.*, 30, 115, 1991.
112. Roos, Y.H. and Karel, M., *J. Food Sci.*, 56, 553, 1991.
113. Roos, Y.H. and Karel, M., *Int. J. Food Sci. Technol.*, 26, 553, 1991.
114. Ablett, S., Izzard, M.J., and Lillford, P.J., *J. Chem. Soc., Faraday Trans.*, 88, 789, 1992.
115. Ablett, S., Clark, A.H., Izzard, M.J., and Lillford, P.J., *J. Chem. Soc., Faraday Trans.*, 88, 795, 1992.
116. Blond, G. and Simatos, D., *Thermochim. Acta*, 175, 239, 1991.
117. Sahagian, M.E. and Goff, H.D., *Thermochim. Acta*, 246, 271, 1994.
118. Chang, B.S. and Randall, C.S., *Cryobiology*, 29, 632, 1992.
119. te Booy, M.P.W.M., de Ruiter, R.A., and de Meere, A.L.J., *Pharm. Res.*, 9, 109, 1992.
120. Izutsu, K., Yoshioka, S., and Terao, T., *Pharm. Res.*, 10, 1232, 1993.
121. Corveleyn, S. and Remon, J.P., *Pharm. Res.*, 13, 146, 1996.
122. Duddu, S.P. and Dal Monte, P.R., *Pharm. Res.*, 14, 591, 1997.
123. Yoshioka, S., Aso, Y., Nakai, Y., and Kojima, S., *J. Pharm. Sci.*, 87, 147, 1998.
124. Yoshioka, S., Aso, Y., and Kojima, S., *Pharm. Res.*, 16, 135, 1999.
125. Yoshioka, S. and Aso, Y., *Chem. Pharm. Bull.*, 53, 1443, 2005.
126. Yoshioka, S. and Aso. Y., *Pharm. Res.*, 22, 1358, 2005.

4 Modulated Temperature Differential Scanning Calorimetry

*Mike Reading, Duncan Q.M. Craig,
John R. Murphy, and Vicky L. Kett*

CONTENTS

4.1 INTRODUCTION

The earlier chapters have described the principles and uses of conventional differential scanning calorimetry (DSC), whereby a linear heating signal is applied to a sample, and the temperature and energy associated with the response are measured. Along with differential thermal analysis (DTA), this method remained the industry standard for approximately 30 years, until the introduction of modulated temperature DSC (MTDSC) in the early 1990s (1–4). The innovation associated with this technique lies in a software modification so as to allow the superimposition of a modulation on the underlying signal, described in more detail in the following text.

At the time of this writing, it is commonly the practice within the pharmaceutical industry and academia to purchase instruments with MTDSC capability. The modulated technique is now an accepted and highly useful supplement to the conventional approach. Although the details of its principles and uses will be described later, it is helpful to outline two important issues at the outset. First, there has been considerable discussion with regard to nomenclature for this family of techniques, with different manufacturers using varying types of modulation and different terminology to describe the outputs. This has been settled to the satisfaction of most parties by referring to all such techniques as *modulated temperature DSC*, or MTDSC (although *temperature-modulated DSC*, or TMDSC, is also sometimes used). In this chapter, we describe the technology and nomenclature associated with the TA Instruments method, but the principles under discussion are applicable to competitor instruments as well. Second, within the pharmaceutical field, the general rule is that the technique is especially applicable to glassy systems, as outlined later, but should be used with great caution for studying melts, as a linear response may well be lost (see ensuring discussion); hence the results can be difficult, or indeed impossible, to interpret correctly. We emphasize this because the study of melting is the most usual application for conventional DSC. There has therefore been some confusion in the past in that workers have initially looked to the technique as a means of assessing issues such as polymorphism, for which the method, in its current state of development, is not well suited (at least not in nonisothermal mode). That said, we describe recent work that has at least partially solved the problem, particularly by the use of quasi-isothermal studies. This approach, as yet largely unexploited in the study of pharmaceutical materials, is described in more detail later.

The chapter is arranged in three sections. First, we outline the principles of MTDSC, including a discussion of the practicalities of running MTDSC experiments. Second, we outline the typical thermal events and systems for which MTDSC is well suited, and finally we review some of the applications of MTDSC within the pharmaceutical sciences to date.

4.2 PRINCIPLES OF MTDSC

4.2.1 THEORETICAL BACKGROUND

MTDSC was introduced by Reading et al. (1–4) and is an extension of conventional DSC. In essence, the technique involves the application of a perturbation to the heating program of a conventional DSC (a sinusoidal wave in most cases, but sawtooth and square waves are also used) combined with a mathematical procedure designed to separate different types of sample behavior. The separation (also called *deconvolution*) procedure can most easily be understood in terms of a simple equation (5,6), that is,

$$dQ/dt = C_p \cdot dT/dt + f(t,T) \tag{4.1}$$

where Q = the amount of heat evolved, C_p = the heat capacity, T = the absolute temperature, t = time, and $f(t,T)$ = some function of time and temperature that governs the response associated with the physical or chemical transformation. The first term on the right-hand side describes the heat flow associated with the sample's heat capacity. By the term *heat capacity* we mean here the energy stored in the molecular motions available to the sample, such as vibrations, rotations, translations, etc. When the temperature is increased by a given amount, more energy is stored in these motions, and when the temperature is reduced over the same temperature interval, exactly the same amount of energy is released by the sample. In other words, this is a perfectly reversible process. When the temperature corresponding to a thermal event such as a chemical reaction is reached, the enthalpy associated with the reaction will also contribute to the heat flow response, and this is expressed by the second term on the right-hand side of Equation 4.1. This is often, though not always, an irreversible process. Conventional rising-temperature DSC measures both components simultaneously and they cannot be separated, even when several different experiments at different heating rates are used; at any point in time in each experiment these two contributions are mixed. At a given temperature, the heat flow from the different experiments cannot be directly compared, as the state of the sample (e.g., extent of reaction) will typically be different.

The advantage of MTDSC is that, within certain limitations, these two components may be measured independently because the sample's response to the modulation is different from its response to the linear component of the heating program. This is significant as many responses may involve two or more overlapping processes, the classic example being the glass transition that occurs when a cross-linking reaction is proceeding in a polymer (discussed in more detail later). The glass transition is essentially a change in heat capacity and is therefore associated with the first term on the right-hand side of Equation 4.1, whereas the reaction is a kinetically hindered event that arises because of the formation of chemical bonds and so contributes to the second term of the right-hand side of Equation 4.1.

The basic operation of the instrument involves the superimposition of a sine wave (or square or sawtooth, depending on the instrument used) on the linear heating signal such that although the underlying heating process may be equivalent to that

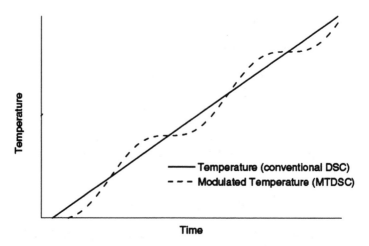

FIGURE 4.1 Schematic representation of temperature as a function of time for conventional DSC and modulated temperature DSC.

of a conventional DSC run, the sample temperature is being oscillated in a sinusoidal manner (Figure 4.1). This will result in the heat flow being modulated. In the simple case in which there is no transition, the heat flow will be modulated in accordance with the modulation in the time derivative of the sample temperature. As all samples will have a nonzero heat capacity, this modulation will always be present whether a transition is occurring or not.

The basis of the separation or deconvolution procedure can be illustrated by a few simple equations. When used with a sinusoidal perturbation, the temperature program for an MTDSC experiment is given by:

$$T = T_0 + bt + B \cdot \sin(\omega t) \tag{4.2}$$

where T_0 = the starting temperature, b = the heating rate, B = the amplitude of the modulated temperature excursion, and ω = the frequency; this is in contrast to the heat flow for conventional DSC, which is given simply by $T = T_0 + bt$. The heat flow equation (Equation 4.1) can therefore be rewritten as:

$$dQ/dt = C_p(b + B\omega \cdot \cos(\omega t)) + f'(t, T) + C \sin(\omega t) \tag{4.3}$$

where $f'(t,T)$ = the underlying kinetic function once the effect of the sine wave modulation has been subtracted, C = the amplitude of the kinetic response (from $f(t,T)$) to the sine wave modulation, and $(b + B\omega \cdot \cos(\omega t))$ = the sinusoidal heating rate. An important assumption that is key to the subsequent analysis is that the response of the sample is linear with respect to the magnitude of the sinusoidal heating signal; this will be the case if the corresponding temperature excursions are small. The heat flow signal will therefore contain a cyclic component that is dependent on the values of B, ω, and C.

The first step in the deconvolution procedure is simply to average the modulated signals over the period of the modulation (or integer multiples of this period). Where the response is linear, this removes the modulation to provide the results that would have been obtained had the modulation not been used, i.e., had only the linear portion of the heating program been used. Signals averaged in this way can be called the averaged, total, or underlying signals.

$$\text{Underlying heat flow} = C_p b + f'(t, T) \tag{4.4}$$

The definition of the $f'(t,T)$ function given earlier can now be modified to become the average of $f(t,T)$ (from Equation 4.1) over the interval of at least one modulation (there are other ways of removing the effect of the modulation besides averaging, but we will consider only this method here). The deconvolution of the heat flow data into the two components can either be a simple deconvolution that does not use the phase lag, or a full deconvolution that does. In most cases, simple deconvolution is perfectly adequate.

Dealing with simple deconvolution first, it can be seen from Equation 4.3 that the heat flow signal is composed of two sets of terms. The first, $C_p(b + B\omega \cdot \cos(\omega t))$, is dependent on the magnitudes of the terms b, B, and ω (all experimental rather than sample parameters) but has no sample dependence other than the value of C_p. Consequently, this first term is a measure of the sample's heat capacity (as defined previously) and, except in the case of the glass transition (see later discussion), can be considered, for the purposes of this discussion, to be an effectively instantaneous sample response. If we focus on the modulation, we see that this is a cosine wave as opposed to the sine wave in the temperature. This means that, when there is no transition, the modulation in heat flow will, in principle, follow this cosine wave (the modulation in heating rate) with zero phase lag when the convention is adopted that endothermic is up. It will be 180° out of phase when the convention is adopted that endothermic is down.

The second component, $f'(t,T) + C \sin(\omega t)$, reflects the heat flow associated with the kinetic or kinetically hindered (because it is not instantaneous) process. The clearest example of such a process is a chemical reaction in which the sample response to a change in temperature will take some time to go to completion. The quantity denoted C is a measure of the degree to which the rate of the kinetic process is modulated along with the temperature (typically, if the temperature is modulated, there will be a corresponding modulation in the heat flow associated with the modulated rate of reaction). This modulation in heat flow follows the sine wave in temperature, and so is 90° out of phase with that associated with the heat capacity, which follows a cosine function. Consequently, the changes in the phase lag are associated with changes in the magnitude of C that only occur during a transition. A key element of the simple deconvolution process is that it is assumed that the magnitude of C is negligible and may be effectively ignored. This is true for many applications, but the reader should be aware that the difference between simple and complete deconvolution rests on whether this assumption is considered to be acceptable for the particular system under study. The quantity C is often negligible except during melting.

Proceeding with simple deconvolution, from Equation 4.2, assuming C is negligible, we can calculate heat capacity from the ratio of the amplitude of the modulated heat flow (A^{MHF}) to that of the modulated heating rate (A^{MHR}) via

$$C_p = (A^{MHF}/A^{MHR}) \qquad (4.5)$$

In reality, the nonideal behavior of the cell requires that we introduce a heat capacity calibration constant K_{C_p} such that $C_p = K_{C_p} \cdot (A^{MHF}/A^{MHR})$. However, in this simple treatment we will assume ideal behavior. Calibration to account for nonideal behavior is discussed later. The amplitudes are obtained by using a Fourier Transform after the average heat flow is subtracted from the raw signal. In effect, using these amplitudes simply relates the magnitude of the temperature change to that of the applied heating signal, as is performed for any heat capacity measurement. If we take Equation 4.4 and Equation 4.5 we obtain:

$$b \cdot (A^{MHF}/A^{MHR}) = \text{the contribution to the underlying heat flow}$$
$$\text{that comes from heat capacity} \qquad (4.6)$$

$$f'(t, T) = \text{underlying heat flow} - b \cdot (A^{MHF}/A^{MHR}) \qquad (4.7)$$

In other words, the value of C_p is calculated from the response to the modulation (Equation 4.5). If this is then multiplied by the linear component of heating rate, b, the heat capacity component of the underlying heat flow equation (Equation 4.6) is obtained. When this is subtracted from the underlying heat flow, the heat flow associated with the chemical reaction is obtained. In this way, these two different contributions to the heat flow are separated.

We now need to consider the question of nomenclature:

$$b \cdot (A^{MHF}/A^{MHR}) = \text{the reversing heat flow}$$

$$\text{underlying heat flow} - b \cdot (A^{MHF}/A^{MHR}) = \text{the nonreversing heat flow}$$

Consequently, it is possible to obtain all three signals, underlying, reversing, and nonreversing, from a single scan. Figure 4.2 summarizes this deconvolution procedure (adapted from [7]).

If we take the example of a chemical reaction, then we could argue that the reversing heat flow is reversible while the nonreversing is irreversible. However, this kind of separation can work for transitions like cold crystallization and other transformations that are reversible by use of large-temperature excursions but that are not reversing at the time and temperature the MTDSC measurement is made. For this reason the terms *reversing* and *nonreversing* have been adopted rather than *reversible* and *irreversible*. This ability to measure heat capacity even when other processes are occurring is one of the main advantages of MTDSC.

It should be noted that the deconvolution process makes two basic assumptions: the first is that the sample response to the modulation is linear (as discussed earlier)

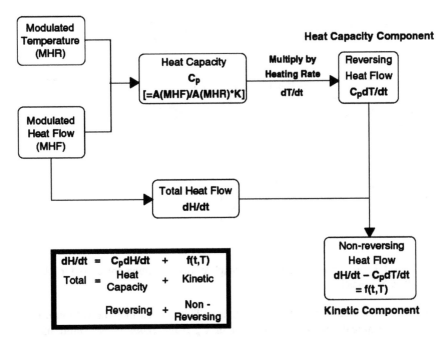

FIGURE 4.2 Flowchart representing the "simple" deconvolution procedure for TA Instruments MTDSC signals. (From Jiang, Z., Imrie, C.T., and Hutchinson, J.M., *Thermochim. Acta*, 387, 75, 2002. With permission.)

and the second is that any changes in the underlying heat flow are slow relative to the time scale of the modulation. In other words, there must be many modulations over the course of a transition (8). When this is not the case, deconvolution cannot work. We would typically expect at least six modulations over the course of that part of the transition where the change in heat flow is most rapid. There are examples In which the requirements for meaningful deconvolution cannot be met, notably in the case of melts of pure materials where the sample cannot increase its temperature above that of the melt until the entire sample has melted. In other words, the sample temperature cannot be modulated and the response is highly nonlinear. However, quasi-isothermal experiments can still provide useful information. This is explored in more detail later.

The simple deconvolution process outlined earlier is the typical methodology used for most MTDSC analysis. However, it is possible to further analyze the signals to take into account the phase lag. Before describing the more complex deconvolution the phase lag itself warrants further explanation.

The phase angle is the angle between the heat flow response and the modulated heating rate. Ideally, the heat flow will be out of phase with the modulated temperature, giving a phase angle of 90° ($\pi/2$ rad) and 180° (π rad) out of phase with the modulated heating rate when the convention is that endotherms go down. When the convention is the endotherms go up, then the phase lags are 90° and 0°, respectively. In reality, this will not be the case because of thermal resistances between the

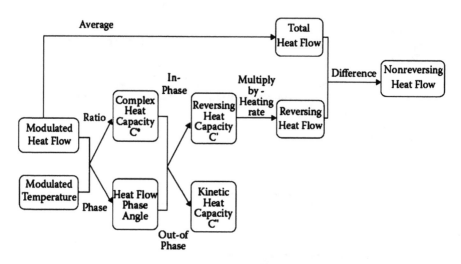

FIGURE 4.3 Flowchart representing the "complete" deconvolution procedure for TA Instruments MTDSC signals.

cell–pan interface and pan–sample interface. This means that the response of the sample is nonideal and the calculation of C_p from dH/dt is not valid unless a correction can be made for these thermal resistances. If the phase angle is plotted with respect to temperature, two features are apparent. First, the phase angle is nonzero, but linear in the regions where no thermal events are observed. Second, there are deviations from this line through the glass transition and recrystallization. The experimental effect mentioned earlier can be corrected for by subtraction of the baseline deviation, using the normal interpolation procedures. Once the baseline has been subtracted, the remaining peak in phase lag can be used to calculate the sine and cosine components to the response to the modulation, as illustrated in Figure 4.3. These peaks may also be helpful in assigning values to glass transitions that occur in "noisy" temperature regions, such as in frozen systems.

The complex deconvolution method therefore bears a greater resemblance to the analysis of dielectric or oscillatory rheological data in which the heat flow response is considered to be vectorial, i.e., either in phase or out of phase with the imposed dT/dt modulation. This "complete" deconvolution procedure is again based on Equation 4.3. By rearranging this equation, it is possible to consider the underlying and cyclic signals (7) via:

$$dQ/dt = BC_p + f'(t, T) \text{ (the underlying signal)}$$

$$+ \omega BC_p \cos \omega t + C \sin \omega t \text{ (the cyclic signal)} \tag{4.8}$$

The contribution of the phase lag, θ, can now be taken into account via

$$C_p = (A^{MHF}/A^{MHR}) \cos \theta \tag{4.9}$$

$$C = A^{MHF} \sin \theta \qquad\qquad (4.10)$$

We can define a new quantity, the kinetic heat capacity $C_{pK} = C/A^{MHR}$. The complete deconvolution then gives rise to a new nomenclature:

$b \cdot (A^{MHF}/A^{MHR})\cos \theta = b \cdot C_p$ = the phase-corrected reversing heat flow

$b \cdot (A^{MHF}/A^{MHR})\sin \theta = b \cdot C_{pK}$ = the kinetic heat flow

underlying heat flow $- b \cdot (A^{MHF}/A^{MHR})\cos \theta =$
the phase-corrected nonreversing heat flow

An alternative nomenclature uses complex heat capacity such that:

$$A^{MHF}/A^{MHR} = Cp^{*2} = Cp'^2 \, iCp''^2 \qquad\qquad (4.11)$$

where C_p' is the real component of complex heat capacity equal to the phase-corrected reversing heat capacity, and C_p'' is the imaginary component equal to the kinetic heat capacity, with "i" being the square root of 1. They may each be described in terms of the phase lag, θ

$$Cp' = Cp^* \cos \theta \qquad\qquad (4.12)$$

$$Cp'' = Cp^* \sin \theta \qquad\qquad (4.13)$$

The interpretation of the signals from the MTDSC then relies on the phase of the modulated signals being used to quantify the components of the complex heat capacity. The real component, C_p, arises from the reversing processes and is in phase with the modulation, whereas the imaginary component, C_p'', arising from the kinetic processes is out of phase with the modulation. The real component of heat capacity (phase-corrected reversing heat capacity) is then multiplied by the heating rate to obtain the reversing heat flow signal. A flowchart representing the complete deconvolution procedure is given in Figure 4.3 (adapted from TA Instruments [8]). For the majority of applications, the simple deconvolution process is used because it can be shown that C_p' and C_p^* are not significantly different during most glass transitions or crystallizations, the principal events analyzed using this technique in this study. The phase lag becomes significant during melting, but in this case the phase lag is difficult to use quantitatively unless considerable pains are taken to calibrate and remove the effects of thermal resistances.

Figure 4.4 shows the raw data and the deconvoluted signals for quench-cooled PET using the simple procedure, illustrating a typical output for the approach. The total heat flow signal shows the glass transition with an accompanying relaxation endotherm, the recrystallization exotherm, and the melting endotherm. On deconvolution, the glass transition is seen in isolation in the reversing signal, whereas the relaxation endotherm and recrystallization peak are seen in the nonreversing signal.

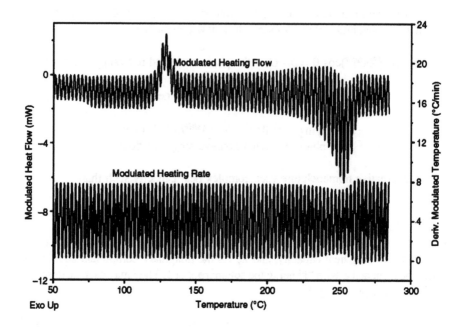

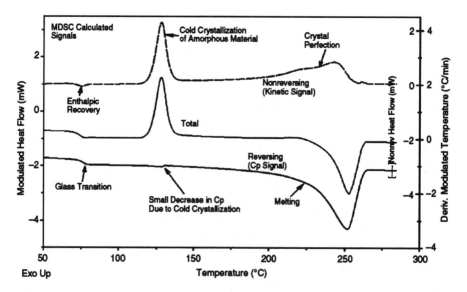

FIGURE 4.4 Raw data (top) and results of simple deconvolution for quench-cooled PET (bottom).

It should be noted that for many polymers the melting response is seen in both deconvoluted signals, despite the breadth of the observed response that, in theory at least, should allow steady-state modulation.

In this case, and in polymer melting generally, there is a distribution of crystallites that have different melting points according to their degree of perfection. The melting transition in these materials is broad because a succession of crystallite populations melts one after the other as the sample temperature reaches their melting temperatures. The amount of energy required to melt these crystallites is fixed, as is their melting temperature. This means that if one wants to melt them twice as fast (i.e., the heating rate is doubled), the rate of energy input must be twice as fast. It follows from this that the heat flow required to melt the crystallites is a linear function of heating rate. Therefore, the enthalpy of melting will be seen in the reversing signal. In a simple case, in principle at least, this type of melting behavior closely mimics heat capacity. However, if there is cooling during the modulation, supercooling may well mean no crystallization occurs; thus, unlike heat capacity, it is not truly reversible, and even the term *reversing* is not strictly applicable. Consequently, the terms *reversing* and *nonreversing* can be misleading in the melt region.

Furthermore, it might be expected, provided conditions are chosen so that cooling never occurs during the heating program, that the relationship between heating rate and heat flow will be invariant (like heat capacity). However, even this is not true, principally because ensuring a uniform temperature within the sample becomes very difficult during melting because of the large amount of enthalpy associated with this type of event. As a consequence of this, the reversing heat capacity is strongly frequency dependent during melting. The result of this is that the quantitative interpretation of the reversing and nonreversing signals during melting is problematic and there is currently no generally accepted procedure for doing this. However, a qualitative assessment can be useful. In Figure 4.4 the nonreversing heat flow is exothermic between the cold crystallization and the end of melting. This is because a process of rearrangement occurs over this temperature interval. Crystallites melt then recrystallize to form crystallites with higher melting temperature that, in turn, melt and so the process continues until the temperature is too high for any crystallization to occur. In conventional DSC, the endothermic process of melting and exothermic process of crystallization cancel each other out to a large extent, and so there is little indication of what is occurring. With MTDSC, it becomes more apparent.

4.2.2 PRACTICAL ASPECTS OF MTDSC STUDIES

4.2.2.1 Advantages and Disadvantages of MTDSC Compared to the Conventional Technique

As can be seen from the previous discussion, MTDSC offers the possibility of analyzing samples with varying degrees of sophistication. However, there are inevitably concomitant disadvantages associated with the technique, and hence it is useful to briefly summarize what we consider to be the main advantages and disadvantages of the method.

In terms of advantages, the ability to deconvolute the signal into the reversing heat flow (which often derives from heat capacity and the nonreversing heat flow, which in turn often comes from kinetically controlled events like a chemical reaction or loss of a volatile such as water) allows much clearer interpretation of events such as glass transitions and curing reactions. This can be seen as the main benefit of the technique for routine applications. Strongly associated with this is the ability to measure glass transitions using the reversing signal that are not greatly affected by thermal history. This makes them easier to identify unequivocally. There is also the benefit of a good signal-to-noise ratio because the modulation component contains a high heating rate and the Fourier transform rejects nonperiodic noise, coupled with high resolution, because the linear component of the temperature program can be very slow. Taken together, these factors have led to MTDSC being very well established, especially for the study of amorphous systems.

The disadvantages of the technique are: First, to obtain reliable results, great care must be taken in both choosing experimental parameters and the subsequent interpretation of the data. It is essential for the sample to undergo at least six modulations through the region of interest for reliable deconvolution to take place, and the frequency must be such that the sample can follow the temperature program. Consequently, it is necessary to use slow underlying rates to account for both these requirements, with 2°C/min being the typical rate employed with a nitrogen purge gas (the effects of different purge gases is discussed later). Similarly, great care is required for calibration, with the temperature, enthalpy, and heat capacity (as opposed to just the first two for conventional DSC) calibration being required. Finally, the technique is by no means suitable for all transitions, the difficulties associated with melting in particular being outlined previously. Indeed, it should always be borne in mind that the use of modulation is not always appropriate. When this is the case, any MTDSC can be used as a perfectly good DSC. In the following subsections, some the aspects mentioned so far will be outlined, particularly in terms of the requirements for obtaining reliable results.

4.2.2.2 Calibration Protocols

Calibration is a potentially complex subject as there can be cell asymmetries in terms of heat capacity, thermal resistances, and other effects such as convection and emissivity. However, here we shall confine ourselves to a simple procedure that will be effective in most cases. We can consider that calibration of an MTDSC consists of three steps: (1) heat flow calibration (by calculation of a cell constant), (2) baseline calibration, and (3) heat capacity calibration (10).

4.2.2.3 Calibration of the Underlying Signal

The underlying heat flow signal is calibrated by the use of standards with known melting temperatures and enthalpies of fusion. A series of such samples is run over the operating temperature range of the instrument. The sample thermocouple has a nominally known relationship between its output and temperature. Any observed differences between measured (by the sample thermocouple) and expected melting temperatures are used to construct a correction curve as a function of temperature (by polynomial fitting), and this correction is then applied in subsequent experiments. Essentially the same approach is adopted to calibrate heat flow. The basic measurement for this quantity is the temperature difference (ΔT) between the sample and reference (here we are restricting the discussion to the most popular form of DSC, the heat flux instrument). The area under the peak in the ΔT measurement is plotted against the known enthalpies of fusion of the standards to construct a calibration graph. This is often done by the manufacturers so the instrument is supplied with a master calibration curve that is then adjusted by experiments carried out by the user. Frequently, a single experiment is used for this adjustment and a constant is applied, called the *cell constant*. Alternatively, several standards can be run, and a correction is applied as a function of temperature. Up to five calibration standards are used, depending on the temperature range of the proposed study, but the most commonly used is the melting point of indium. It is important to use the same heating rate and experimental conditions in the calibration as in the subsequent experiments.

For the underlying signal, baseline calibration is carried out in essentially the same manner as in a conventional DSC; the baseline (obtained with empty matched pans) underlying heat flow can simply be subtracted from the underlying heat flow for the sample. An alternative is to mathematically fit a polynomial through this baseline and subtract this fitted curve from subsequent runs. The benefit of the latter approach is that it reduces noise.

4.2.2.4 Calibration of the Reversing Signal

For MTDSC, it is also essential to calibrate for the reversing heat capacity to allow quantification of the deconvoluted results. There are several approaches by which this may be achieved, with more accurate methods requiring greater sophistication and more time; hence, a decision needs to be made with regard to how important accurate heat capacity data are to the objectives of the study. For most pharmaceutical applications, fairly simple calibration procedures such as those about to be outlined are usually sufficient. However, for more accurate work it is essential to use more detailed approaches such as that described in Reference 11. In this summary, we outline only the simple approaches, but readers should be aware of the availability of more complex methods that yield more reliable results.

4.2.2.5 One-Point Calibration

The heat capacity constant may be simply determined by dividing the literature heat capacity of a calibration standard at a given temperature by the measured heat capacity. Artificial sapphire is often used for this purpose and is available as a standard from the instrument manufacturers (either as a powder or compacted into a disk). This is clearly a simple and rapid approach but is limited by the assumption that the constant does not change over the temperature range of the study of interest. This assumption becomes increasingly problematic when studying systems in which several transitions, over a wide temperature range, are being examined. It has also been shown that the $K_{C_p} \cdot (C_p)$ value changes rapidly below 300 K, so that over a large temperature range the use of a single calibration point could result in errors as large as 15 to 20% (12). When this is the case, the method outlined next is preferable.

4.2.2.5 Multipoint Temperature-Dependent Calibration

The accuracy of the results over the entire temperature range can be improved by transferring the uncalibrated data to a spreadsheet, and calibrating each temperature point separately with its own heat capacity constant. Artificial sapphire is often used as the standard because an equation (Equation 4.14) giving its change in heat capacity across a given temperature range has been determined.

$$C_p = 8.6446 - (0.4929T) + (8.2336 \times 10^{-3}T^2) - (3.5739 \times 10^{-5}T^3) +$$
$$(7.556 \times 10^{-8}T^4) - (8.033 \times 10^{-11}T^5) + (3.4219 \times 10^{-14}T^6)$$

(4.14)

Comparison of the empirical with the literature values gives the required calibration constant at each temperature. The heat capacity constant K_{Cp} is not now assumed to be constant with temperature but is assumed to be constant from experiment to experiment.

This reversing heat capacity calibration is usually performed as follows. The same average heating rate to be used in the experiment must be used in the calibration. Empty pans of the same type to be used in the subsequent experiments are used in the baseline calibration (mass-balanced to ± 0.1 mg to allow compensation of mass imbalance across the cell). If the modulated heat flow and modulated heating rate signals are in phase, the baseline reversing heat capacity is added to all subsequent runs; if not, the heat capacity is subtracted as appropriate. Some workers prefer to deliberately bias the results by using a reference pan with a lower mass than the sample pan so that it is clear that the baseline (empty pan) must always be subtracted, not added. It has been found that long period times and large modulation amplitudes give the most accurate heat capacity measurements (13). Data from the uncalibrated files for the sample, the aluminum oxide, and for the empty pans are saved as data tables and imported into a spreadsheet program. Each data table will contain columns for temperature, underlying heat flow, and reversing heat capacity.

The heat capacity constant (K_{Cp}) is determined by first subtracting or adding the baseline (empty pan) values as appropriate from the sample and calibration reversing heat capacity data. The baseline-corrected measured reversing heat capacity for the aluminum oxide is compared to the expected heat capacity, determined from the literature heat capacity data as follows:

$$K_{Cp} = Cp_{lit}/Cp_{meas}$$

(4.15)

where Cp_{lit} is the literature heat capacity, Cp_{meas} is the measured reversing heat capacity (after baseline correction), and K_{Cp} is the heat capacity constant at each temperature. In this way, the reversing heat capacity constant, K_{Cp}, is then determined as a function of temperature. This calibration constant is then applied to the sample data from which the baseline (empty pan) results have already been subtracted.

It is also possible, if desired, to use the same data set to calibrate the underlying signal in terms of heat capacity. To do this, the baseline (empty pan) underlying heat flow is subtracted from this signal for the aluminum oxide and the sample experiments. The underlying heat flow is then converted to a heat capacity by dividing it by the linear component of the heat rate, and this value is compared to the known values to calculate an underlying heat capacity calibration constant K_{CpU}, viz:

$$K_{CpU} = Cp_{lit}/((UHF_{AlO} - UHP_{baseline})/b) \qquad (4.16)$$

where UHF_{AlO} is the underlying heat flow for the aluminum oxide, and $UHP_{baseline}$ is the underlying heat flow obtained from the baseline (empty pan) experiment. The sample underlying heat flow is then converted to heat capacity by subtracting the baseline heat flow, dividing by b, and multiplying by K_{CpU} at each temperature. In this way, both the reversing and underlying signals can be expressed as heat capacities.

4.2.2.6 Choice of Modulation Parameters

When preparing to run a sample under MTDSC conditions, certain protocols need to be followed to ensure that meaningful data are gained from the experiment. More details of the choice of parameters may be obtained from the manufacturers and from a number of texts (e.g., Reference 12 to Reference 16). The period and amplitude of the modulation as well as the underlying heating rate have a profound effect on the measurement. The underlying signal is equivalent to the heat flow that would have been obtained had modulation not been used. The modulation is suppressed typically by simply averaging over the period of one or more modulations. For this signal to be undistorted, the underlying signal must only change slightly over the course of a modulation. When a transition has the form of a peak, TA Instruments recommends choosing a period such that there are six modulations over the width of the peak at half height. When the transition has the form of a step change, this can be expressed as a peak by taking the derivative and then the same rule applies. Clearly, there is some latitude with this guideline and judgment must be exercised.

The number of modulations that occur over any given temperature interval can be changed by varying either the period of the modulation or the underlying heating rate. Increasing the frequency (decreasing the period) so as to allow sufficient modulations through a transition entails the danger that steady state might be lost, i.e., the sample may not follow the temperature changes caused by the modulation. In practice, periods of 30 to 60 sec are typically used. Using too low an underlying heating rate can make the experiment overly long and the signal-to-noise ratio of the underlying heat flow, which decreases as the linear component of heating rate decreases, may become unacceptable. As mentioned earlier, the most rapid underlying heating rate used is typically 5°C/min, with 2°C/min being the most commonly used rate.

The modulation amplitude must be specified. A compromise is required between avoiding distortions of the response to the modulation and achieving a reasonable signal-to-noise ratio. If the amplitude used is too low, then the reversing heat capacity signal becomes extremely weak and difficult to distinguish from the baseline, whereas too high a value can result in a nonlinear response, i.e., the sample response to a sinusoidal modulation will not be an accurate sine wave. In practice, values between approximately 0.2 and 0.5°C are typically used. A further consideration is whether the choice of amplitude, period, and underlying heating rate leads to heat only or a heat–cool cycle. If, for example, one uses a large amplitude and a slow underlying heating rate, the sample may actually decrease in temperature during the cycle. Conversely, a small amplitude coupled with a relatively rapid heating rate will result in a heat-only program with the minimum heating rate being greater than zero.

There is the special case where the minimum heating rate is set to zero. When studying chemical reactions and glass transitions, there is typically no problem with having a heat–cool program. When studying melting that occurs over a wide temperature range, such as polymer melting, it is typically undesirable to have a cooling component. This is because material that melts on heating will not crystallize on cooling, and the response is nonlinear. In these cases, it is best to set the minimum heating rate to zero. The combination of frequency and amplitude that will ensure zero or heat-only conditions may be easily predicted from

$$A^{MHR} \leq B \cdot P/2\pi 60 \qquad (4.17)$$

where P is the modulation period in seconds.

4.2.2.7 Choice of Purge Gas

As with conventional DSC experiments, a flowing purge gas is used in MTDSC. A flowing atmosphere reduces temperature gradients and carries away any evolved gases that might react with and damage the cell. An additional benefit of this is that any moisture or oxygen present in the cell is evacuated by the gas, improving the lifetime of the cell (9,13).

There are two purge gases commonly in use, nitrogen and helium. Nitrogen is economical, allows a wide temperature range of operation (–150 to 725°C), and provides good sensitivity. The effect of flow rate of the gas on the cell constant is negligible with nitrogen, and TA Instruments recommends a flow rate of 40 to 60 ml/min. Helium has the advantage over nitrogen of having a higher thermal conductivity, allowing the sample to respond more quickly to the temperature program because it is more effectively coupled to the furnace. This, in turn, facilitates the use of shorter modulations and higher underlying heating rates. The disadvantages of helium compared to nitrogen are that it is more expensive, and the flow rate has a marked effect on the cell constant, with the value increasing as flow rate increases. This is because it is very difficult to establish an atmosphere of 100% helium in the cell, because of the diffusion of air. This then means that the cell contains a mixture, and the percentage of helium depends on the flow rate, which then impacts on the cell constant. TA Instruments advises the use of a nitrogen purge at 50 ml/min. It is essential that the flow rate be controlled by a mass flow controller or an equivalent device. Simply using a needle valve usually means the flow rate will change gradually over the course of a day and from day to day; this would change the calibration.

4.2.2.8 Choice of Sample Pans

As in conventional DSC, the choice of sample pan can drastically affect the results of the MTDSC experiment. There are three basic types of pan: open, closed, and hermetically sealed (also included in this category are pinholed hermetically sealed pans that allow vapor loss in a reproducible manner). It must be noted that there is a great deal of variation between manufacturers, particularly in terms of the size and temperature resilience of hermetically sealed pans (13,17). Overall, the considerations regarding pans are more demanding with MTDSC because the reversing signal is more affected than the underlying signal (or ordinary DSC signal) by poor thermal contact between the pan and the sensor. It is essential that the bottom of the pan be as flat as possible. Hermetically sealed pans frequently distort with increasing pressure and the resulting increase in internal pressure. There may be no easy solution to this problem, and so the quantitative value of the reversing signal can become compromised. This may not be a serious problem, however, if all that is necessary is to detect transitions (the accuracy of temperature measurement would not usually be greatly affected) rather than, for example, to quantify heat capacity changes accurately.

4.2.2.9 Sample Mass

When filling the sample pan, it is important to recognize that the response of the sample to the applied temperature program depends on the thermal contact between the sample and the pan. A thin layer, compressed on to the bottom of the pan, will provide the best response. A thick sample (one of large mass) will provide a broader response owing to thermal gradients within the sample (11), as the middle of the sample (or top if the sample has no lid) cannot keep up with the temperature modulation and lags behind the bottom of the sample, which has a better thermal contact with the pan. This results in the measured reversing heat capacity being lower than it should be and can lead to apparent broadening of transitions with increasing sample mass. A sample mass of 10 to 15 mg is recommended for polymer samples. Heavier masses are allowed for inorganic or metal alloy samples (8,11) because of their higher thermal conductivity. It is sometimes necessary to use larger masses with biological samples because the transition are weak; however, their poor conductivity can lead to problems with quantification.

4.2.2.10 Lissajous Analysis

Lissajous analysis is another method of presenting MTDSC data that allow the operator to monitor whether the sample is in steady state or otherwise. In the case of MTDSC, the Lissajous figure is obtained by plotting the derivative of the mod-

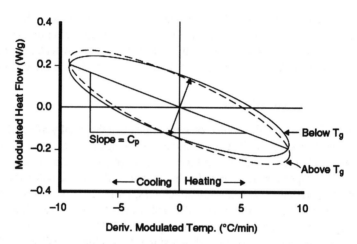

FIGURE 4.5 Schematic of a typical Lissajous figure (after TA Instruments). (From Jiang, Z., Imrie, C.T., and Hutchinson, J.M., *Thermochim. Acta*, 387, 75, 2002. With permission.)

ulated temperature against the modulated heat flow. Information concerning the stability of the MTDSC system can be obtained from the figure because a stable MTDSC system will produce an ellipse that retraces. Any change in the sample properties will result in a change of the Lissajous figure. This gives another way of looking at transitions by MTDSC and can provide information on steady state and whether linearity is achieved, because when this is not the case the shape of the Lissajous figure is distorted. Figure 4.5 gives a schematic of a typical Lissajous figure, as applied to MTDSC.

4.2.2.11 Measurement of the Phase Angle

The interpretation of the complete deconvolution has been the subject of intense discussion, particularly in terms of the phase angle. Although the analysis outlined in an earlier section points out that the heat capacity measured via examination of the ratio between the modulated heat flow to the modulated heating rate is in fact the complex heat capacity, with a contribution associated with the kinetic response, it is now well established that the measured phase angle is also associated with heat transfer processes within the sample and the cell; such processes will be time dependent and will therefore result in an out-of-phase contribution to the measured heat capacity. A number of approaches have been described whereby this effect may be corrected for (e.g., Reference 18 through Reference 21). The simplest approach, a linear or sigmoidal baseline under the peak, is also the most robust. It remains true that, in principle, the phase correction should be applied because the phase-corrected signals provide a more accurate separation of the reversing and nonreversing components. However, in practice the phase correction often results in a very small change to the reversing signal while introducing unwanted noise. Consequently, it is often neglected. The phase angle by itself can be useful as an additional

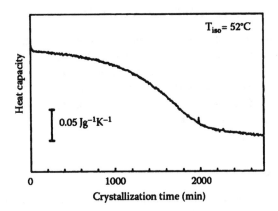

FIGURE 4.6 Heat capacity evolution of the crystallization of pregelatinized waxy corn starch at 52°C. (From De Meuter, P., Rahier, H., and Van Mele, B., *Int. J. Pharm.*, 192, 77, 1999. With permission.)

means of delineating a transition (because it exhibits a peak). However, currently its importance in pharmaceutical applications remains uncertain.

4.2.2.12 Stepwise Quasi-Isothermal MTDSC

Stepwise quasi-isothermal MTDSC involves running the instrument with an underlying heating rate of zero at a set temperature for a set time period, then heating or cooling the sample by an incremental amount (usually 1°C) to the next temperature of interest, and so on until a complete data set at a range of temperatures has been obtained. It is necessary to run the experiments for a sufficiently long period so as to allow equilibration at the set temperature, and it is also necessary to bear in mind that, because of the finite heating rate between runs, the method is not truly isothermal but a close approximation to it. That said, the method has attracted interest owing to the possibility of directly monitoring time-dependent processes. For example, De Meuter et al. (22,23) have used the approach to monitor the recrystallization of starch in real time (Figure 4.6). These authors argue that because of the small enthalpic change associated with the slow crystallization of starch, an alternative approach would be to study the change in heat capacity via the reversing heat flow signal. The authors compared MTDSC to XRD, dynamic mechanical analysis, and Raman spectroscopy, showing that MTDSC produced data that compared favorably with the more established techniques.

Other studies using the quasi-isothermal approach include that of Hohne and Kurelec (24), who modeled the quasi-isothermal response of ultrahigh molecular mass polyethylene in the premelting region to obtain the time constants and activation energy of the associated relaxation processes. In addition, Pak et al. (25) have proposed that quasi-isothermal measurements are an effective means of obtaining absolute values of the complex heat capacity; the heat flow lags that invariably occur when measuring in non-isothermal mode need not be corrected for.

4.3 TRANSITIONS STUDIED USING MTDSC

There is now a wealth of literature associated with MTDSC, and it is not feasible within the scope of the current chapter to review all the theoretical and application developments that have taken place in the 12 years since the introduction of the technique. Instead, the following discussion is intended as an introduction to some of the main areas of activity, with a view to considering how such investigations may be of relevance to pharmaceutical studies. A more detailed exposition on these topics is available in (11).

4.3.1 THE GLASS TRANSITION

The characterization of the glass transition represents the classic use of MTDSC, and indeed represents one of the main applications of the technique. The theoretical aspects of the glass transition have been discussed in a previous chapter (Chapter 1), and hence the emphasis here will be on the benefits of by the modulated technique in comparison to conventional methods.

MTDSC has several significant practical advantages for studying glass transitions. The first is that the limit of detection is increased. The effect of using the Fourier Transform to eliminate all responses not at the driving frequency of the modulation is to decrease unwanted noise. As will be discussed in the next section, the ability to detect subtle glass transitions has several important pharmaceutical applications. They range from the characterization of materials that exhibit only a small change in heat capacity through the transition and are thus difficult to detect using conventional DSC, through to the assessment of small quantities of amorphous material on otherwise crystalline samples. A second related advantage is that of increased sensitivity. Although the underlying heating rate may be comparatively slow (which in itself improves resolution), the high rate of temperature change afforded by the modulation results in the generation of a high heat capacity signal. Third, the ability to deconvolute the signal into the heat capacity and kinetic components allows the operator to assign glass transitions with considerably more confidence than is possible using conventional DSC. When a glass transition is weak and set against a rising baseline owing to the gradual increase in heat capacity of other components, the presence of a relaxation endotherm can give the impression of a melt or some other endothermic process rather than a glass transition. A clear step change in the reversing signal makes a correct assignment unequivocal in most cases. A fourth advantage is that quantification of amorphous phases is made more accurate, particularly in terms of assigning a value to T_g, which would otherwise be difficult due to the presence of the relaxation endotherm and also in terms of assessing the heat capacity change through the glass transition, either by direct observation of the reversing heat flow or by analysis of the derivative of the reversing heat capacity with respect to temperature. This approximates very well to a Gaussian distribution, and numerical fitting procedures can be used to quantify multiphase systems with overlapping transitions (11,26–31).

The ability to visualize the reversing and nonreversing signals separately has many concomitant advantages in terms of more sophisticated analysis such as the

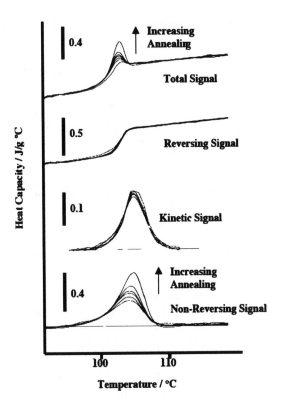

FIGURE 4.7 Results for MTDSC experiments on polystyrene annealed for different lengths of time up to 45 min. The inset bars are a guide to the scale for each signal.

calculation of relaxation times from the magnitude of the endotherm seen in the nonreversing signal as a function of annealing time. However, it is essential to have a reasonable grasp of the principles and limitations associated with the measurement to avoid substantial errors when treating the data quantitatively. This is exemplified by Figure 4.7 depicting the effects of annealing on polystyrene that has been annealed for time periods up to 45 min. It can be seen that, as expected, the total signal is the same as that observed for a conventional DSC experiment. As annealing increases, the characteristic endothermic peak at the glass transition increases. In contrast, the reversing and kinetic signals are largely unaffected by annealing; thus, the nonreversing signal shows an increasing peak with annealing time.

However, some caution is required in interpreting such data as the frequency dependence of the glass transition. It is well known that the temperature of the glass transition is frequency dependent, from measurements made with dynamic mechanical analysis and dielectric thermal analysis (32). This same frequency dependence is seen in MTDSC (33). Figure 4.8 shows the results for the glass transition of polystyrene run at a variety of modulation frequencies. T_g is seen to occur at lower temperatures as the frequency decreases, with a lower apparent signal seen for the T_g measured in the total heat flow, which is of course not obtained from the modulation but from the underlying signal. The situation is in many ways analogous to

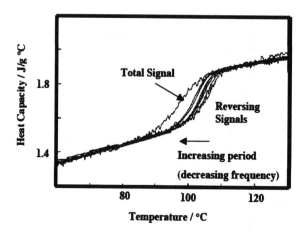

FIGURE 4.8 MTDSC results for polystyrene cooled at 1°C/min using different modulation periods from 10 to 100 sec.

the cooling rate dependence of the glass transition seen with conventional DSC. When we consider a cooling experiment with MTDSC we have both a cooling rate, β, and a frequency (the frequency of the modulation, ϕ) to take into account. The result seen in the reversing signal is largely independent of the cooling rate once it is slow enough to ensure many modulations, over the transition region (bearing in mind that when there are not many modulations the result is meaningless in any case). Thus, as we keep the frequency of the modulation constant and vary the underlying cooling rate, T_g seen in the average signal changes whereas that seen in the reversing signal remains the same.

Similarly, as shown in Figure 4.8, if we keep the cooling rate the same and vary the frequency, the underlying signal remains constant, whereas the reversing signal changes. In an MTDSC experiment, the average signal will always give a lower T_g than the reversing signal because the underlying measurement must, in some sense, be slower (i.e., on a longer time scale) than the reversing measurement. As a consequence of this, as the sample is cooled there is a peak in the nonreversing signal that is clearly not related to annealing but is a consequence the difference in effective frequency between the average measurement and that of the modulation; this is sometimes termed the *frequency effect* and must be accounted for when calculating relaxation times from the endothermic peak. The frequency effect may also be seen on inspection of Figure 4.7; here we see a peak in the non-reversing signal that is due to the annealing process but also contains a contribution from the difference in T_g calculated from the reversing and the total heat flow signals. We can consider these effects, to a first approximation, to be additive; hence, the non-reversing signal gives a measure of the enthalpy loss on annealing with an offset due to the frequency difference.

We can consider a simple model for the glass transition that consists of only a single relaxation process. For the response to the modulation, it has been shown that the reversing (real) and kinetic (imaginary) heat capacities (C_{pR} and C_{pK}) may be approximated as follows (11,30,34):

$$C_{pR} = \Delta Cp/(1 + \exp(-2\Delta h*/(T - T_g)/RT_g^2))   \quad (4.18)$$

$$C_{pK} = \Delta Cp \exp(-2\Delta h*/(T - T_g)/RT_g^2)/(1 + \exp(-2\Delta h*/(T - T_g)/RT_g^2))   \quad (4.19)$$

where ΔC_p = the heat capacity change at the glass transition, $\Delta h*$ = the apparent activation energy, and T_g = the glass temperature.

It should be borne in mind that these calculations are based on a very simple model of the glass transition and cannot be expected to accurately model real systems. However, it is adequate for our purposes here. The fact that C_{pR} is time scale dependant in this case can be made explicit by using the notation $C_{p\phi}$ for the response to the modulation and $C_{p\beta}$ for the response to the underlying heating or cooling rate. Thus, Equation 4.20 can be considered to model an experiment in which a single frequency of modulation is used, and the sample is cooled from above the glass transition at β; some annealing is carried out, and then the sample is heated at β, viz

$$dQ/dt = \beta C_{p\beta} + \langle f(t,T) \rangle \ldots \text{ the underlying signal}$$
$$+ C_{p\phi}B\phi \cos \phi t + C \sin \phi t \ldots \text{ the response to the modulation} \quad (4.20)$$

where $C_{p\phi}$ = the reversing heat capacity at the frequency ϕ, $C_{p\beta}$ = the reversing heat capacity implied by the cooling rate β, $C = C_{pK}B\phi$ where C_{pK} is given by Equation 4.19, and $\langle f(t,T) \rangle$ = some function that represents the recovery of enthalpy lost below the glass transition after cooling at β.

The form of C_{pK} in Equation 4.19 is very different from the case of an Arrhenius-type process. However, it is still basically a manifestation of the kinetics of the glass transition and, thus, the concept that this signal (and thus C from Equation 4.20) is a measure of the kinetics of the transition, remains valid. Even without annealing, the underlying signal on heating at β (after cooling at β) will not be identical, and so some hysteresis will always be present. In this formulation, this hysteresis is the minimum contribution from $\langle f(t,T) \rangle$, which will increase with annealing.

We can now express the nonreversing heat flow as:

$$\text{Nonreversing heat flow} = \beta(C_{p\beta} - C_{p\phi}) + \langle f(t,T) \rangle \quad (4.21)$$

There is a contribution from the differing "frequencies" of the cooling rate and the frequency of the modulation expressed in $\beta(C_{p\beta} - C_{p\phi})$, as well as a contribution from the effect of annealing expressed as $\langle f(t,T) \rangle$. This then invites the question of how we define $C_{p\beta}$ and relate it to $C_{p\phi}$. One approach is to define $C_{p\beta}$ as the result obtained on cooling at β. This means there is no contribution from $f(t,T)$ except on heating, when it embodies the effects of annealing and reheating. This contribution would tend to some relatively small value, as the annealing time was reduced. An approach to defining the relationship between $C_{p\phi}$ and $C_{p\beta}$ is to give a modulation frequency that gives the same glass transition temperature as a given cooling rate. This glass transition temperature equivalence could be defined as either the temperature at which the sample is 50% devitrified or at which the derivative of the heat capacity reaches a maximum, or the intersection of the extrapolated enthalpy lines. Each measure would give a slightly different equivalent frequency. It has already been observed that the underlying signal on cooling is not the same in shape as the reversing signal (33,34); thus, a simple frequency equivalence is not strictly accurate, and distribution of frequencies would be required as a function of extent of vitrification. Whether one of these alternatives or another approach is taken is largely a matter of convention.

In summary, the important concepts embodied by this discussion are as follows. The glass transition, as measured by the reversing heat capacity, is a function of frequency. T_g as measured by the total heat capacity ($= <dQ/dt>/$) on cooling is a function of cooling rate. Broadly, there is an equivalence between these two observations because changing both the frequency of the modulation and the cooling rate changes the time scale over which the measurement is made. This means that there is always, in the nonreversing signal, a contribution from $\beta(C_{p\beta} - C_{p\phi})$ that is present, regardless of annealing (for example, it is present when cooling). Aging below the glass transition produces enthalpy loss that is recovered as a peak overlaid on the glass transition. However, this aging does not have a great effect on the reversing signal, and this is intuitively satisfactory as the aging effect is not reversible on the time scale of the modulation. This means that the nonreversing signal includes a contribution from the different "frequencies" of the cyclic and underlying measurements plus a contribution from annealing expressed as $f(t,T)$ in Equation 4.21. This implies that the relationship between the enthalpy loss on annealing and the area under the nonreversing peak should be linear. In reality, at higher degrees of annealing, the reversing signal is significantly affected, thus this simple relationship breaks down giving rise to overestimates as high as 20%. However, the increase in the area underneath the nonreversing peak still increases monotonically with annealing; thus, ranking samples is still possible. Alternatively, a correction can be applied (11), but discussion of this is beyond the scope of this chapter.

As mentioned earlier, there is an extensive literature available on the use of MTDSC for the study of the glass transition. For further information the interested reader is referred to Reference 35 through Reference 43, among others.

4.3.2 FURTHER THERMAL EVENTS CHARACTERIZED BY MTDSC

4.3.2.1 Phase Separation

A highly important issue within the polymer sciences, and one that has great relevance to pharmaceutical systems such as drug or plasticizer-loaded polymeric matrices and films, is that of phase separation. Many binary systems are partially miscible, the extent of phase separation showing both compositional and temperature dependence. The phase separation of polymers in the liquid state may be characterized by a lower or upper critical solution temperature (LCST and UCST) on heating or cooling, respectively, these values usually being identified using optical microscopy (see Figure 4.9). Although thermal analysis has been used as a supplementary

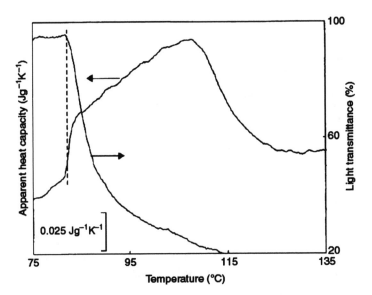

FIGURE 4.9 Apparent heat capacity and optical microscopy measurement (% light transmission) of a 75/25 poly(oxyethylene)/poly(ether sulfone) blend. (From Dreezen, G., Groeninckx, G., Swier, S., and Van Mele, B., *Polymer*, 42, 1449, 2001. With permission.)

technique to monitor these processes, such studies are hampered by the small demixing enthalpy and the slow rate of the diffusion-controlled process.

Dreezen et al. (44) investigated the use of MTDSC as a means of studying the phase separation of poly(ethylene oxide) with poly(ether sulfone) and poly(3,4'-diphenylene ether isophtaloyl amide). The authors were able to show that by measuring the heat capacity from the MTDSC reversing heat flow signal, it was possible to monitor the onset of the demixing process (Figure 4.10). In addition, by performing isothermal experiments at two different temperatures, the authors were able to directly monitor the demixing process (Figure 4.11), ascribing the decrease in heat capacity to a combination of time-dependent evolution of the composition of the coexistent phases to a value commensurate with that calculated from the additivity of the heat capacities of the two components in the case of the studies at 103°C, whereas the data obtained at 138°C also reflect devitrification of the PES-rich phase.

4.3.2.2 Curing and Chemical Reactions

There has been an extensive body of work, particularly from the group of Van Mele, on the use of MTDSC to study curing processes (e.g., Reference 11, Reference 45 to Reference 47). The group has examined the use of MTDSC for both isothermal and nonisothermal curing processes, with emphasis on examining the devitrification process. When a material cross-links during curing, the glass transition increases until it becomes equivalent to the curing temperature. At this point, the system enters the glassy state, and the rate of the cure process decreases. Van Assche et al. (47) examined the curing of a range of thermosetting systems and were able to demon-

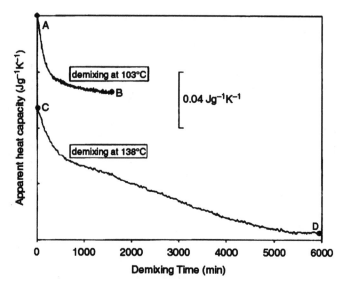

FIGURE 4.10 Quasi-isothermal experiments showing the apparent heat capacity of a 75/25 poly(oxyethylene)/poly(ether sulfone) blend at two temperatures as a function of demixing time. (From Dreezen, G., Groeninckx, G., Swier, S., and Van Mele, B., *Polymer*, 42, 1449, 2001. With permission.)

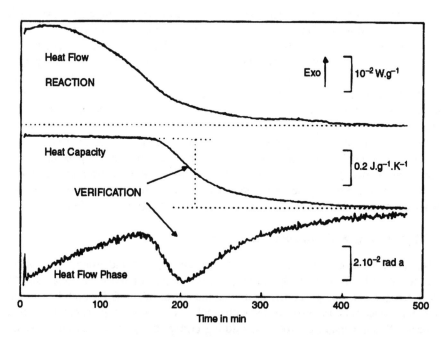

FIGURE 4.11 Quasi-isothermal cure and vitrification of an epoxy-anhydride resin at 80°C at a frequency of 1/60Hz. (From Van Assche, G., Van Mele, B., and Saruyama, Y., *Thermochim. Acta*, 377, 125, 2001. With permission.)

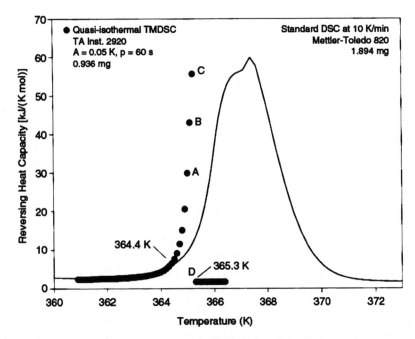

FIGURE 4.12 Reversing heat capacity of *n*-pentacontane by standard DSC and quasi-isothermal MTDSC. (From Pak, J., Boller, A., Moon, I., Pyda, M., and Wunderlich, B., *Thermochim. Acta*, 357–358, 259, 2000. With permission.)

strate that the change in heat capacity associated with the glass transition could be distinguished from the reaction exotherm (Figure 4.12). This important effect is not visible using conventional DSC, thereby demonstrating the advantage of using the modulated technique for this application.

4.3.2.3 Melting

As mentioned previously, nonisothermal measurements during melting processes tend to show responses in both the reversing and nonreversing signals because of the difficulties associated with truly modulating a sample during a melting process, often overlaid on a complex process of melting and crystallization. This is unfortunate as one of the most common uses for conventional DSC within the pharmaceutical field is the study of polymorph melting; hence, the method is apparently inapplicable to one of the most important potential applications of the technique. In fact, it is possible to study melting processes using MTDSC, even for low-molecular-weight materials that have narrow melting points, by using the method in quasi-isothermal mode. This method has been pioneered by the group of Wunderlich and involves measuring the reversing heat flow of the sample at a series of temperature increments through the transition (11). By adapting this quasi-isothermal approach (outlined in Subsection 4.2.2.12), it is possible to determine the reversibility of the process and to obtain the isothermal temperature dependence of the melting transition. Interestingly, the group has found that low-molecular-weight systems melt reversibly as long as crystal nuclei are still present on which crystal growth may occur within the time frame of the modulation, whereas high-molecular-weight materials tend to melt irreversibly because of the time scales required for nucleation and growth being long in the context of the modulation (48). Nevertheless, reversible melting for polymers has been observed (11).

The approach is illustrated for *n*-pentacontane in Figure 4.12. The standard DSC response is shown via the solid curve, and the quasi-isothermal response is shown as a series of filled circles. As the sample is heated through the melt, it may be seen that the process is in fact complete (point D) at a temperature well below that indicated by conventional DSC. In addition, it is possible to summate the endotherms and exotherms for the complete data set (including those associated with heating the sample between successive temperature increments), the value being in good agreement with the total heat of fusion measured using conventional DSC. However, the approach allows the operator to then examine the temperature range over which the majority of the melting takes place. In the present case, it was noted that although the melting occurs over a narrow temperature range of about 1 K, the majority of this process (66%) occurs over a range of 0.1 K from 365.25 to 365.35 K. Similarly, by examining the equivalent heat flow responses as a function of time (Figure 4.13), it may be seen that up to region D the signals are symmetrical and rapidly reach steady state, with the majority of the associated heat of fusion being related to the incremental advances between isothermal runs. Consequently, up to and including

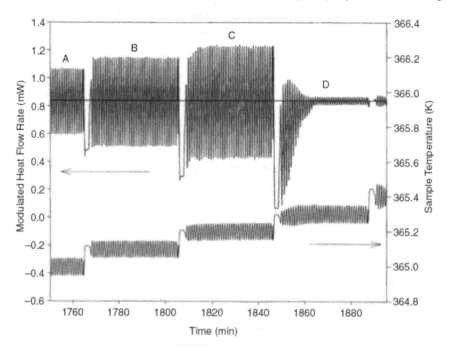

FIGURE 4.13 Modulate heat flow rate and sample temperature plotted vs. time for n-pentacontane by quasi-isothermal MTDSC. (From Pak, J., Boller, A., Moon, I., Pyda, M., and Wunderlich, B., *Thermochim. Acta*, 357–358, 259, 2000. With permission.)

measurement C, the melting and crystallization are reversible in each modulation cycle. At point D, however, more melting occurs in the first few cycles than does crystallization; hence, steady state is lost until the system finally settles in to the reversible heat flow characteristics of the liquid.

4.3.2.4 Crystallization

Crystallization processes are simpler to study using MTDSC than melting, because of the kinetics of the nucleation or growth process being such that the enthalpic change over the course of a single modulation may be sufficiently small so as to render the system effectively nonreversing, with only the heat capacity component seen in the reversing signal. Furthermore, as crystallization is usually governed by Arrhenius-type kinetics, the signals may be analyzed via

$$dQ/dt = \beta C_{pR} + <Hf(\alpha)Ae^{-E/RT}> + C_{pR}B\beta\omega \cos \omega t + C \sin \omega t \qquad (4.22)$$

where $C = Bf(<x>) \cdot d(HAe^{-E/R<T>})/dT$.

The first two terms on the right-hand side refer to the underlying signal and the last two to the response to the modulation. The inclusion of the expression $Hf(\alpha)Ae^{-E/RT}$ simply refers to the kinetic model used to describe the crystallization process.

Scherrenberg et al. (49) used quasi-isothermal measurements to study the crystallization of polyethyelene, including in their investigation an examination of the effects of experimental parameters on the crystallization process. In particular, they emphasize the necessity of establishing conditions whereby the change in the degree of crystallinity during a single modulation is small, so as to ensure that the reversing heat flow is a function of the underlying heat capacity rather than the kinetic events associated with the recrystallization. Similarly, the authors point out that the crystallization process should not be influenced by the modulation parameters applied. The possibility of this being a significant influence on the data obtained is shown in Figure 4.14, whereby the increased amplitude leads to a more rapid crystallization process. Indeed, the authors strongly recommend that the system be checked to ensure that it is in steady state, possibly by using Lissajous analysis, as otherwise misleading information may be obtained.

4.4 PHARMACEUTICAL APPLICATIONS OF MTDSC

4.4.1 Basic T_g Measurements

Several examples are now available whereby the glass transitions and relaxation endotherms of amorphous drugs have been deconvoluted to the extent that this has now effectively become routine practice. One of the earliest such studies was that of Royall et al. (50) in which the glassy behavior of amorphous saquinavir was examined. In this investigation, the authors demonstrated the effectiveness of MTDSC as a means of visualizing the glass transition of a low-molecular-weight drug in isolation from the relaxation endotherm, as indicated in Figure 4.15. In addition, the paper describes the effect of modulation parameters on the data obtained

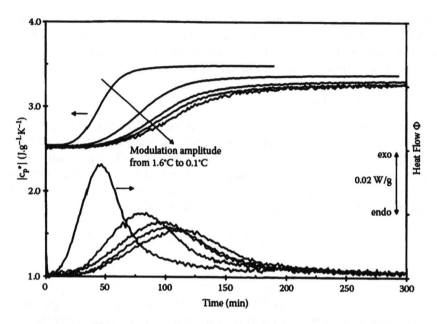

FIGURE 4.14 The effect of modulation amplitude on the recrystallization behavior of linear polyethylene. (From Scherrenberg, R., Mathot, V., and Van Hemelrijck, A., *Thermochim. Acta*, 330, 3, 1999. With permission.)

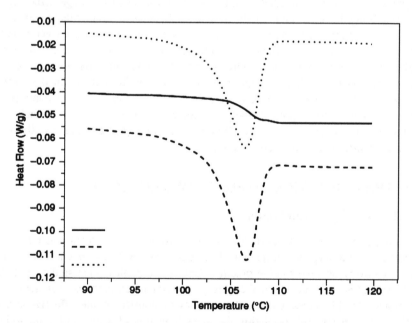

FIGURE 4.15 MTDSC response of amorphous saquinavir on heating. (From Royall, P.G., Craig, D.Q.M., and Doherty, C., *Pharm. Res.*, 15, 1117, 1998. With permission.)

and the potential use of quasi-isothermal studies to obtain T_g values at minimal heating rates.

The authors used the accurate measurement of T_g to study the plasticization effects of water (51). In particular, the study demonstrated that it was possible to build up a theoretical diagram using which the stability of a glassy drug could be predicted as a function of storage temperature and sorbed water levels. This analysis was based on the concept that storage above T_g is likely to result in recrystallization, whereas storage more than 50°C below T_g will maintain the drug in the amorphous state; intermediate storage temperatures would result in there being some risk of recrystallization. However, as most drugs are to some extent hydroscopic, one must consider the T_g of the plasticized material as well as of dry glass.

The authors suggested that the critical water content w_c at which the T_g of the plasticized glass falls below that of the storage temperature (T_S) may be predicted from

$$w_c = \left[1 + \frac{Tg_2\rho_2(T_S - 135)}{135(Tg_2 - T_S)}\right]^{-1} \tag{4.23}$$

where T_{g2} and ρ_2 are the glass transition and density of the unplasticized glass, respectively. Such a diagram is shown in Figure 4.16 for saquinavir, thereby allowing the operator to have some idea of how stable a material is likely to be as a function of water content.

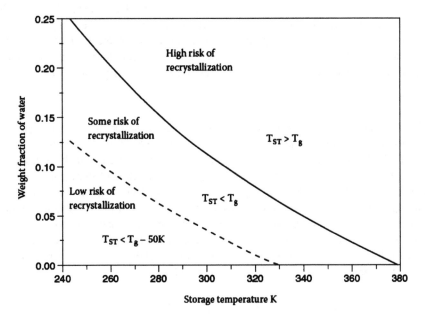

FIGURE 4.16 Predicted storage stability of saquinavir as a function of temperature and water content. (From Royall, P.G., Craig, D.Q.M. and Doherty, C., *Int. J. Pharm.*, 190, 39, 1999. With permission.)

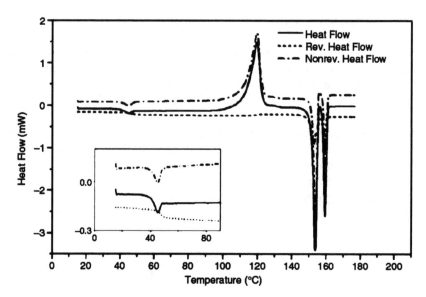

FIGURE 4.17 MTDSC response of amorphous indomethacin, with details of the glass transition region shown inset. (From He, R. and Craig, D.Q.M., *J. Pharm. Pharmacol.*, 53, 41, 2001. With permission.)

In a later study, He and Craig (52) used MTDSC to identify the glass transition temperature of amorphous indomethacin, with Figure 4.17 showing T_g and the relaxation endotherm in the reversing and nonreversing signal. The recrystallization is seen in the nonreversing signal only, with only the decrease in heat capacity (increase in reversing heat flow) seen in the reversing signal through this transition. The melting response is seen in both signals for the reasons given earlier.

4.4.2 RELAXATION BEHAVIOR

MTDSC may also be used to assess the relaxation behavior of glassy drugs. This should be considered in the context of the existing body of work in which conventional DSC has been used to measure the relaxation time of a drug via the assessment of magnitude of the endothermic relaxation peak as a function of storage temperature and time (e.g., Reference 53 through Reference 55). In brief, the magnitude of the enthalpy change is considered to represent the relaxation of the glass to the extrapolated liquid state; in other words, the relaxation process reflects the volume relaxation back to that of a supercooled liquid equivalent to that directly extrapolated from the liquid curve. At the temperature of storage (T_S) the maximum corresponding enthalpy change is given by $\Delta H_\infty {}_{(TST)}$, estimated via

$$\Delta H_{\infty(T_S)} = (Tg - T_S)\Delta C_p \tag{4.24}$$

where T_g is the (fictive) glass transition temperature, and ΔC_p is the change in heat capacity between the glass and the liquid at T_g. The measured enthalpic recovery ΔH at temperature (T_S) and time (t_S) is then given by the empirical equation

$$\Delta H(t_S T_S) = \Delta H_\infty(T_S)[1 - \Phi(t_S)] \qquad (4.25)$$

where $\Phi(t)$ is the relaxation function, which may be related to the relaxation time via the Williams–Watt expression

$$\Phi(t) = \exp[-(t/t_c)^\beta] \qquad (4.26)$$

where β is a power law function that is associated with the distribution of relaxation times, and t_c is a characteristic time that is usually considered to be equivalent to the average relaxation time. When using conventional DSC, the approach involves measuring the magnitude of the relaxation peak as a function of time at a series of storage temperatures from which the relaxation time as a function of temperature may be ascertained; this method has been used by Hancock et al. (56) to measure relaxation behavior below T_g for pharmaceutical materials. MTDSC potentially offers some advantages over the conventional technique; the relaxation endotherm may be measured in isolation in the nonreversing signal and the value of ΔC_p may be easily obtained (given the provisos stated earlier). This was demonstrated for amorphous lactose (57), whose relaxation time was measured as a function of temperature (Figure 4.18). However, the frequency effect mentioned earlier needs to be taken into account. A later study by Van den Mooter et al. (58) used a similar approach for amorphous ketoconazole, and Six et al. (59) also used this method to

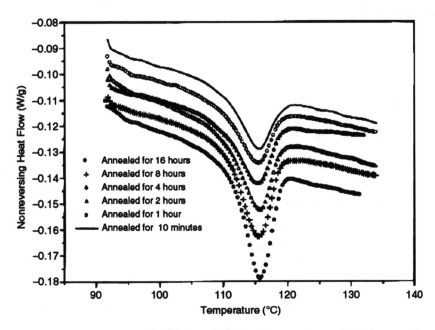

FIGURE 4.18 Nonreversing heat flow as a function of storage time at 90°C for amorphous lactose. (From Craig, D.Q.M., Barsnes, M., Royall, P.G., and Kett, V.L., *Pharm. Res.*, 17, 696, 2000. With permission.)

estimate the relaxation time of a series of structural analog of itraconazole (micon-
azole and ketoconazole), reporting a relationship between molecular complexity and
the relaxation time.

4.4.3 POLYMERS AND DRUG DELIVERY SYSTEMS

MTDSC has also been used to characterize drug delivery systems, particularly those
based on polymeric materials. Hill et al. (60) studied progesterone-loaded polylactide microspheres, demonstrating plasticization of these materials by the drug up
to concentrations of 30% w/w, at which the drug was present as a separate amorphous phase on the surface of the spheres. On increasing the drug loading to 50%
w/w, the external layer was crystalline, reflected by an absence of a plasticization
effect and the melting of two polymorphs of crystalline progesterone. Passerini and
Craig (61) studied the glassy behavior of cyclosporin, an undecapeptide used as an
immunosuppressant. The authors were studying the encapsulation of this material
into polylactic acid microspheres and noted that the drug as received showed a small
transition at about 120°C that had not been previously reported. These authors (62)
also examined the potential plasticization effect of water on PLA microspheres,
demonstrating that residual water resulting from the manufacturing process could
markedly reduce the T_g of the spheres (Figure 4.19). The effect of water plasticization was also studied by Steendam et al. (63), who used the Couchman–Karasz
approach to model the decrease in T_g as a function of water content. These authors
also studied the relaxation behavior of the PLA, using axial expansion and thermomechanical analysis.

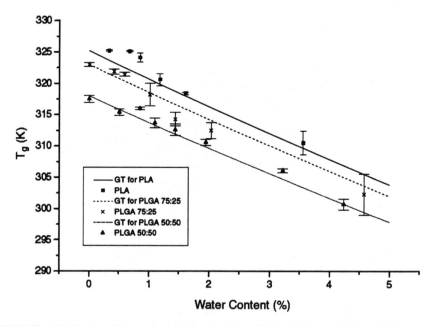

FIGURE 4.19 Gordon Taylor relationships for PLA, PLGA 75:25, and PLGA 50:50.

An area in which MTDSC could perhaps be more extensively used is in the field of biomaterials and biodegradable polymers, as the physical characteristics of these materials may be intimately associated with their functionality. An example of such a study is that of Cesaro et al. (64), in which poly(hydroxybutyrate), a biodegradable polymer that is a substrate for soil microorganisms, was studied in relation to the effect of preparation conditions on the degree of crystallinity. The authors used MTDSC to demonstrate that the glass transition could be clearly visualized for the quench-cooled systems but was also present for unquenched samples, indicating that there was an amorphous component to the polymer, irrespective of conditioning.

There has been a persistent interest in the development of polymeric matrices for slow or fast release of drugs. Wang et al. (65) studied ternary solid dispersions of PVP VA64, Myrj 52, and itraconazole, prepared by a solvent evaporation technique. The authors used MTDSC in conjunction with XRD to demonstrate that the drug was miscible with the PVP VA64 in the solid state although, interestingly, the presence of the Myrj 52 resulted in phase separation, with concomitant implications for drug release behavior. Similarly, Tatavati et al. (66) studied the release of two weakly basic drugs (papaverine hydrochloride and verapamil hydrochloride) from hydrophilic bases in the presence and absence of a methacrylic (Eudragit L100) and acrylic acid (Carbopol 71G), using MTDSC and FTIR to detect an interaction product between the Eudragit and verapamil HCl. Hussain et al. (67) used MTDSC in combination with atomic force microscopy to image and characterize cyclosporine A film dispersions in HPMC, noting that the two components were phase-separated on a nano- or microscale, depending on the proportions used. The authors also noted that MTDSC was able to confirm phase separation via measurement of T_g of the two components, which was found to be unchanged on preparing the dispersions.

A further area in which more studies could be usefully performed is the investigation of gelling systems using MTDSC. Tan et al. (68) studied the glass transition of maize starch during the gelatinization process, noting that the T_g overlapped with the gelatinization peak for all maize starch formulations studied. The authors proposed an hypothesis for the gelation process based on the side-chain liquid crystalline model for starch that incorporated both the observed gelation and glass transitional behavior. Although starch gelation is of clear importance for pharmaceutical applications in its own right, this paper is of particular interest because of the wider implications for the study of polymeric-controlled release systems.

4.5 CONCLUSIONS

The chapter has attempted to outline the theoretical and practical principles of MTDSC as well as give an indication as to the manner in which the technique may be used within the pharmaceutical sciences. The approach has now become something of an industry standard, with MTDSC capability now supplied as a matter of course for many instruments. Clearly, the method has great potential and can usefully be further explored within the pharmaceutical context. However, it is also essential to be aware of the limitations of the technique and the problems for which it is best suited, so that artifacts can be avoided. Nevertheless, the method provides a highly useful supplementary tool to conventional DSC and is likely to remain an integral part of any modern physical characterization laboratory.

REFERENCES

1. Seferis, J.C., Salin, I.M., Gill, P.S., and Reading, M., *Proc. Acad. Greece*, 67, 311, 1992.
2. Gill, P.S., Sauerbrunn, S.R., and Reading, M., *J. Therm. Anal.*, 40, 931, 1993.
3. Reading, M., Elliot, D., and Hill, V.L., *J. Therm. Anal.*, 40, 949, 1993.
4. Reading, M., *Trends Polym. Sci.*, 1, 248, 1993.
5. Reading, M., Elliott, D., and Hill, V., *Proc. 21st NATAS Conf.*, 145, 1992.
6. Reading, M., *Thermochim. Acta*, 238, 295, 1994.
7. Jiang, Z., Imrie, C.T., and Hutchinson, J.M., *Thermochim. Acta*, 387, 75, 2002.
8. Aubuchon, S.R. and Gill, P.S., *J. Therm. Anal.*, 49, 1039–1044, 1997.
9. Verdonck, E., Schaap, K., and Thomas, L.C., *Int. J. Pharm.*, 192, 3, 1999.
10. *Modulated Temperature Differential Scanning Calorimetry: Theory and Practical Applications in Polymer Characterisation*, Vol. 6: *Hot Topics in Thermal Analysis and Calorimetry*, Eds., Reading, M. and Hourston, D.J., Springer-Verlag, Dordrecht, the Netherlands, 2006.
11. Varma-Nair, M., Wunderlich, B., Balogh, J.J., and Aldrich, H., *Proc. 23rd NATAS Conf.*, 26, 1994.
12. Hill, V.L., Craig, D.Q.M., and Feely, L.C., *Int. J. Pharm.*, 192, 21, 1999.
13. Sauerbrunn, S.R. and Blaine, R.L., *Proc. 23rd NATAS Conf.*, 57, 1994.
14. Sauerbrunn, S.R., Gill, P.S., and Foreman, J.A., Modulated DSC: the effect of period, *Proc. 23rd NATAS Conf.*, 51, 1994.
15. Jones, K.J., Kinshott I., Reading M., Lacey A.A., Nikolopoulos C., and Pollock H.M., *Thermochim. Acta*, 305, 187, 1997.
16. Cser, F., Rasoul, F., and Kosior, E., *J. Therm. Anal.*, 50, 727, 1997.
17. Varma-Nair, M. and Wunderlich, B., *J. Therm. Anal.*, 46, 879, 1996.
18. Jiang, Z., Imrie, C.T., and Hutchinson, J.M., *Thermochim. Acta*, 336, 27, 1999.
19 Reading, M. and Luyt, R., *J. Therm. Anal. Calorimetry*, 54, 535, 1998.
20. Lacey, A.A., Nikolopoulos, C., and Reading, M., *J. Therm. Anal.*, 50(1–2), 279, 1997.
21. Wunderlich, B., Jin, Y., and Boller, A., *Thermochim. Acta*, 238, 277, 1994.
22. De Meuter, P., Rahier, H., and Van Mele, B., *Int. J. Pharm.*, 192, 77, 1999.
23. De Meuter, P., Amelrijckx, J., Rahier, H., and Van Mele, B., *J. Polym. Sci.: Part B: Polym. Phys.*, 37, 2881, 1999.
24. Hohne, G.W.H. and Kurelec, L., *Thermochim. Acta*, 377, 141, 2001.
25. Pak, J., Boller, A., Moon, I., Pyda, M., and Wunderlich, B., *Thermochim. Acta*, 357–358, 259, 2000.
26. Song, M., Hammiche, A., Pollock, H.M., Hourston, D.J., and Reading, M., *Polymer*, 36, 3313, 1995.
27. Song, M., Hammiche, A., Pollock, H.M., Hourston, D.J., and Reading, M., *Polymer*, 3725, 5661–5665, 1996.
28. Song, M., Pollock, H.M., Hammiche, A., Hourston, D.J., and Reading, M., *Polymer*, 383, 503, 1997.
29. Hourston, D.J., Song, M., Hammiche, A., Pollock, H.M., and Reading, M., *Polymer*, 381, 1, 1997
30. Song, M., Hourston, D.J., Pollock, H.M., Schafer, F.U., and Hammiche, A., *Thermochim. Acta*, 304/305, 335, 1997.

31. Song, M., Hourston, D.J., Reading, M., Pollock, H.M., and Hammiche, A., *J. Therm. Anal. Calorimetry*, 56, 991, 1999.
32. Reading, M. and Haines, P.J., Thermomechanical, dynamic mechanical and associated methods, in *Thermal Methods of Analysis Principles, Applications and Problems*, Haines, P.J., Ed., Blackie Academic and Professional, Glasgow, U.K., 1995.
33. Reading, M., Jones, K.J., and Wilson, R., *Netsu Sokutie*, 22, 83, 1995.
34. Jones, K.J., Kinshott, I., Reading, M., Lacey, A.A., Nikolopoulos, C., and Pollock, H.M., *Thermochim. Acta*, 304/305, 187, 1997.
35. Swier, S., Van Assche, G., and Van Mele, B., *J. Appl. Polym. Sci.*, 91, 2798, 2004.
36. Hutchinson, J.M. and Montserrat, S., *Thermochim. Acta*, 377, 63, 2001.
37. MacDonald, I., Clarke, S., Pillar, R., Giric-Markovic, M., Marisons, J., *Z. Therm. Anal. Calorimetry*, 80, 781, 2005.
38. Boller, A., Schick, C., and Wunderlich, B., *Thermochim. Acta*, 266, 97, 1995.
39. Boller, A., Okazaki, I., and Wunderlich, B., *Thermochim. Acta*, 284, 1, 1996.
40. Hutchinson, J.M. and Montserrat, S., *J. Therm. Anal.*, 47, 103, 1996.
41. Hutchinson, J.M. and Montserrat, S., *Thermochim. Acta*, 286, 263, 1996.
42. Wagner, T. and Kasap, S.O., *Philos. Mag. B*, 74, 667, 1996.
43. Thomas, L.C., Boller, A., Okazaki, I., and Wunderlich, B., *Thermochim. Acta*, 291, 85, 1997.
44. Dreezen, G., Groeninckx, G., Swier, S., and Van Mele, B., *Polymer*, 42, 1449, 2001.
45. Van Assche, G., Van Hemelrijck, A., Rhanier, H., and Van Mele, B., *Thermochim. Acta*, 286, 209, 1996.
46. Van Assche, G., Van Hemelrijck, A., Rhanier, H., and Van Mele, B., *Thermochim. Acta*, 304/305, 179, 1997.
47. Van Assche, G., Van Mele, B., and Saruyama, Y., *Thermochim. Acta*, 377, 125, 2001.
48. Wunderlich, B., Okazaki,I., Ishikiriyama, K., and Boller, A., *Thermochim. Acta*, 324, 77, 1998.
49. Scherrenberg, R., Mathot, V., and Van Hemelrijck, A., *Thermochim. Acta*, 330, 3, 1999.
50. Royall, P.G., Craig, D.Q.M., and Doherty, C., *Pharm. Res.*, 15, 1117, 1998.
51. Royall, P.G., Craig, D.Q.M. and Doherty, C., *Int. J. Pharm.*, 190, 39, 1999.
52. He, R. and Craig, D.Q.M., *J. Pharm. Pharmacol.*, 53, 41, 2001.
53. Cowie, J.M.G. and Ferguson, R., *Macromolecules*, 22, 2307, 1989.
54. Brunacci, A., Cowie, J.M.G., Ferguson, R., and McEwen, I.J., *Polymer*, 38, 3263, 1997.
55. Richardson, M.J., Susa, M., and Mills, K.C., *Thermochim. Acta*, 280, 383, 1996.
56. Hancock, B.C., Shamblin, S.L., and Zografi, G., *Pharm. Res.*, 12, 799, 1995.
57. Craig, D.Q.M., Barsnes, M., Royall, P.G., and Kett, V.L., *Pharm. Res.*, 17, 696, 2000.
58. Van den Mooter, G., Craig, D.Q.M., and Royall, P.G., *J. Pharm. Sci.*, 90, 996, 2001.
59. Six, K., Verreck, G., Peeters, J., Augustijns, P., Kinget, R., and Van den Mooter, G., *Int. J. Pharm.*, 213, 163, 2001.
60. Hill, V.L., Craig, D.Q.M., and Feely, L.C., *J. Therm. Anal.*, 54, 673, 1998.
61. Passerini, N. and Craig, D.Q.M., *J. Pharm. Pharmacol.*, 54, 913, 2002.
62. Passerini, N. and Craig, D.Q.M., *J. Controlled Release*, 73, 111, 2001.
63. Steendam, R., van Steenbergen, M.J., Hennink, W.E., Frijlink, H.W., and Lerk, C.F., *J. Controlled Release*, 70, 71, 2001.
64. Cesaro, A., Navarini, L., and Pepi, R., *Thermochim. Acta*, 227, 157, 1993.
65. Wang, X., Michoel, A., and Van der Mooter, G., *Int. J. Pharm.*, 303, 54, 2005.
66. Tatavati, A.S., Mehta, K.A., Augsburger, L.L., and Hoag, S.W., *J. Pharm. Sci.*, 93, 2319, 2004.

67. Hussain, S., Grandy, D.B., Reading, M., and Craig, D.Q.M., *J. Pharm. Sci.*, 93, 1672, 2004.
68. Tan, I., Wee, C.C., Sopade, P.A., and Halley, P.J., *Carbohydr. Polym.*, 58, 191, 2004.

5 Thermogravimetric Analysis: Basic Principles

Andrew K. Galwey and Duncan Q.M. Craig

CONTENTS

5.1 INTRODUCTION

The terms *thermogravimetry* (TG), *differential thermogravimetry* (DTG), and *thermogravimetric analysis* (TGA) describe any experimental method whereby changes in mass are used to detect and to measure the chemical and, less frequently, the physical (e.g., sublimation) processes that occur on heating a reactant under investigation. This laboratory technique has been extensively and successfully applied in studies of diverse reactions that occur on heating substances, including a wide variety of pure compounds, natural materials, mixtures, etc., many of which are initially solid. TG observations have been particularly fruitful in measurements of reaction rates. Data obtained by this method are often more accurate than those from other techniques but usually require the support of complementary chemical and structural characterizations to identify and to confirm the reaction stoichiometry. To a lesser extent, TG has been used to study the thermal behavior of liquid reactants and gas–solid reactions and to measure some thermodynamic quantities. In the pharmaceutical sciences, usage of the technique is extensive but arguably narrow, most studies using the method to measure the temperature range in which dehydration occurs and the quantity of water lost from solid drug or excipient systems. This topic is reviewed in Chapter 6. A key objective of this chapter is therefore to outline in some detail the possibilities that the technique affords, particularly, with reference to studies that, although outside the immediate medical or pharmaceutical arena, demonstrate principles that are applicable to systems in which most readers of this book would be interested. Furthermore, it is hoped that this account will aid pharmaceutical scientists in understanding accounts of the use of the TG method appearing in the wider literature and appreciate the untapped potential of the technique in their own field.

TG methods have been profitably employed in studies of reactions yielding one or more volatile products for a most extensive range of diverse reactants. These include decompositions of simple inorganic compounds, losses of water from all types of hydrates and the evolutions of various ligands from complex salts, together with other types of thermal processes. Decomposition studies using TG methods have been applied to most classes of organic reactants, including pure substances, salts of organic acids, organometallic compounds, and synthetic materials of high molecular masses, such as polymers. Useful information about the thermal reactions and reactivities of natural materials have included TG examinations of clays, coals, wood, wool, and many foodstuffs, in which mass loss measurements can be used to determine or to compare stabilities. In favorable systems, partial chemical analyses can be deduced from the measured magnitude of a mass change that occurs at a temperature known to be characteristic of a particular reaction. It is difficult, even pointless, to attempt to summarize all the diverse types of reactions that are suitable for study by TG methods, but representative examples are given throughout this chapter.

TG measurements are possible for reactants heated under very wide ranges of conditions. The following descriptions give some indications of the versatility of the technique.

Temperature intervals across which TG measurements are practicable extend between about 120 and 2600 K (the accuracy achieved obviously varies within different ranges).

The *sample temperature* can be held constant throughout the experiment (isothermal reactions) or systematically varied according to a specified program of change with time, or other suitable parameter such as reaction rate. A constant rate of temperature increase has been widely used, but sometimes its variation is controlled to achieve a constant reaction rate, $d\alpha/dt$ (where α is the fractional reaction).

The reaction can be completed in a maintained vacuum, during which volatile products are continually evacuated from the vicinity of the substance undergoing chemical change (*reactant environment*). These conditions can be used to measure the sublimation rate of a solid or the evaporation rate of a liquid. Alternatively, reactions can be studied in the presence of a controlled (often constant) pressure of product or of a volatile reactant, a method for investigation of gas–solid reactions.

This chapter is primarily concerned with research applications of TG. These are directed toward providing fundamental scientific information of value in elucidating chemical and (to a lesser extent) physical characteristics of thermal processes, including reaction stoichiometry, absolute and relative levels of reactivity with their controls and reaction mechanisms. Such investigations contribute toward the confirmation and extension of the scientific theory and differ from empirical studies. Work of the latter type is undertaken to obtain specific information of often considerable value in the design and development of manufacturing or preparative processes, determining stabilities (including pharmaceuticals), quality control, etc., but requires no support from theory.

Interpretation of TG data in fundamental research is frequently aided by confirmation from appropriate complementary measurements (though such information is not always provided by reports in the literature). Limitations of the TG method are mentioned throughout this discussion, but it may be helpful to mention first the following aspects as having general significance:

- A measured mass loss does not necessarily identify reaction stoichiometry completely and additional analytical information may be required to confirm the chemical change.
- For historical reasons, and because many reactants are initially solid, TG kinetic studies often proceed with rate data analysis assuming that the reaction proceeds in the solid phase. This is not necessarily correct, but physical changes, most notably melting, are not detected by TG measurements. It is important, therefore, in describing a reaction adequately to establish whether it proceeds in the liquid or in the solid state.
- Reaction kinetics, particularly for many solid-state endothermic and reversible rate processes, are sensitive to reaction conditions [the "procedural variables" (1)]. Because of the participation of secondary controls, the reaction rate measured is not necessarily that of the limiting chemical (or rate-determining) step.
- The literature concerned with nonisothermal rate data interpretation, including TG, uses approximate rate equations in which apparent values of kinetic parameters (Arrhenius activation energy, E_a, preexponential factor, A, etc.) vary with the particular mathematical procedure used in the calculations. Analysis of data must proceed with careful consideration of all assumptions, including term definitions and units in the methods used (3). Aspects of the theory of reaction kinetics in this field remain unresolved. Considerable care must be exercised in ascribing chemical significance during interpretation of experimentally measured data and in using values reported in the literature.

A representative range of sources selected to include, particularly, accounts of TG are found in Reference 4 through Reference 16.

5.2 THERMOGRAVIMETRIC APPARATUS AND MEASUREMENTS

5.2.1 GENERAL CONSIDERATIONS

Before considering the functioning of the various constituent parts of the equipment used in TG investigations, it is useful to make some general points.

Stoichiometric conclusions for thermal reactions require confirmation by chemical analyses or structure determinations and should not be based solely on the mass changes measured. Chemical changes do not invariably conform to expectation or necessarily fulfill the requirements of an anticipated stoichiometric equation. Residual products may be mixtures arising through different participating reactions, and many solids exist in more than one crystallographic form. Secondary reactions, parallel or consecutive, may contribute to the overall changes that occur; this becomes more probable when investigating larger and more complex reactant molecules.

The interpretation of TG observations frequently requires support from complementary measurements. Important methods to confirm the identities of rate processes investigated by TG often usefully include enthalpy measurements (DSC and DTA), evolved gas analysis (EGA), structure determinations (x-ray diffraction), microscopy (textural changes of solids), etc. In particular, within the pharmaceutical arena, there is often a tendency to assume that mass losses are exclusively due to water; it is recommended that such an assumption should be made with care.

Some dissociations of solid reactants are reversible. The kinetic characteristics of *reversible reactions* (based on measured apparent values of α, the fraction reacted) are frequently sensitive to pressures of all gases present (including particularly products) in the immediate vicinity of the reaction zone. Little is known about the magnitude of the contribution made by this effect in many rate processes. However, for some specific reactions [e.g., calcium carbonate dissociation (17) and nickel oxalate dihydrate dehydration (18)] it has been shown that, to measure the rate of the forward (i.e., dissociation) step only, very low pressures of product (and other gases that act as an effective barrier to product removal) are required. Measured rates, and the derived kinetic parameters, for such reactions are sensitive to the procedural variables (1), including reactant mass and packing, particle sizes, product pressure, heating rate, etc. Reaction rates observed may be influenced by controls other than, or in addition to, an interface rate-determining step. These controls may depend on the overall rate of product removal, through its ease of diffusive escape through intercrystalline channels in which inhomogeneous distributions and compositions of gas may arise within an assemblage of reactant particles. Such effects are greatest for rapid reversible reactions proceeding within relatively large reactant samples and in the presence of higher pressures of gases in the reaction zone (including inert gases in the atmosphere).

Reactant temperature within the reaction zone may be appreciably different from that measured for the controlled furnace reaction vessel because of local self-cooling or self-heating as a consequence of the reaction enthalpy. The significance of self-cooling in dehydrations has been discussed by Bertrand et al. (19). L'vov et al. (20) have developed a computer model to represent the effect with reference to the endothermic dehydration of $Li_2SO_4 \cdot H_2O$. Not all research reports discuss the possible consequences of reaction enthalpy in influencing reactant temperature.

Mass loss measurements do not detect melting or other phase changes. TG studies do not detect fusion. Melting may be a significant mechanistic feature of reaction and of importance in the interpretation of observations. TG has been widely used to study reactions that occur on heating reactants that are (at least initially) solid (21); it must be remembered that one of the possible consequences of heating is melting.

Characteristics of equipment must be noted. During the formative years of the subject, equipment was often manufactured "on site" and adapted to the specific requirements of the systems to be investigated. The commercially available equipment now generally used is more sophisticated, less easily modified, and frequently used as supplied. It must be remembered, however, that this equipment has usually been manufactured for use in a variety of applications and inevitably contains some design features that represent compromises between conflicting or irreconcilable

requirements (8). A researcher intending to make measurements of the highest
achievable accuracy must satisfy himself that the equipment selected for this use is
capable of fulfilling his specific requirements. To optimize performance it may be
necessary to calibrate the reaction temperature across the reaction zone and to
confirm the absolute accuracy of measurements within the reactant container for the
range of conditions of interest. It is also important, particularly for studies of
(possibly) reversible rate processes, to investigate the sensitivity of the observations
to changes of reaction conditions (the procedural variables) such as reactant mass,
pressure and flow of gas within the reaction vessel, etc. The relevance of software
supplied with commercial equipment for interpretation of the measured data must
be carefully considered and any potential limitations addressed.

It is always important to confirm reproducibility of behavior. It is not always
apparent from many published articles that the authors have made quantitative
investigations of the accuracy achieved in repeated experiments and that all quantitative results reported (e.g., Arrhenius parameters) include realistic error limits.

5.2.2 Balance Mechanism

The principal components and ancillary connections required for the operation of
most types of TG apparatus are summarized and shown schematically in Figure 5.1.

Several alternative geometric arrangements, involving different relative dispositions of the components, have been successfully used in thermobalances, in which
the essential feature is that the sample is suspended in the temperature-controlled
zone. Typical designs of apparatus capable of making mass measurements of heated
samples are described in References 4–8, 10, 12, and 16.

5.2.2.1 Beam Balances

These often consist of two identical arms of equal length, one supporting the reactant
sample suspended in the heated zone and the other bearing a countermass that can
be varied to offset changes in reactant mass. An electrically induced magnetic field,
operating on a metal bar suspended from the beam, can be used, by variation of the
current supplied, to operate the balance by maintaining zero displacement, the null
mode. Beam movement can be detected by an attached shutter that interrupts a light
beam focused on a photocell whereby the imbalance is detected. This response is
offset by a control mechanism that causes the necessary change in current generating
the magnetic field and can be calibrated to measure mass change.

The equipment must include all the features that are characteristic of a sensitive
analytical balance to measure, with maximum accuracy, the mass changes that occur
during reactions of small reactant samples. The fulcrum and suspension points at
the beam ends consist of sharp, durable knife edges, supported on similarly hard
surfaces. These instruments are mounted on a vibration-free plinth, sometimes
including an elastic layer capable of absorbing small disturbances. The balance is
housed in a container to shield it from external air movements and must be insulated
from any heating effects or convection currents arising from the furnace. The beams
are fabricated from materials of low coefficient of expansion. In some designs, the

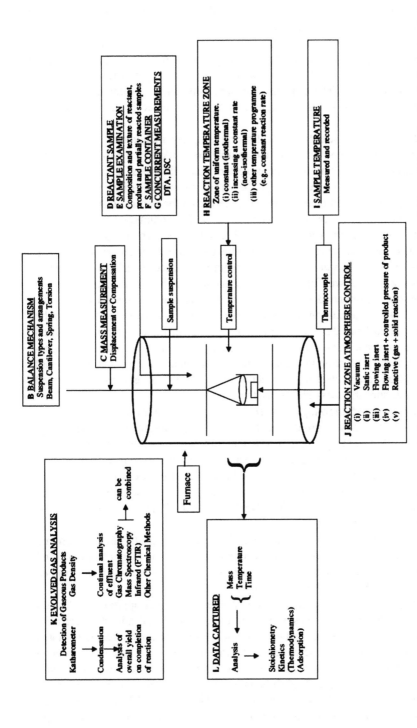

FIGURE 5.1 The thermobalance — diagrammatic representation of component parts and summary of methods used. Letter headings refer to descriptions in Section 5.2 of the text.

sample suspension wire is extended through a fine, cooled tube to eliminate the effects of convection and radiation on the balance mechanism (the wire must not touch the walls of the tube in which it is located).

Variations from this basic design include single-pan balances, in which there is a single, possibly damped, countermass, and provision may be made for the addition or subtraction of calibrated masses resting on a pan acting on the same side of the fulcrum as the sample. There are also balances using arms of unequal lengths, giving a known leverage ratio. Some design features are discussed in the following text:

Helical spring: A sensitive helical spring is rigidly anchored at its upper end, and the lower end supports the sample suspended in a furnace beneath. Again, the mounting must be protected from vibrations, draughts, thermal variations, etc. The extension of the spring is usually in direct proportion to sample mass, and changes are measured from the vertical displacement of a pointer observed by a traveling microscope. Null point operation is possible (as for the cantilever beam described later) through the provision of a (measured) magnetic support for a metal bar, also suspended from the spring.

Wire or ribbon torque: The balance beam is attached (usually at the center) to a horizontal wire or ribbon firmly anchored at both ends. Any change in the mass suspended from one end of the balance arm results in a torque that twists the suspension wire. Again, the force causing the twist can be measured from the displacement of a pointer or the countermass required to restore the balance to the zero position. Alternatively, a magnetic opposing torque of sufficient magnitude to maintain the balance at the null position can be electrically generated and the current required is calibrated to measure the mass change of the sample.

Vibrating reed thermogravimetric analyzer: Another design utilizes a mounted reed, supported toward the center, and vibrated by an electromagnet (22). The sample is in a container attached to one end of the reed, and changes in mass are detected by variations in the resonance frequency of reed vibration, which also has the beneficial effect of constantly mixing the sample. Mass resolutions of better than ±1.5% are claimed for (relatively large) samples, 0.5 g. Advantages claimed are low cost, easy operation, and robustness.

Cantilever beam: A horizontal beam, of a flexible material, is firmly anchored at one end and is deflected (bent) in proportion to the mass of sample (and container) suspended from the free end. The change in mass is determined from the change in beam deflection.

Quartz crystal microbalance: A different experimental approach, but again measuring mass changes, is used in the quartz crystal microbalance (23). A thin disk of the reactant crystal, having a planar upper face, is mounted on a thermostat; temperature control may be improved by contact through a metal eutectic. About 20 mm above the reactant surface is positioned a quartz crystal held at 78 K throughout the experiment, with the intervening gap maintained at a very low pressure (below 4×10^{-5} Pa) by continual pumping. The resonance frequency of the quartz crystal, excited through an external circuit, is changed by condensation of gaseous products released from the heated reactant. It has been claimed (24) that weight losses investigated through this resonance frequency change can be measured to within $\pm 10^{-11}$ kg and that temperature control is accurate to 0.1 K.

The method can only be used for product gases that are condensed quantitatively under reaction conditions; these include water, carbon dioxide, and ammonia. Reactions can be studied only at low pressures and significant investigations of dehydrations using this technique have been published (23–25). This technique can alternatively use a quartz crystal with the reactant sample deposited on or otherwise attached to it and the mass loss monitored through resonance frequency changes as described earlier.

Furnace position: Positioning the furnace in the most obvious situation, below the sample suspension point, can be undesirable, particularly for high-temperature studies. The convection currents developed and the radiant energy dissipated can result in unequal heating of balance parts, which may introduce uncertainties into observations. These are reduced, but not necessarily eliminated, by the use of insulators and baffles. To minimize such errors, balances have been constructed to locate the sample either above or beside the balance arm.

Balance container: To achieve maximum accuracy and sensitivity and, at the same time, control the reaction environment, thermobalances must be designed to avoid the effects of spurious heating from the furnace and to permit control of the gaseous environment within which reactions proceed. For reactions in air at atmospheric pressure, it may not be necessary to mount the balance beam (helical spring, cantilever arm, etc.), suspension wires, sample, and container in a closed envelope. However, a sealed container is essential when working in vacuum, at high pressure or in an atmosphere of controlled or inert composition. In many types of commercially supplied apparatus all moving parts are accommodated within a containing vessel, which can be connected to pumps, pressure gauges, and supplies of a gas that is directed to flow through the heated chamber in which the reactant is located. The balance container must be mounted on a vibration-free foundation and protected from draughts, temperature variations (including sunlight), etc.

A *dynamic (flowing gas) atmosphere* within which the reaction takes place usually exerts a force on the reactant container that changes the apparent mass of the suspended reactant. This may be constant or may be determined by comparative calibration under static and dynamic conditions.

A *condensable reaction product* may accumulate on balance parts, including any cool regions of a suspension wire or even the beam, cantilever, or spring (26). Elimination of any spurious contribution to sample mass can be achieved for normally volatile substances by suitable management of gas flows. For less volatile products, it may be necessary to heat the possible condensation surfaces or to ensure that all such materials are contained in a hot gas stream leaving the furnace.

A *corrosive gas*, introduced into the thermobalance as a reactant or generated as a product, must be kept away from contact with all metal or reactive parts of the apparatus. Reaction with balance components will (at best) cause error in measurements and (at worst) could ultimately destroy expensive equipment. For experiments involving corrosive compounds, it may be necessary to use balances manufactured from materials that do not react with the particular gases concerned or to include purging flows of inert gas that oppose undesirable contacts; for example, a maintained flow across the balance beam is directed toward the sample.

High-pressure thermogravimetry equipment capable of TG studies at high pressures, 3 MPa (4,5) and above (8,27) has been described.

5.2.3 MASS MEASUREMENT

In the different types of apparatus described in the previous section, mass changes may be measured through the amount of countermass variation required to restore the balance to the null position or through calibration of the distance of displacement from the zero position. For small movements, the displacement distance is usually directly proportional to the mass change (and always requires confirmation). Alternatively, both methods can be used in combination, the larger changes compensated by added calibrated masses and the smaller changes determined by a displacement measurement.

Most of the widely employed types of commercial equipment use a magnetic force generated by an electric current acting on a magnetic material incorporated into the beam, or suspended from it, to restore the balance to the null position. The current required can be accurately measured and calibrated to determine the total mass and the mass changes of the reactant. This method requires a detector sensor on the beam to identify any departure from the balance (null) position, often through changes in the amount of light passing through an attached shutter interposed in the light beam. This sensor can be used to operate a sensitive feedback electrical response that imposes just sufficient force to restore the null alignment. In automatic equipment the magnitude of this restoring force is measured in a form that has been quantitatively calibrated to give reactant mass or mass change, and these values are recorded at appropriate time intervals.

These principles were applied in an early electrobalance, often associated with Cahn and Schultz (26). A conventional balance beam is supported by a taut torsion wire, or ribbon, that passes through it at right angles and is fixed rigidly at both ends. Imbalance from a preset, null, position causes, through a response circuit, a change in the current supplied to a torque motor surrounding the balance beam or suspension wire to offset the sample mass change. The change of sample mass is recorded at intervals corresponding to known reaction times.

5.2.4 REACTANT SAMPLE

During quantitative TG investigations, the reactant mass is measured with maximum accuracy at a series of specified times for a sample heated to constant, or programmed, temperature in a controlled atmosphere. The magnitudes of (m, t, T) values are recorded at prespecified intervals. In the present account we use the convenient and conventional definition of the fractional reaction α, where

$$\alpha = (m_t - m_f)/(m_o - m_f) \qquad (5.1)$$

Sample masses are m_o at the start of reaction, m_t at time t, and m_f, the final mass, on completion of the reaction (*infinite time*).

The quantity of reactant used must be sufficiently large to permit each measured mass loss to be determined with the precision necessary for its intended use. For kinetic studies, it is desirable that values should be accurate (where possible) to at least ±0.1% of the total mass change. The most accurate TG investigations are achieved for reactions that involve the loss of a high proportion of the reactant mass. The required precision of measurement may be difficult to realize if the reactant mass is small, whereas the use of a larger sample introduces kinetic uncertainties because of the possibility of increased contributions from reversible reactions, temperature variations within the sample, etc. Sample temperature deviations from the value measured due to reaction enthalpy [self-cooling or self-heating (28)] are relatively less in small sample masses, particularly, when a powdered material is spread rather than aggregated or compacted tightly.

In many, perhaps most, TG investigations the most reliable measurements are obtained through experiments using the smallest practicable sample masses. By minimizing the thermal inertia of the sample, the time required for the reactant to reach furnace temperature at the start of reaction is diminished, and, in programmed temperature experiments, the thermal lag is least. During vacuum studies, sample heating is by radiation only, which is not efficient for a thermal furnace. When a gaseous atmosphere is present, heat transfer is increased by convection, but this may diminish the rate of volatile product escape by diffusion, thereby increasing the probability of secondary reactions. The location of the reactant sample within an environment of low thermal conductivity introduces error and uncertainty into quantitative TG measurements.

To obtain data for use in kinetic analysis it is essential that the set of α values examined should refer to a single rate process and that the limiting mass values are correctly measured. Values of m_o must exclude contributions from initial (precursor) rate processes due to rapid breakdown of unstable surface species or volatilization of adsorbed materials (both from solid reactants) or contributions resulting from the completion of a previous reaction. The possibility that the reaction of interest might already have commenced during initial heating, outgassing, or previous aging may also require consideration. Measurement of m_f can be unreliable because of onset of a subsequent rate process. For this reason, it is often undesirable to increase the temperature toward the end of reaction to accelerate its completion (i.e., to reach m_f more quickly), because the change of conditions sometimes introduces unforeseen errors.

5.2.5 SAMPLE EXAMINATION

TG measures mass change resulting from a chemical process and, for many reactions, provides probably the most accurate method of measuring its progress. However, because mass determinations alone reveal relatively little about the processes involved, complementary observations are required to characterize the nature of the changes that have occurred to reactant constituents, compositions, structures, textures, products formed, etc. Concurrent measurements (e.g., TG + DTA) and examinations of partly reacted samples and products through chemical analyses, microscopy, etc. are capable of revealing much about the chemistry of the processes of interest. Such investigations are not always included and described in published reports. However, because so much can be learned from these observations, especially for reactions of solids, the value of complementary studies is particularly emphasized here.

In all quantitative TG investigations, it is necessary to characterize the reactant, and this may require both chemical (elemental) analysis, together with structure determination for those solids that exhibit polymorphism (where reactivity can vary with crystal form). The reaction identified must be consistent with both mass change and compositional evidence (which also confirms the balance calibration). Several aspects of sample examination and reaction characterization that merit consideration are mentioned in the following paragraphs:

Consecutive reactions: Kinetic analysis of measured rate data must be concerned with a single rate process, and the rate equations used are based on the assumption that α-time data are calculated to refer to only that one chemical change. Reactions proceeding through a sequence of consecutive steps may require individual stoichiometric confirmation, and, certainly, kinetic analyses must consider each single step individually.

Reactant particle size distributions: Kinetic characteristics of some reactions of solids depend sensitively on reactant particle sizes (29). Ideally, reactants to be used in kinetic studies should be composed of crystallites of identical (known) sizes and shapes, to which the geometry of interface advance can be related quantitatively. This is not, however, always (or easily) achieved in experimental studies, and most powder samples contain particles of disparate sizes; for example, sometimes crystals are mixed with fine powder. The kinetic model giving the best apparent fit to the data then may not accurately represent the reaction. Dependencies of rate on particle size are only rarely investigated. The state of subdivision of a solid reactant is most frequently described in literature reports only by qualitative terms, such as *single crystals* or *crushed powder.*

Reactant and product textures: Textural examinations of solid reactants and their products can provide information of great value in understanding reactivities and reaction controls and mechanisms. Microscopic observations are frequently capable of identifying the phase in which a chemical change proceeds. Melting of a solid may be detected by optical examinations, although recognition is not always as straightforward as might be anticipated, particularly, for a liquid phase occurring only during partial or local/temporary intracrystalline melting. Fusion during the decomposition of copper(II) malonate (30) was ascribed to the intervention of acetate and was detected by the formation of intraparticular bubbles. It was concluded that reaction proceeded in melt retained within a resilient "skin" formed during the early reaction that maintained the individuality and shapes of the reactant particles. In contrast, the generally similar decomposition of silver malonate was identified (31), again by microscopic examinations of partially reacted crystals, as proceeding by a nucleation and growth solid state reaction (29). These studies did not involve TG, but this method could, in principle, be used to measure these reaction kinetics. These examples illustrate mechanistic behavior for thermal chemistry of crystalline reactants.

5.2.5.1 Intermediates in Solid-State Reactions

The possibility that reactive/transient intermediates participate in solid-state rate processes has not been widely accepted or investigated. This contrasts with homogeneous reactions in which the identification of any intermediate species participating in the conversion of reactant into product is a routine part of mechanistic studies. One possible reason for this difference in attitude is the general acceptance of the view that, during interface reactions in solids, all chemical changes are completed within a thin active reaction zone (probably of molecular dimensions). During such processes, the total amounts present of any intermediates participating would be very small, or undetectable. In consequence, attention in solid-state chemical studies is sometimes directed towards the probable structure of an assumed transition state generated by activation of a reactant component.

Nevertheless, there are examples of stable, well-characterized intermediate phases that have been identified during the stepwise decompositions of solids. Perhaps, the most obvious examples are found during dehydrations (32), in which a sequence of lower hydrates may intervene between a highly hydrated salt and the anhydrous product of a sequence of thermal reactions. For example, depending on conditions (33), the dehydration of $MgSO_4 \cdot 7H_2O$ may proceed through hydrates containing 6, 4, 2.5, 2, and 1 H_2O (and finally to the anhydrous salt). One particular step, the dehydration of the tetrahydrate to the dihydrate, was used (19) in a study of Smith-Topley behavior (29).

The sequence of intermediates formed during a stepwise dehydration may vary significantly with changes in the availability of water vapor. Nickel sulfate hexahydrate is converted directly to the monohydrate in a good vacuum (34) whereas, in the presence of water vapor product, the tetra- and dihydrates have been characterized as intermediate phases by x-ray diffraction (35,36). Another investigation of this reaction (37) suggested that three main processes may contribute to the overall dehydration: (1) detachment of the H_2O molecule from the crystallographic site, (2) diffusion of this volatile product through the water-depleted residue, and (3) crystallization of this solid to the product phase. If water elimination and product-phase crystallization are distinct and separated processes, the (amorphous or zeolitic) material between these represents a type of reaction intermediate.

In decompositions of other reactants, initially solid, molten intermediates have been recognized or suggested. Examples (28) include the interventions of acetate and copper(I) in copper(II) malonate decomposition, of CrO_3 in the thermal reactions of ammonium dichromate, and of NO_2ClO_4 in NH_4ClO_4 breakdown. It appears that the identification of intermediates in thermal reactions of solids in which melting is not evident have not been widely attempted. One example, which now seems to be agreed (39), is that $K_3(MnO_4)_2$ is an intermediate in the decomposition of $KMnO_4$, which was originally regarded almost as a model (one-step) solid-state decomposition (29).

This approach to intermediate identification is not generally applied in solid-state chemistry. The topic is, however, mentioned here to contribute toward advancing chemical understanding of thermal reactions through reconsideration of an aspect of interpretation that seems not to have been adequately addressed. Development of this viewpoint might be capable of recognizing patterns of behavior in a subject in which classification of reaction types has not been successful in ordering the thermal chemical properties of diverse and different reactants (29,32). A wider view of the identities and functions of reaction interfaces may be useful in formulating reaction mechanisms and in helping to understand the chemistry of solids. The detection and measurement of amounts of participating reaction intermediates will usually require chemical analyses or structural (x-ray) investigations (or both) of samples partially reacted to known extents.

5.2.5.2 Microscopy and Solid-State Decompositions

Kinetic analyses of the thermal reactions undergone by initially solid reactants is usually based on the fit of yield–time or rate–time data to rate equations (kinetic models) derived from geometric representations of interface advance (29). A short introduction to this topic is given in Section 5.4 of this chapter. Conclusions from such kinetic analyses, identifying the most probable pattern of interface development, can be confirmed by microscopic observations if the product phase can be reliably distinguished from the reactant. When such reactions yield a volatile product and are, therefore, suitable for TG investigation, the residual material usually does not fully occupy the reactant volume from which it was derived. Such residual product phases may then be microscopically recognizable because these materials are often, characteristically, finely divided, are permeated by cracks, may be opaque, and may be bounded by distinct surfaces. Microscopic examinations are most productive when the reactant and product phases can be unambiguously distinguished. Examination of samples of reactant and partially reacted material can then be used to establish whether or not the reaction proceeds at an interface and, in favorable systems, to confirm the form of the reaction geometry development inferred from the kinetic analysis.

For samples that have undergone small extents of reaction (low α), nuclei may be identified on crystal faces by the appearance of patches of characteristic texture (often with symmetrical shapes), each of which increases in size as α increases. This is a feature of nucleation and growth processes, and microscopy has enabled the kinetic rate laws for nucleation to be deduced (29).

Reactions that are initiated rapidly across all (or some) surfaces immediately on attaining the reaction conditions with establishment of an interface that continually progresses inward thereafter are referred to as *contracting envelope* processes. Such behavior is most reliably confirmed by microscopic examination of sections of partially reacted crystals (24). After reaction to known extents, single-crystal samples are cleaved through the center, or powder is lightly crushed, and the sections are examined to determine the interface distribution and disposition. The most reliable conclusions will result from consistent behavior patterns observed from microscopic examinations of external and internal features of several samples reacted to different α values.

The most significant result from microscopic work is probably the provision of positive evidence that the reaction takes place at an interface. When progressive variations of interface distributions with time are expressed quantitatively, this enables the reaction geometry to be characterized. Other useful observations may include evidence of melting, particle size changes (fragmentation and sintering), and local or general cracking.

5.2.6 SAMPLE CONTAINER

The reactant container must be sufficiently robust to support the sample in the heated zone but of low total mass so that the measured changes are predominantly due to the reaction of interest. Ideally, the thermal inertia of the reaction vessel should be small, and the thermal conductivity of its walls high, so that the heat supplied maintains the reactant at the controlled furnace temperature with rapid responses to programmed variations.

Sample containers must be inert. Reaction vessels must be thermally stable and inert across the temperature interval of use; these are fabricated from a variety of materials, including metals and refractory oxides such as silica, alumina, and ceramics. It is important that the material of the container does not interact chemically with the sample; for example, some oxides (e.g., SiO_2) react at high temperatures with the residual alkaline compounds (oxides or carbonates) formed by decompositions of sodium, potassium (and other) salts of organic acids. Platinum and other metals are capable of catalyzing the breakdown of many organic materials, so the reaction investigated may not be that intended or may contain a proportion of some promoted parallel process [possibly unrecognized (40)]. Secondary reactions between primary products adsorbed on the active surface of the metal sample container may change the compositions of the volatile products identified by EGA during the TG experiments. In addition to such changes of reaction stoichiometry, products such as acids or oxygen-containing molecules may be capable of reacting with sample vessels made of oxidizable metals such as nickel. The reduction of Re_2O_7 with H_2 proceeded differently in silica and platinum crucibles (41), which was ascribed to "spillover" of dissociated hydrogen from the active platinum metal into the reactant.

Even small traces of oxygen in a purge gas passing through the reaction zone may introduce errors into the mass changes measured by oxidizing the metal of a sample container or by participating, to a perceptible extent, in the rate processes being studied (42). Similarly, the container itself must not react with corrosive gases used in studies of gas–solid reactions; the possibility of such processes can be tested in blank experiments. In studies of magnetic materials, contributions to apparent mass change from magnetic fields (43) must be eliminated, or complete allowance made in data interpretation.

The shape of the container can significantly influence the composition of the gaseous environment in which reaction proceeds. Under frequently used conditions, the flow of an inert gas past the reactant-containing vessel diminishes markedly the locally maintained pressure of the volatile products. It may then appear that the reaction is independent of the gaseous atmosphere, although this requires careful experimental verification, particularly, for reversible reactions. Under other controlled experimental conditions, a known pressure of the product (or other reactive gas) is maintained in a flow of gas entering the reaction chamber. This approach can be used to investigate the influence of the pressure of a selected additive on reaction rate. Another method, which has been used to study reversible processes, including the thermal reactions of $MgNH_4PO_4 \cdot 6H_2O$ (44), is to design reactant containers that vary the ease of diffusive escape of the volatilized product. In this comparative study, reactant was spread thinly on several open trays, thereby promoting the rapid removal of volatile products. Other containers used included crucibles either open or loosely closed with a lid. To delay departure of evolved gases, a labyrinthine crucible composed of a series of concentric covers was devised, necessitating extended diffusion through the escape channels to ensure that reaction proceeded in a self-generated atmosphere. Usually, such comparative studies are limited to examination of the effects of the presence or absence of a lid on the containing vessel. A loose lid or a cover of metallic gauze may be required to prevent reactant losses when samples froth, sputter, or disrupt vigorously on heating.

5.2.6.1 Sample Container with Temperature Sensors

Specialist reaction vessels include those incorporating temperature sensors and heating elements in equipment designed for concurrent measurements of TG, together with DTA or DSC. Such apparatus cannot give the most accurate TG data, but there are advantages in combining techniques to enable concurrent measurements to be made by other complementary TA methods. The use of a single sample, reacting under constant conditions, eliminates the possibility of variations of characteristics due to the procedural variables (1,15). The same sample (particle sizes, packing density, etc.) reacts under the same conditions (temperature, atmosphere, etc.), so that the loss of precision between the different measurements is offset by the greater reliability of the effects being measured in the single reaction. Successive, nominally identical, reactions may proceed appreciably differently because of the inherent variations between the slightly different samples and conditions used.

5.2.7 Concurrent Measurements

Mass determinations alone usually provide insufficient information to characterize fully the chemical change or changes from which they arise. Therefore, any investigations requiring a knowledge of the specific reactions involved must interpret observations in the context of appropriate complementary measurements. TG provides a reliable method of determining the temperatures and rates of reactions, but their stoichiometry must be supported by chemical analyses of the products, residual phases, and compounds volatilized. TG observations are usefully complemented by techniques capable of detecting phase changes, particularly melting, by concurrent measurements for the same sample though heat flow, e.g., DTA and DSC. Other thermal techniques that are sometimes used, though requiring separate experiments, include thermodilatometry, thermomechanical analysis (determination of mechanical properties under tension), thermomagnetometry, etc. (4,5). Chang et al. (45) discuss the advantages of coupling TG with thermo-Raman spectroscopy in identifying intermediates during the thermal decomposition of $CaC_2O_4 \cdot H_2O$.

The rates of many reactions of solids are sensitive to properties of the sample (particle sizes and the types and numbers of imperfections in the sample crystals) and, for reversible reactions, the composition of the atmosphere surrounding the sample. This possibility of variation of behavior between nominally identical experiments makes it desirable (for some purposes) to complement thermochemical studies through concurrent measurements by some alternative sensor. A most successful combination with TG has been DTA, sometimes referred to as the *derivatograph* (10) and associated with the names J. Paulik and F. Paulik (12). In one form of combined equipment, two sample holders accommodating thermocouples measuring reactant and inert reference material temperatures have balance suspensions containing the thermocouple leads. In an alternative design, the reference pan is positioned beside the reactant. The incorporation of the additional sensor and electrical contacts inevitably diminishes the performance and sensitivity of the balance. However, these complementary measurements refer to the same reactant and the same reaction conditions. In many studies, such results are of greater value than more accurate data from separate instruments that inevitably refer to (very slightly) different reactants and to not quite identical reaction conditions.

5.2.8 REACTION ZONE TEMPERATURE

The heated zone, within which the reactant is maintained throughout each experiment, must be uniform and controlled at a known temperature, which may be constant or varied according to a specified program. The constant temperature zone may be smaller in a null point instrument than where deflection measurements require some sample movements. Separate sensors are often used for the control of furnace temperature and the measurement of the sample temperature. This avoids the possibility of the heating controls exerting any influence on the recorded sample temperature (10). The furnace, which should be noninductively wound to avoid interference with these sensors, may be regulated in response to outputs from a resistance thermometer or a thermocouple.

The sample temperature, particularly during nonisothermal experiments, will always experience some delay compared with furnace temperature. This lag will be greatest for larger samples of a poorly conducting material heated more rapidly in a vacuum or in a high flow rate of a gas atmosphere that has not been preheated. Ideally, the furnace will, without initial delay, rapidly reach the reaction temperature and accurately maintain this value thereafter or respond quickly to programmed variations. For high-temperature work, all heated components must be capable of withstanding (with a suitable safety margin) the most severe conditions expected, including possible interactions with reactive atmospheres introduced and all volatilized products. Ranges of suitable working conditions are usually specified for manufactured equipment. It must be ensured that metal parts of the balance do not catalyze secondary reactions between primary products, particularly if the initial products are to be analyzed. There is also the possibility that catalytic reactions on or near a thermocouple junction (at the metal surface) might introduce error into the apparent sample temperature recorded. Atmosphere movements and radiant heat from the furnace must not be allowed to influence the balance mechanism and may be prevented by baffles inside or outside (or both inside and outside) the suspension envelope.

In addition to heating by electrical resistance furnaces, a number of other methods have been used, including infrared radiation focused on the sample and microwaves (4,5,8).

5.2.8.1 Furnace Temperature Control Regimes

The different temperature control regimes used and their possible variations illustrate the range of experimental conditions that provide particular types of information. The regimes that are widely used are discussed in the following text:

Determination of the temperature interval at which each contributory reaction occurs: TG can provide a rapid and effective method of qualitatively establishing a pattern of reactivity and determining whether an overall change proceeds as a single process or through a sequence of consecutive steps. Mass losses at each step, as a fraction of the total, can be used as a qualitative indication of reaction stoichiometries and the temperature interval recorded. Such preliminary conclusions should be regarded as requiring analytical or structural confirmation, EGA, or x-ray diffraction.

Stepwise isothermal analysis: This has been described (46,47) as a method whereby reaction temperatures and mass loss steps can be accurately characterized by a complementary combination of simultaneous TG and DTA methods. The sample is heated at a slow, constant rate until a reaction is detected and, on reaching a specified (low) rate of mass loss, the temperature is maintained constant. When this rate process has been completed, indicated by diminution to a second specified (relatively much lower) rate, the slow rate of temperature increase is recommenced. Thus, each successive rate process is completed isothermally and both the mass loss (evidence of stoichiometry) and the temperature of that step are determined precisely.

Constant temperature: A large number of important kinetic studies using isothermal TG measurements have been reported. The influence of the gaseous products present on apparent rates of reversible dissociations must be investigated and possible contributions from sample self-cooling or self-heating considered. (This refers to any change of temperature within the reaction zone that arises as a consequence of an endothermic, or exothermic, reaction taking place.) The influences sample buoyancy and the flow of a gaseous atmosphere on sample mass measurements should be determined by suitable calibration.

Nonisothermal TG measurements: Mass–time measurements during a constant rate of temperature rise are widely used (29) in kinetic studies. Each set of mass (α)–time temperature data should refer to a single-rate process only, identified as a particular stoichiometric chemical reaction. Again, the influences of gaseous products on the rate of the chemical change of interest and any variations with conditions must always be investigated (particularly for reversible processes). Rates of reactions should be sufficiently slow to eliminate the effects of self-cooling or self-heating.

Controlled (often constant) rate thermal analysis (CRTA) (48–50): An alternative to the previous technique is based on kinetic measurements of mass changes, which are maintained at a constant (usually slow) rate by variations of reactant temperature ($d\alpha/dt$ = constant). An advantage of the method is that the influence of volatile products on reaction rate is probably constant throughout reaction (though not necessarily zero). In this respect, vacuum conditions maintained during reaction differ from those proceeding with product accumulation where the contribution from the "reverse" step changes to an unknown extent as reaction progresses. It is also probable that the effects of reactant self-cooling (or self-heating) remain constant throughout the rate process. This method can be used to determine the reaction activation energy by the "rate jump" method (48,51). From the temperature difference required to achieve a predetermined increase in the (constant) rate of reaction, the Arrhenius parameters can be calculated, together with any variation with α.

5.2.9 SAMPLE TEMPERATURE

Sample temperature is usually measured by a thermocouple placed as closely as is practicable beneath (52) or beside the sample holder, but it must not touch the suspended container. It may be possible to achieve closer proximity during null point balance operation because the sample holder is effectively stationary. It is alternatively possible to incorporate the sensor into the holder, even in direct contact with the sample, but the electrical leads (perhaps incorporated into the suspension) must not interfere with the working of the balance. The extension of TG measurements across wide temperature intervals (up to 120 to 2600 K) has been discussed by many authors (8,52,53).

Errors between the reactant temperature recorded and that existing within the reactant are greatest for larger, and poorly conducting, reactant masses, particularly during rapid heating. Alves (54) has discussed errors in temperature measurement, including methods of determining the corrections necessary. The possibility of reactant self-cooling (or self-heating) during chemical change must also be considered. The cooling due to the flow of gas past the sample holder should be determined by calibration under experimental conditions. Although potentially the most convenient method, temperature calibration using the melting points of pure compounds cannot be performed because there is no mass loss. An exception is the use (4) of mass supports, fashioned from fusible metal; on reaching its melting point, the metal drops a mass, which is detected. A convenient alternative, to calibrate the instrument under working conditions, is to use the Curie points of a series of ferromagnetic materials (55). At the Curie temperature, known for a number of metals (e.g., nickel, iron) and alloys (e.g., alumel), the magnetic properties change. If a sample of suitable metal is held in the balance pan in a suitably oriented magnetic field, there will be an apparent mass change at the characteristic temperature, which can be compared with the sensor reading. Brown et al. (56) have discussed the use of rapid exothermic reactions as a method of calibrating TG apparatus.

5.2.10 REACTION ZONE ATMOSPHERE CONTROL

The presence of gas within the reaction zone usually influences the apparent reactant mass, and corrections are often necessary. A flowing gas atmosphere will exert a constant force on the balance pan that may be determined by calibration with and without flow.

At low gas pressures, below about 300 Pa, any temperature gradient within a balance suspension or support will result in a (thermomolecular) flow of the gas present, from hot to cold. This will exert a force on the balance components, registering as a mass. The effect can be counterbalanced by the use of a symmetrical balance, similarly heated on both sides. Alternatively, the necessary correction can be determined by blank experiments, or a small pressure of an inert gas (above 300 Pa) can be introduced to eliminate the effect.

At higher gas pressures, a buoyancy correction must be made to allow for the volume occupied by the sample; this increases with pressure. The mass of gas displaced is given by MPV/RT, where P and V are pressure and volume of gas of molar mass M, and R is the gas constant. During reactant decompositions, the volume of the residual sample changes; for example, the nickel metal product formed on heating nickel oxalate dihydrate occupies only about 10% the volume of the reactant from which it was formed. Other corrections include the contributions from any gas flow in and around the balance mechanism and the effects of convection that increase with pressure. Some specific issues are discussed in the following text:

Reversible reactions: For many endothermic decompositions, the amount of volatile product present within the reaction zone exerts a significant control on the rate at which the overall change proceeds. This may be due to changes in equilibria at the reaction site, through which the overall frequency of the successful dissociation step is diminished or the products recombine in reacted material, or both. The effects on measured kinetic results can be large and in mechanistic studies cannot be ignored. The consequences of inhibiting product escape can be characterized by measuring the influences of the procedural variables (1).

Vacuum: Volatile products released from the reactant are lost most easily during reactions in vacuum. No corrections are required for sample buoyancy or atmosphere flow. However, the absence of convective heating reduces the rates of heat movement and may introduce uncertainties into measured reactant temperatures. The effect of product availability on rates of reversible reactions is reduced, but, at least in some of these reactions, to eliminate such effects entirely, pressures must be maintained throughout at very low values, e.g., below 10^{-2} Pa. The sensitive variation of the apparent measured value of the activation energy with pressure has been demonstrated for the dehydration of $NiC_2O_4 \cdot 2H_2O$ (18) and the decomposition of $CaCO_3$ (17). It is not known how many other reactions of solids show similar behavior, but, if the rate of (only) the "forward" dissociation step is to be determined, comparative kinetic measurements should be made at various pressures, including the lowest attainable. For these rate measurements, the quartz crystal microbalance is a useful technique.

Static inert atmosphere: These conditions maximize contact between reactant, residual products, and all volatilized products, gas escape is inhibited by the immobile surrounding atmosphere, and reaction proceeds in a self-generated environment. The effective pressures of product will systematically change during the course of the reaction, as a function of the relative rates of generation and of diffusive/convective removal within the volume available. Gas compositions, both within and beyond the sample, will be at least temporarily inhomogeneous. For irreversible reactions, an inert gas enhances thermal equilibration within the reactant and with the surroundings. Mass measurements require buoyancy but not gas flow corrections; at higher pressures, the effects of convection may require consideration. Care must be taken to avoid condensation of products on the cooler parts of the balance or suspension (57).

Flowing inert atmosphere: The effect of gas flow on apparent reactant mass must be determined by calibration. The dynamic atmosphere in the reaction zone can be used to remove volatile products and may prevent condensation on the balance suspension. The local, and possibly inhomogeneous, product pressure maintained in and around the reactant will be determined by the reaction rate and by the flow pattern in the vicinity of the sample (usually contained in a bucket-shaped pan). If the gas flow rate is increased, there is the possibility that a fine-powder sample will be ejected from the reaction vessel, but if it is too slow, more product gas will be temporarily retained within interparticular channels. The conditions existing within the sample depend on several (often uncharacterized) parameters, including particle sizes of reactant and product, sample mass, diffusion coefficients of gases concerned, etc. In many systems, the effective pressure of product within the reactant is not known, is probably inhomogeneous, and will vary during the course of reaction. Furthermore, impurities in gases may be sufficient to cause detectable reactions (42). For example, there is a small proportion of oxygen in "pure" nitrogen, and other gases supplied may contain traces of water vapor. The total amount of reactive additive introduced in a flowing, inert gas during a prolonged experiment can be larger than might be expected. The possible effects of impurities on the phenomenon being studied should always be critically appraised.

Flowing inert atmosphere containing controlled amount of volatile product: One method for investigating the influence of product pressure on reaction rate is to measure reaction kinetics by TG studies in flowing atmospheres that contain controlled amounts of the product of interest. This approach is useful in studies of, for example, the Smith–Topley effect where dehydration rates of many crystalline solids show the same characteristic pattern of variation with pressure of water vapor in the vicinity of the reactant (28,29). The corrections required are the same as those in the previous paragraph.

Gaseous reactant: Many gas–solid reactions are studied by TG mass determinations during heating of a nonvolatile reactant in a flowing or maintained atmosphere of the volatile reactant (e.g., 58). This method is suitable for investigation of metal oxidations and other corrosion processes, including measurements of the changes in rate with gaseous reactant pressure. The corrections required are as in previous paragraphs, together with consideration of the possibility that impurities in the gas flow introduced may influence behavior.

5.2.11 EVOLVED GAS ANALYSIS

All suitable and practicable methods for the qualitative and quantitative detection and analysis of gases emerging in the gas flow from the reaction zone have been applied in characterizing the volatile products formed during TG measurements. The simplest examinations of effluent gases provide qualitative evidence that a reaction has occurred. More sophisticated techniques of EGA are capable of identifying a range of different products, measuring their yields, and determining the rates at which some or all of these are evolved and their variations with α. Many aspects of this topic have been described in the literature, the following accounts merit mention: References 14, 59–64.

Following release, the volatile products from a reaction within a TG apparatus will be transported by a flowing atmosphere from the reactant surface out from the furnace exit port after which these may be analyzed. During the short time taken to escape from the reactant particles, the reaction vessel, and the furnace, there will be diffusive mixing of products with the carrier gas. Thus, the yield–time response curve determined from measurements in the effluent stream must differ to some extent from that of the reaction taking place (65). To minimize errors in kinetic measurements based on gas analyses, the time taken to transport the products to the sensor is reduced as far as practicable by the use of fast carrier gas flow rates, in narrow tubes across the least practicable distances. These are conditions that tend to decrease the errors and uncertainties in TG measurements. It is also important that there is no condensation of any less volatile product on the cool parts of the apparatus. After formation, there is always the possibility of secondary reactions between primary products in the hot gas stream, on the walls of the apparatus or during the temperature decrease after leaving the heated zone. The compounds detected as products are not necessarily those generated in the chemical controlling step. Furthermore, there is also the possibility of interactions with the inert carrier gas or any reactive impurities contained in it.

5.2.11.1 Evolved Gas Detection

The presence of an unidentified product leaving the furnace can be detected through a change in the physical properties of the flowing gas. By using a low molecular weight carrier gas of high thermal conductivity (hydrogen or helium), the appearance of (higher molecular weight) products is detected as a change in thermal conductivity (katharometer), gas density, or other sensors of the types that are used in gas chromatography (GC), including the flame ionization method that is useful for detecting organic compounds. This approach differs from gas chromatography in that the products flow directly from the reaction vessel. Mixtures are not separated, and components are not identified; the occurrence of reaction is demonstrated by a change in a (usually physical) property of the effluent gas. Some (highly qualitative) information can be obtained about product volatilities by interposing suitable cold traps between reactant and detector. Such evidence of condensation temperature may distinguish between two possible products (e.g., CO_2 and CO) or provide some confirmation of an expected product.

5.2.11.2 Evolved Product Analysis (Completed Reaction)

Most reaction products can be separated from excess carrier gas (inert, with low boiling point, e.g., helium and nitrogen) by chilling the product-containing stream leaving the reaction zone, perhaps to 78 K. The accumulated yields of all volatile substances formed on completion of the reaction, and retained in the cold trap, can then be analyzed by any suitable technique. The analytical methods that can be used will depend on the complexity of the mixture expected or as indicated by preliminary experiments. Many of the reactants selected for TG studies, those offering results of greatest interest for mechanistic studies, are relatively simple compounds. These give a single, or a small number of low molecular weight products in thermal reactions. Dehydrations frequently give water as the predominant, or only, product, perhaps in several steps, and in most reports no other volatilized compound is mentioned. Some reactions, however, yield a proportion of acid, formed by hydrolysis. Carbonates usually yield CO_2, and many oxalates also give CO_2 or a mixture of CO and CO_2, depending on the electropositive character of the cation. The thermal reactions of organic compounds and of complex salts that contain large organic ligands are capable of giving product mixtures that, for some systems, will be composed of several, or many diverse, compounds. The analysis of such mixtures offers challenges to the researcher, and to the handling methods, so that chemical or physical techniques applied must be selected specifically to address the qualitative and quantitative analytical problems anticipated.

The simplest approach to qualitative or semiquantitative analysis of products that are gases at ambient temperature (e.g., CO_2, SO_2, some nitrogen oxides, perhaps H_2O, etc.) is through retention with a refrigerant in a vacuum apparatus. Measurements are then made of the pressure exerted on subsequent volatilization of the condensed gases present within a vessel of known volume. By the use of several cold traps, estimations of the compositions of a limited range of simple mixtures are sometimes possible.

Other methods of quantitative analysis, which are much more accurate and applicable to wider ranges of mixtures, include GC, mass spectrometry (MS), the GC–MS combination, in which products separated by GC are detected by MS, infrared spectroscopy, specific sensors to measure water, sulfur dioxide, etc., and all other appropriate techniques, which may include "wet chemical" analytical methods.

To characterize the less volatile products, the substances condensed from the gas stream can be dissolved (quantitatively) in a suitable solvent, the mixture separated, and the solutes identified by any of the techniques normally used in inorganic and organic chemistry. Again GC, MS, and infrared analyses, etc., may provide suitable analytical methods. (A similar approach to the characterization of reaction intermediates can be used by dissolving or extracting a sample of partially reacted material in a suitable solvent and identifying the dissolved substances.)

Completed analysis of the product mixture, taken with the compositions of the reactants and of the residue should, through comparisons of the elemental distributions, provide a reliable stoichiometric description of the overall chemical reaction that has occurred. In particular, the volatile products detected must be expected to account for the total mass loss. The products detected may, however, alternatively have arisen through secondary interactions between the primary products in the flowing gas or on any active surface available. Furthermore, instances in which the course of chemical change varies during its progress are known; this is probable where there are two consecutive and overlapping reactions. Alternatively, it sometimes results from temperature changes during nonisothermal experiments. These possibilities should be considered in data interpretations.

5.2.11.3 Evolved Gas Analysis during the Progress of Reaction

Mass loss, TG, and kinetic measurements may be complemented with concurrent EGA, which can be continuous, frequent, or intermittent, depending on the method. These complementary observations can be used to confirm the TG rate data, characterize the rate of evolution of one or more selected volatile products, and resolve the components in a complex process in which the courses of the contributing and overlapping reactions are different.

Equipment intended to obtain kinetic data by EGA methods must minimize delay between evolution and detection by reducing, as far as possible, the distance traveled and diffusive dispersal (65). Specific detectors for an identified product may operate continuously, e.g., electrolysis cells for water or sulfur dioxide (64), infrared absorption responding to a particular bond, MS operating to detect a particular mass, etc. Alternatively, MS can be used to scan repeatedly a selected range containing several products. GC necessarily analyzes samples at time intervals dictated by the longest-retained component, and the output can use MS as a detection method.

The value of MS in analyses of gases evolved from polymer degradation during concurrent TG studies as was discussed by Raemaekers and Bart (66). The sensitivity of the method was sufficient to provide useful kinetic information about volatile product evolution. Szekely et al. (14) conclude that MS observations are useful in determining the sequence of product evolutions and to resolve TG curves. The limitations of the technique must, however, be remembered in undertaking quantitative analyses, and in kinetic studies. MS has also been used to measure aspirin stability and in identifications of its degradation products (67). Product analyses using TG–MS methods were applied to investigate the mechanism of thermal decomposition of α-tetralylhydroperoxide (68).

5.2.12 DATA CAPTURE

The data collected in each thermogravimetric experiment is a set of measured (m, t, T) values resulting from the chemical changes undergone by a reactant sample heated under constant or programmed temperature conditions. These are discussed as follows:

   Mass: Mass measurements can be used to investigate stoichiometry and kinetics of reaction. Evidence concerning stoichiometry may be obtained by comparisons of overall mass loss on completion of the rate process with the original reactant mass, $(m_o - m_f)/m_o$, and the reactant formula. The conventional method of expressing extent of reaction for *kinetic analysis* is through the fractional reaction $\alpha = (m_o - m_t)/(m_o - m_f)$, where m_o, m_t, and m_f, are the reactant masses at zero time, at time t, and the final mass on completion of reaction. Data for a kinetic analysis, i.e., sets of (α, t, T) values, must always refer to a single identified rate process only (unless specialized methods are used). Interpretations using conventional rate expressions (29) of data incorporating two or more concurrent or overlapping rate processes are unreliable, usually meaningless, and not necessarily applicable to any of the contributing reactions. Mass measurements may (where applicable) require correction for buoyancy and the effects of gas atmosphere flow or thermomolecular flow.

   Temperature: The sensor is located as near the sample as is practicable, and the reliability of measurements is increased by calibration, such as in the Curie point (magnetic) method. There may be delays in the transfer of heat to the sample due to conductivity and thermal inertia of both reactant and container, a temperature programmed change, the gas atmosphere flow, and reactant self-cooling or self-heating during the reaction.

   Time: This is probably the parameter that is measured most accurately through the use of automated data collection equipment. The most significant errors arise in identifying the time when reactant reaches reaction temperature; during rising temperature experiments, there may be a time delay between the value recorded and the value representative of the sample.

   Reproducibility: It is always necessary to confirm that the (m, T, t) data measured is characteristic of the reaction of interest only and that any conclusions reached from these are based on reliable, reproducible observations. Comparisons between successive, nominally identical, experiments (and for different prepared samples of the reactant) enable the accuracy of the methods used to be assessed and meaningful error limits to be attached to reported parameters, e.g., activation energy (E_a). Repeated experiments also allow the identification (and reconsideration) of the occasional inconsistent result (was it because of unrepresentative sample, malfunction of equipment, or a programming bug?). Experiments specifically performed to establish reproducibility are not mentioned in every report, but it is always possible for unanticipated errors to occur. An appropriate proportion of replicated tests are profitably incorporated in every research program reported.

5.3 THERMOGRAVIMETRY: APPLICATIONS AND INTERPRETATIONS

5.3.1 STRENGTHS AND WEAKNESSES OF THERMOGRAVIMETRY

Thermogravimetry is an attractive experimental technique for investigations of the thermal reactions of a wide range of initially solid or liquid substances, under controlled conditions of temperature and atmosphere. TG measurements probably provide more accurate kinetic (m, t, T) values than most other alternative laboratory methods available for the wide range of rate processes that involve a mass loss. The popularity of the method is due to the versatility and reliability of the apparatus, which provides results rapidly and is capable of automation. However, there have been relatively few critical studies of the accuracy, reproducibility, reliability, etc. of TG data based on quantitative comparisons with measurements made for the same reaction by alternative techniques, such as DTA, DSC, and EGA. One such comparison is by Brown et al. (69,70). This study of kinetic results obtained by different experimental methods contrasts with the often-reported use of multiple mathematical methods to calculate, from the same data, the kinetic model, rate equation $g(\alpha) = kt$ (29), the Arrhenius parameters, etc. In practice, the use of complementary kinetic observations, based on different measurable parameters of the chemical change occurring, provides a more secure foundation for kinetic data interpretation and formulation of a mechanism than multiple kinetic analyses based on a single set of experimental data.

5.3.1.1 Solid-State Reaction Mechanisms

TG measurements record mass losses only, and characterization of the chemical significance of TG data requires complementary or confirmatory observations, sometimes by equipment incorporated into the thermobalance (e.g., DTA) added beyond the termobalance (e.g., EGA) or by residue analysis (e.g., x-ray diffraction). Before discussing the interpretation of TG observations generally, it is useful to consider some specific applications of the technique to the thermal decompositions of solids, for which numerous TG studies have been reported (21). It is believed that many of these reactions are completed within a thin, advancing interface and that these are acceptably represented by the advancing interface model, expressed mainly by geometric rate equations. The rate equation most closely representing the (α, T, t) data set for a particular reaction is identified through comparisons of the closeness of fit with each of those several kinetic models that comprise the set that is applicable to solid-state decompositions. Satisfactory agreement (often described as the *best fit*) with one equation from this set is often accepted (on a statistical criterion, e.g., a correlation coefficient) as sufficient evidence to conclude that the reaction proceeds according to the geometric or diffusion criteria used in the derivation of that preferred equation (Section 5.4). Confirmation of this interpretation by supporting observations, including microscopic examinations, is highly desirable but not always provided. Detailed information about the nature of the changes occurring is generally experimentally inaccessible in thin interfaces, within which reactions are believed to occur preferentially (329).

The uncritical acceptance of a best-fit result from a kinetic analysis, restricted to comparisons involving solid-state reaction rate models only, can fail to identify features of the reaction mechanism that can be of great interest and are accessible to the study. There is always the possibility that an additional reaction model, unconsidered in comparisons restricted to equations within the set examined, might be capable of providing a more satisfactory or realistic portrayal of behavior. The equations most usually considered in this type of kinetic analysis (71) were originally based on patterns of behavior observed microscopically, sometimes during reactions in progress. This theory was developed during the formative years of the subject. The range of compounds now of interest, however, has been very much extended and some of these reactants, initially solid, undergo melting, perhaps local and temporary, introducing additional factors that must be considered and incorporated into the interpretation of rate data.

5.3.1.2 Melting and Intermediates

Probably the most usual consequence of heating a solid is that it melts. This possibility cannot, therefore, be excluded from consideration in studies of thermal processes involving reactants initially crystalline and yielding products that are solid when cooled for inspection and analysis after reaction. By exclusively basing the kinetic analysis of diverse thermal chemical changes on solid-state reaction models, mechanisms involving melting are excluded from consideration, and this appears to have been implicitly accepted during many recent studies in this field. If, however, it could be agreed that a fundamental feature of the description or mechanism of any chemical reaction is characterization of the phase within which the chemical change occurs, then this must be a priority in the interpretation of all such thermal investigations. However, because this aspect is so often ignored within current accepted practices, it is necessary to recognize this particular limitation of accepted methodology by developing, in future, a more imaginative and open-minded approach toward establishing the range of possible mechanisms of thermal reactions undergone by diverse, initially solid, reactants.

There is evidence (38) that many decompositions proceed relatively more rapidly in a melt than as a solid, because the constraining and stabilizing forces of a crystal structure are relaxed. In some reactions, liquefaction may arise through eutectic formation (sometimes composed of reactant together with one or more products) or the formation of a liquid intermediate. Melting of some reactants may be comprehensive and is not detectable by TG but readily recognizable from a characteristic response form in DTA or DSC measurements. In other compounds, fusion during the progress of reaction may be partial, local, and temporary and can be much more difficult to recognize (72). Kinetic behavior shown by a reaction that involves the participation of a liquid phase may generally resemble that characteristic of the decomposition of a solid, but specific equations have not been developed to identify and to distinguish rate processes occurring in these mixed phases because of the considerable variability of behavior that is possible. The different, and possibly changing, proportions of liquid participating are difficult (or almost impossible) to measure. The selected examples given later illustrate some of the solid or melt reaction types that have been reported. These are mentioned here to emphasize the diversity of chemical behavior that has been characterized for thermal rate processes that resemble, and has even formerly been regarded as, solid-state reactions but have since been shown to involve at least some melting. Appropriate microscopic examinations and chemical analyses of partially reacted samples have, for these systems, confirmed the contribution of fusion and recognized participating intermediates. Such textural and analytical information is required to formulate detailed reaction mechanisms that involve a number of distinct reactions through which the reactant is transformed into product, requiring complementary observations from several techniques. TG was not used in the illustrative examples described later, but certainly the method would be capable of contributing useful kinetic data in such mechanistic studies. Wider appreciation of the essential complexity of thermal decompositions is expected to introduce a more chemical approach to these investigations, which may be capable of recognizing rate processes proceeding in a homogeneous melt, constituting perhaps only a small proportion of the total reactant mass.

The following thermal decompositions were formerly (and perhaps uncritically) regarded as proceeding in the solid state. However, subsequent detailed kinetic and analytical investigations have provided evidence that there is at least partial melting, together with contributory reactions involving identified intermediates. For example, microscopic observations demonstrated that the decomposition of ammonium dichromate proceeds within zones of local melt (73). It was concluded that the several reactions contributing to the overall change involved the participation of molten CrO_3 together with ammonia oxidation, thus accounting for the intervention of nitrate and nitrite, detected in earlier studies. Although the kinetic measurements, again not obtained by TG, gave a (somewhat asymmetric) sigmoid shaped α-time curve, the analytical evidence shows that the mechanism is too complicated to be characterized completely by yield–time determinations. The thermal decomposition of ammonium perchlorate also may proceed with the participation of fused intermediates, and the chemical changes occur in a molten phase (74).

Other decompositions, which had previously been accepted as simple reactions proceeding in the solid state, have subsequently been shown to be more complicated than was discerned from overall kinetic data. The thermal breakdown of potassium permanganate exhibits almost symmetrical sigmoid curves, now regarded (39) as proceeding with the intermediate formation of $K_3(MnO_4)_2$ by at least two, possibly consecutive, reactions. Dehydration of calcium oxalate monohydrate proceeds (75) with the loss of H_2O molecules from two different types of site by two concurrent reactions that proceed at slightly different rates.

5.3.2 REACTION STOICHIOMETRY

The mass change measured on heating a characterized reactant (usually through a recorded temperature interval) is frequently accepted as evidence of the occurrence of an anticipated chemical change, often represented by a balanced equation. Numerous reports in the literature base stoichiometric conclusions on mass changes alone. Some of these deductions refer to a single process; others identify two or more rate processes that may be individually resolved or overlap to a greater or lesser extent. It is probable that many of these conclusions are correct. However, although the mass change must be consistent with that indicated by the equation representing it, there is sometimes the possibility that an alternative reaction can equally satisfactorily explain the measured change. Experimental error and the possibility of non-stoichiometry of reactants and products must be considered in the data analyses. Balanced rate expressions were developed to describe homogeneous processes, where interactions involve individual species. Their application to chemical changes of solids is not invariably acceptable where alternative structures are not normally represented, and it may be necessary to accommodate compositional variations, together with the participation of crystal imperfections.

Conclusions based on TG evidence alone are not invariably reliable, and further information may be required to establish unambiguously the identities of the reactant and products (compositions and structures). The following aspects of TG data interpretation require consideration:

Concurrent or overlapping consecutive reactions may not be resolved by TG into individual steps. Modification of experimental conditions, particularly a change of heating rate during nonisothermal investigations, may be capable of separating the successive contributing processes. However, where one reaction remains uncompleted before the onset of that following, its overall yield corresponding to completion may require estimation by extrapolation, thereby introducing error into the value of m_f determined (and α values calculated for use in kinetic analysis). Investigation of stoichiometry in stepwise processes has been improved by a temperature programming procedure (47).

The chemical change that occurs may depend on conditions within the reaction zone. For example, in vacuum the dehydrations of $MgSO_4 \cdot 7H_2O$ (76) and of $NiSO_4 \cdot 6H_2O$ (34) proceed directly to the monohydrate. However, in a gas atmosphere that opposes the diffusive escape of the H_2O evolved, these dehydrations proceed stepwise with the intervention of intermediate lower hydrates (33,35). Characterization of each of these by mass loss may or may not be possible, but additional tests, such as x-ray diffraction, are required in some systems to confirm the identities of the intermediate hydrates (35,36).

The occurrence of concurrent or substantially overlapping reactions will not be characterized by TG data alone. This has recently been shown by kinetic studies of the dehydration of $CaC_2O_4 \cdot H_2O$ (75) and of the decomposition of $KMnO_4$ (39). Both reactions were accepted as single-rate processes, but recent more detailed examinations have identified more complicated behavior.

Compound identification (reactant or product) based on chemical analysis must be precise to define composition reliably. Many solids deviate from stoichiometric expectation, there is also the possibility that precursor processes can precede the reaction of interest, resulting in loss of adsorbed or included solvent, breakdown of superficial unstable material, or a limited surface initial process (77).

Thus, to establish reliably and to characterize completely the chemical reaction resulting in a measured mass change, it may be necessary to include consideration of (1) initial reactions (such as superficial processes [77]) or nonstoichiometry of participants, (2) the possibility that the overall change may be composed of two or more contributory (but distinct and concurrent or overlapping) rate processes, and (3) the fact that one reaction is completed and its mass change can be reliably determined before the onset of a subsequent step.

For chemical changes that are simple, and have already been characterized, a measured mass change may be regarded as an adequate confirmation that the reaction anticipated has occurred as, for example, in the dissociation of calcite (1,17) or the dehydration of nickel oxalate dihydrate (18). However, when investigating novel systems or reactions in which different changes occur under varied reaction conditions, TG data alone must be regarded as insufficient to determine stoichiometry. High-temperature decompositions of complex salts containing large ligands can be confidently expected to involve secondary decompositions or interactions between primary products, and, without supporting analytical data, products can be incorrectly identified. (For example, the molecular weights of $2H_2O$ (36.03) and HCl (36.46) are close, and mass changes must be very accurately measured for these possible products to be reliably distinguished on this evidence alone.)

5.3.3 REACTION KINETICS

Many, perhaps most, measurements of reaction rates are undertaken either to obtain kinetic data for a practical purpose (*empirical kinetic studies*) or to investigate the fundamental chemical characteristics of the reaction (*fundamental kinetic studies*).

5.3.3.1 Empirical Kinetic Studies

Reaction rates are determined for a specific practical objective. Representation of data through kinetic expressions and the Arrhenius equation provides a useful method of summarizing results (empirically) and perhaps enables useful extrapolations of the observations to be made beyond the range of conditions experimentally measured. This systematization of results is directed primarily toward characterizing levels of reactivity and patterns of behavior. Such correlations of data are not, however, intended to advance theory, establish insights into the chemistry of the processes, or formulate reaction mechanisms.

Empirical studies are usually directed toward specific practical objectives, including process design, where behavioral variations through the influences of temperature, pressure, etc., may be used to explore methods of maximizing the overall efficiency of product manufacture in an industrial process. Empirical measurements can also be applied to determine shelf lives of unstable compounds, including pharmaceuticals. Rates of degradation can be investigated under realistic storage conditions (temperature, humidity, additives including packaging, etc.). Results from accelerated testing, in which shorter times are required through the use of slightly elevated temperatures, can be extrapolated to the lower values expected to be encountered, for example, in tropical climates.

Empirical experiments undertaken for an identified purpose must mimic, as closely as possible, the conditions believed to apply during the manufacturing process, storage, etc. Thus, providing that all contributory reactions behave as in the situation of practical interest, it is not necessary to identify the part played by each or the roles of such factors as reaction reversibility, phase changes, and self-heating. Also, it may be useful to determine the effects of the procedural variables (1) in controlling reaction rates without establishing why and how these operate. It may not be necessary or useful to identify a rate-limiting step or speculate about the significance of the activation energy E_a.

5.3.3.2 Fundamental Kinetic Studies

These are investigations directed toward extending or developing chemical theory, generally or specifically, to establish the factors that control reactivity, to elucidate the mechanisms of particular reactions, to compare and to explain the behavior patterns characteristic of similar or related rate processes, etc. Studies of this type are intended to provide fundamental insights into, and provide models for all or specific aspects of the chemistry, of the target reactions while perhaps also contributing to the advance of theory more generally. Such work is expected to introduce and to extend order and theory of the subject and to increase the reliability and range of useful predictions that can be made for hitherto untested systems.

5.3.3.3 Kinetics of Homogeneous Reactions

A simplistic view of the methodology of homogeneous reaction kinetics is that the rate equation (e.g., reaction order) most satisfactorily representing the data is identified. This may enable mechanistic deductions to be made (e.g., reaction molecularity). From measurements made across a range of temperatures, calculated rate constants are correlated with the Arrhenius equation. The activation energy (E_a) and the reaction frequency factor (A) are then calculated and, from the transition state theory, identified with the magnitudes of the energy barrier to reaction in the rate-limiting step and the frequency of occurrence of the reaction situation, respectively. This general approach, sometimes suitably adapted, has been applied most successfully to innumerable homogeneous reactions in the gas phase and in solution.

5.3.3.4 Kinetics of Condensed-Phase Reactions

The range of reactions that are capable of being investigated by TG do not include most homogeneous rate processes that involve gaseous reactants or species in solution (unless involving an appreciable mass change). Those most usually investigated by TG methods can be classified into three groups:

1. Reactions of solids, including decompositions, sublimations, interactions with gases, and any other process involving a mass change.
2. Reactions of solids accompanied by melting, partial or complete. The occurrence of melting gives no TG response, and its specific detection requires additional observations.
3. Reactions of liquids, melts, or other liquid systems that contain little or no solvent but are accompanied by a mass change.

It must be stressed that the theory of reaction rates available for consideration of the kinetic characteristics of these condensed-phase reactions is unsatisfactory. At the present time, suitable theories and reaction models for reactions of crystals are not generally available. Accordingly, the present account adheres to the accepted practice of using theory derived for homogeneous rate processes, suitably modified where possible. This is unsatisfactory in many important respects, a theoretical limitation that must be emphasized here, but the approach reflects the content of the recent literature. Many superficial parallels exist between homogeneous kinetic theory and the rate characteristics of heterogeneous rate processes, most notably, the applicability of the Arrhenius equation. However, the collisional activation of gaseous molecules cannot apply to chemical changes in a crystal and, indeed, little is known about the interface conditions that result in preferred reactions within these zones.

5.3.4 REACTION REVERSIBILITY

Kinetic measurements undertaken to determine a reaction mechanism are often intended to characterize a rate-limiting step, for which the magnitude of E_a is regarded as providing a measure of the energy barrier to the bond redistribution step, and A is the frequency of occurrence of the reaction situation. This approach, successfully applied to homogeneous reaction kinetics, is frequently applied directly to solid-state rate processes. However, it cannot always be assumed that the overall reaction rate is controlled exclusively by a single "slow" step. This has been particularly demonstrated for reversible decompositions of solids that are sensitive to the procedural variables (1); these include reactant mass, particle size and packing of a powder, heating rate, pressure of gases present (particularly including the volatile products), etc. Kinetic characteristics are often sensitive to reaction conditions, most notably, those which influence the ease of volatile product removal from the reaction zone. Within a large reactant mass, it is probable that the distribution of the evolved gas may become inhomogeneous and vary throughout the progress of reaction. The partial pressure of product is probably greatest between the most densely packed particles in the center of the reactant aggregate and less at the peripheries, from which escape is easier. Gas readsorbed does not contribute to the apparent overall rate measured and may participate in interface equilibria, diminishing the frequency of the dissociation step, with a consequent reduction in the rate of product release.

A theoretical problem, probably hindering the advance of this subject, is that it is not known whether the features described as follows are representative of reversible solid-state reactions generally or are atypical, applying only to the specific reaction mentioned. More work is undoubtedly required to establish the generality of these patterns of behavior, but their existence means that there is the possibility, even probability, that other reactants exhibit similar considerable sensitivities of kinetic properties to reaction conditions. It certainly cannot be assumed, without appropriate experimental support, that reported values of calculated Arrhenius parameters can be taken to be applicable to a rate-limiting step for the overall chemical change observed. Characteristic properties of some reversible solid-state dissociations are illustrated by the examples following:

Dissociation of Calcium Carbonate: A number of studies of this decomposition have concluded that *Ea* is similar to the dissociation enthalpy for equilibrium conditions, about 173 kJ mol^{-1}. A study of the reaction in vacuum, below 10^{-2} Pa (17), however, measured the activation energy as 205 kJ mol^{-1}, or about 30 kJ mol^{-1} greater than the enthalpy of the dissociation process; these values have been supported by subsequent studies (78). This study investigated the reaction between 934 and 1013 K through measurements, by a quartz microbalance, of the constant rate of mass loss during inward interface movement from a single flat reactant crystal face. No reduction of rate, attributable to impedance by increasing thickness of the product CaO layer, was detected. Beruto and Searcy (17) suggested that, under higher pressure conditions, different from those applied in their work, CO_2 pressures developed within the sample approach equilibrium values. Diffusion rates are proportional to gradients of partial pressure and, consequently, the enthalpy and E_a values are similar. Measurement of the rate of the chemical-controlling step, by which the volatile product is released, requires removal of contributions from all secondary processes participating in reaction zone equilibria.

Dehydration of Nickel Oxalate Dihydrate: A TG study of this reaction (18) used reaction conditions designed to remove, very rapidly and efficiently, the water vapor evolved from the dehydrating salt. A vacuum was maintained throughout the reaction by constant pumping through wide bore tubing at a rate capable of achieving 10^{-4} Pa. Rate constants, identified as being proportional to the velocity of interface penetration, were obtained from the linear initial reaction rate. To minimize the influence of water vapor, rates were measured for small reactant masses, 3 to 0.2 mg. The activation energy, 129 ± 5 kJ mol^{-1} between 358 and 397 K, was calculated from rates extrapolated to zero mass. This measured E_a is significantly greater than other values reported (e.g., 42, 76, 95 kJ mol^{-1} [18]) for the same reaction, studied in higher-temperature intervals and under conditions that did not ensure water removal so stringently. It was further shown that, at 383 K, the presence of a water vapor pressure of only 5 Pa diminished the dehydration rate × 0.04. The dehydration rate of this reactant was shown to be highly sensitive to the availability of (product) water vapor.

Dehydration of Nickel Sulfate Hexahydrate: In a vacuum, the dehydration of $NiSO_4 \cdot 6H_2O$ proceeds directly to the monohydrate in a single step (34). However, in a static atmosphere, during nonisothermal heating at 8 K min^{-1}, water is lost stepwise, with the intervention of the tetra- and dihydrates. The identities of these intermediates, formed at appreciably higher temperatures than the single-step reaction ($-5H_2O$), were confirmed by x-ray diffraction (35). Water vapor present within the reactant particle assemblage because escape is impeded by the atmosphere enables formation of the intermediate phases, and dissociation is possible only at higher temperatures than in the vacuum reaction (0.1 Pa above 308 K).

Dehydration of Lithium Sulfate Monohydrate: In a discussion of the dehydration of $Li_2SO_4 \cdot H_2O$, another dehydration for which rate characteristics vary with reaction conditions (79), L'vov et al. (28) resolve the kinetic inconsistencies and distinguish two types of behavior. Reactions in the *equimolar mode* proceed in the absence of volatile products, whereas in the *isobaric mode* the (constant) partial pressure of a volatile reaction product significantly exceeds its equilibrium dissociation pressure. This accounts quantitatively for the influence of water vapor in diminishing this dehydration rate.

Dehydration of Chrome Alum: In contrast with the previous example, it appears that the dehydration of $KCr(SO_4)_2 \cdot 12H_2O$ is insensitive to the presence of water vapor, though experiments did not extend to the lowest pressures (80–82). A possible explanation is that dehydration is promoted by retained intranuclear water. The rate of interface advance at 290 K increased by about one quarter at 100 Pa water vapor compared with that in vacuum.

These examples demonstrate conclusively that the availability of volatile product in the immediate vicinity of the site of a reversible dissociation can markedly influence the apparent kinetic characteristics of crystolysis reactions. Some systems are highly sensitive to such effects but others, such as chrome alum dehydration, are markedly less affected. The kinetics of $CaCO_3$ dissociation vary considerably with reaction conditions (83), and here the pattern of reaction rates is also influenced by heat flow during the endothermic, reversible reaction. It follows that it cannot be assumed, without examination of the influence of the procedural variables, that measured kinetic parameters are determined by a slow rate-limiting step.

5.3.5 REACTIONS OF SOLIDS WITH MELTING, PARTIAL OR COMPLETE

Probably the most usual consequence of heating a solid is melting, a change that cannot be directly detected by TG. The possible role of fusion in thermal behavior of solids is conveniently introduced by considering some highly hydrated inorganic salts. Melting points are listed for alum (365 K, with $-9H_2O$) and for chrome alum (362 K), which appears inconsistent with kinetic studies reported (81) for the solid-state (partial) dehydrations of these reactants at much lower temperatures, in vacuum below 310 K. However, if large samples of alums are (relatively) rapidly heated in air, under conditions that do not favor water escape, the "melting" point is reached when the salt effectively dissolves in its own water of crystallization. Thus, the changes observed on heating vary significantly with prevailing conditions. This occurrence of alum melting is an alternative behavior pattern to the different kinetic characteristics, in vacuum or in air, described earlier for $NiSO_4 \cdot 6H_2O$ dehydration. Water loss can be extensive ($-5H_2O$) or stepwise and again is a consequence of product water availability. These examples concern hydrates and have been selected because the properties of this group of solids have been extensively studied (32), and provides well-characterized examples.

The solvent properties of water on heating highly hydrated salt are well known, but, for other diverse reactants, a wide range of different possibilities exist (e.g., other solvent ligands, eutectic formation, and a molten intermediate). Unless specifically sought, the participation of melting during a solid-state decomposition may not be recognized. In the formulation of a reaction mechanism, it must be accepted as of central importance to establish the phase in which the chemical steps that are significant in transforming the reactant into product take place. In general, it is probable that the kinetic characteristics of reaction in a solid will be different from that in a liquid (38). A complete description of reaction must, therefore, require the determination of both; this is experimentally difficult (even effectively impossible) where the amount of liquid present would require its detection and measurement in a reactant undergoing temporary local or progressive fusion. There is some evidence that a number of chemical changes take place relatively more rapidly in a melt, in which the stabilizing, attractive forces of the crystal structure are relaxed and there is greater stereochemical freedom for the precursors to change to adopt the conformation most favorable for the reaction to proceed. The following examples include some different types of decompositions in which an essential role of melting has been identified:

Decomposition of copper(II) malonate: Examination of fracture sections of cooled, solidified particles of partly reacted salt, by electron microscopy, revealed the intraparticular generation of a frothlike material composed of bubbles with rounded surfaces (30). This texture was ascribed to the evolution of gaseous products within a fluid matrix. The predominantly curved surfaces were ascribed to surface tension control, and no regularly aligned features or flat surfaces, characteristic of crystalline materials, were found. It was concluded that the formation of copper acetate, together with copper reduction to the monovalent state, resulted in the formation of a molten intermediate or a eutectic. The progressive formation of melt, as reaction advanced, meant that no sharp endotherm was given, though a small exotherm toward completion of the first reaction stage was ascribed to solidification. This is consistent with the asymmetric sigmoid-shaped α-time curve and the early increases in the amount of the intraparticle frothlike material. The occurrence of fusion was not, however, obvious. Apparently, on heating, the reactant formed a relatively stable superficial layer that inhibited crystallite coalescence, thereby maintaining the individuality and shapes of the original crystals, and the product particles, therefore, remained pseudomorphic with the reactant. It was only by examination of internal textures, revealed by crushing of partly reacted salt, that participation of local, probably temporary, melting could be identified.

d-*Lithium potassium tartrate decomposition:* The sigmoid-shaped isothermal α-time curves for the thermal decomposition of d-$LiKC_4H_4O_6$ dehydrated, single crystals, 485 to 540 K, fitted the Avrami–Erofeev equation during the greater part of reaction, between $0.04 < \alpha < 0.96$ (84). From microscopic observations, it was concluded that the reactant did not undergo comprehensive melting, but, during reaction, droplets of brown liquid appeared, adhering to surfaces and penetrating cracks in the white reactant crystals. The amount of liquid increased and the proportion of solid decreased as reaction advanced, and no intermediate compounds could be detected. It was concluded that the anion breakdown reactions were completed at, or close to, the crystal/molten phase contact. These kinetic studies used gas accumulatory measurements, which are similar to TG mass loss; neither includes contributions from the condensed-phase processes whereby carboxylate in the anion is converted to alkali carbonate product. The usual approach to interpretation of the kinetic data would identify this as a nucleation and growth process, but microscopy recognized that the "nuclei" are composed of a viscous liquid, a possibility not normally included in solid-state reaction theory.

Comment. These examples (see also 73,85) have been described at greater length here than is customary in many surveys to emphasize that thermal rate processes exhibiting sigmoid α-time curves do not invariably or necessarily proceed by a solid-state reaction mechanism. There is the possibility of fusion, enabling reaction to occur in a liquid-phase that may be local and/or temporary. During such processes, it may be possible to identify intermediates that are an essential feature of the sequence of chemical changes, a knowledge of which may be necessary to enable the reaction controls and the contributions to the overall mechanism to be understood. However, data relating to the measured chemical change may be insufficient to characterize all the factors controlling product formation. A complete kinetic description may require investigation of the rates of production and loss of all significant participating intermediates. Melting is a frequent consequence of heating, and, in at least some rate processes, there may be greater freedom for chemical change and enhanced reactivity in the liquid phase (38). The view that the demonstration of "best fit" of date to a particular solid-state kinetic model is sufficient to characterize the mechanism cannot be sustained. Reactions in a condensed phase are not necessarily completed within a narrow interface zone; the thermal behavior of solids is more complicated than is envisaged in the theory often applied.

5.3.5.1 Kinetic Equations for Reactions Proceeding with Partial Melting

The principles underlying the formulation of rate equations applicable to the decompositions of solids are presented in Section 5.4. In summary, these result in the replacement of the concentration terms generally applicable in homogeneous rate processes by geometric or diffusion parameters. It is possible, in principle, to formulate a further set of kinetic models that describe concurrent reactions proceeding in a solid reactant as well as in a molten phase. However, the unavoidable incorporation of additional variables (at least two different rate constants for the two rate processes together with a factor expressing their relative proportions, which probably changes as reaction progresses) makes kinetic analysis unreliable in the absence of additional specific rate data. Quantitative information about the amount of liquid phase present, and its variation with α, would facilitate data interpretation. However, such studies do not appear to have been undertaken and are not obviously or easily accessible experimentally. The incorporation of additional parameters, not independently measurable, into the kinetic analysis of rate studies does not offer any immediate prospect of resolving these difficulties. At present, the qualitative demonstration of melting and the identification of intermediates appear to be worthwhile investments of effort directed toward the elucidation of mechanisms of condensed-phase reactions. The following limiting/particular situations merit consideration.

Chemical changes are completed within a thin layer (perhaps only a few molecular layers thickness) of liquid situated between the reactant solid and one or more solid products. This represents local enhancement of mobility of the reaction participants within an active advancing interface. Systematic changes in the reaction rate will be determined by the geometry of interface advance, and the kinetic characteristics will be identical with those of the solid-state rate processes (see Section 5.4), although the interface mechanisms and controls will be different. A related pattern of behavior, also exhibiting solid-state reaction kinetic characteristics, is the nucleus' retaining some (81) or a significant amount (85) of a liquid product, which promotes and/or participates in the interface process.

Reactions occurring in a liquid, the amount of which increases in proportion to the extent of reaction, results in rate acceleration throughout the range for which this control applies. Such a rate processes is expected to fit the exponential rate law (30).

5.3.5.2 Reactions with Complete Melting

In these reactions, there is comprehensive melting before the initiation of reaction conditions. These are homogeneous rate processes, and are therefore expected to be represented by rate laws based on reaction order, the amount of reactant remaining. However, if the reactant is not inspected during the progress of change, fusion may not be detected. This observation should be undertaken at reaction temperature to avoid the possibility that the reactant has solidified on cooling, after removal from the apparatus. Studies of dehydrations have shown that the kinetic characteristics during reactions of melts are sometimes approximately zero-order and a more-than-single-rate process may be involved (86,87). Participants in reactions proceeding close to a melting point may undergo a more-than-single-phase change (i.e., melting and solidification) before completion. This behavior is expected to vary with reaction temperature, and thus kinetic and mechanistic characteristics are likely to be complicated.

5.3.6 Self-Cooling or Self-Heating During Reaction of a Solid

Heat is initially supplied to the reactant sample to raise it to reaction temperature, at which, ideally, it should be maintained without change thereafter in isothermal studies or at the specified range of values during a programmed sequence of variations. However, in addition, account must be taken of the heat absorbed in endothermic processes or evolved during exothermic reactions resulting in reactant self-cooling or self-heating to an extent that will depend on reaction rate, enthalpy, thermal conductivity, etc. This may be significant in causing an appreciable difference between the temperature within the reactant particles and, more specifically, inside the reaction zone, and that measured for the controlled reaction environment. This difference is difficult to measure. Many solids are poor conductors of heat and, when held in a vacuum, there is the possibility that there can be a significant difference between reactant temperature and that recorded.

In the early studies of solid-state dehydrations, often using large single crystals, self-cooling as source of error was recognized (80,88). The extent was measured for these endothermic processes and due allowances made in the calculations of E_a values. More recent kinetic studies appear to have largely ignored the significance of contributions from such temperature deviations. However, recent computer modeling for the dehydration of $Li_2SO_4 \cdot H_2O$ revealed (28) a substantial reduction in the rate of water evolution, compared with that expected, and the effect was greater in powders than in crystals. This control was capable of accounting for the main features of the Smith–Topley effect that has been regarded as characteristic of many dehydrations (29). At low water vapor pressures, the rate of water release diminishes markedly. Subsequent increases in water vapor pressures result in a rise of dehydration rate to a maximum value that later falls to zero at the dissociation equilibrium. In an earlier detailed study of this behavior pattern (19), it was concluded that the interface conditions included a discontinuity at the reaction front, involving the equilibration of movements of heat and water. The abnormal behavior was shown to occur in a well-defined relative pressure/temperature range for several different salts. In commenting on this earlier work, L'vov et al. (28) pointed out that the model discussed had not developed in the subsequent two decades, possibly because the temperature differences indicated appeared to be unacceptably large. However, the recent revival of consideration of the kinetic consequences of self-heating, supported by computer modeling, should reopen a topic that has not always received the attention it deserves. The dependence of dehydration rates on water vapor pressure is a further example of the complexity of kinetic characteristics of reversible processes and the ability of the volatile product to control the apparent dissociation rates.

5.3.7 Thermodynamics, Reaction Heats, and Adsorption

Although the movement of heat (thermodynamics) cannot be directly measured through mass changes, TG alone or in combination with complementary observations can often provide thermodynamic data. Again, the accuracy, convenience, and reliability of the technique commends its use, and examples of many ingenious applications are to be found. The following examples demonstrate and illustrate the types of work that have successfully incorporated TG measurements.

5.3.7.1 Dissociation Equilibria

For a reversible process, the dissociation enthalpy of a solid reactant may be determined from the variation of equilibrium decomposition pressure, p_e with temperature, for example, the dissociation of calcium carbonate (17):

$$CaCO_3 \rightleftarrows CaO + CO_2$$

and the equilibrium can be written in the form

$$\ln(p_e/p) = -\Delta G^\circ/RT \tag{5.2}$$

This is applicable because the two crystalline participants do not form a solid solution, which would introduce a further term into the energy balance equation. Such measurements may be made by heating the reactant in a known pressure of the volatile product, here CO_2, and TG can be used to detect the cessation of reaction, when the equilibrium pressure and temperature are measured. It is essential that there is sufficient reactant to ensure that both solid phases are present at equilibrium, which should be approached from both directions.

The applications of thermodynamic functions in thermal decomposition kinetics have been discussed in detail by L'vov (89,90). The Hertz–Knudsen–Langmuir equation expresses the rate of monotonic evaporation, J, (which can be measured using TG) by

$$J = N_A P/(2\pi MRT)^{1/2} \tag{5.3}$$

where N_A is the Avogadro number, P is the hypothetical equilibrium partial pressure for the evaporated product of molar mass M. Consideration of the subsequent processes involving these participants then may give information concerning the reaction controls and mechanisms. Other aspects of TG studies of sublimation, based on a similar theoretical approach, have been discussed (91). The sublimation enthalpies of the six isomers of dihydroxybenzoic acid, measured by TG in a study of matrix assisted laser desorption/ionization-mass spectroscopy, were compared with literature values.

5.3.7.2 Surface Adsorption Processes and Surface Characterization

Provided that the surface area, and therefore the quantity of gas adsorbed, is sufficiently large, mass changes can be used to investigate surface properties. TG was used to measure the nitrogen adsorption isotherms at 77 K on the mixed oxides: SiO_2/Al_2O_3, SiO_2/TiO_2, SiO_2/MgO, and Cr_2O_3/Al_2O_3 (92). It was concluded that "vacuum balances are efficient in determining the textural properties of catalysts" and "provide a more or less complete, clear pattern of the surface acidity of solid catalysts." In a different TG study, Goworek et al. (93) investigated the porosity of silica gel by thermal desorption of benzene for various heating programs. The pore size distributions were similar but not identical with those found from low temperature nitrogen adsorption. Properties of water films on surfaces of porous silica were investigated (94) by high resolution TG and DSC, to measure amounts adsorbed and the associated heat changes. Information was obtained about the mechanism of silica hydration and the adsorption capacity, including the influence of pore structures on the freezing and melting of adsorbed water, which occurred at temperatures different from those of bulk water. TG and DSC measurements have been used (95) to characterize both mass change and heat absorbed during zeolite dehydrations.

5.3.7.3 Vapor Pressure and Boiling Points

TG measurements were used to determine the boiling points of binary mixtures of short-chain triglycerides (96) and, from these data, the effective heats of vaporization were calculated. Boiling points were found from mass loss measurements across a wide temperature range enabling extrapolations to be made both at low (slow mass losses) and higher temperatures (more rapid mass losses). Where these two tangents intersect is the onset temperature. It was concluded that the TG measurements, described in detail, provide rapid, accurate vapor pressure (studied between 100 and 2.6 kPa) and boiling point data (to about 670 K) for binary mixtures of some triglycerides (substitute diesel fuels).

The heat of vaporization of liquids can be measured (97) by TG–DSC to determine both mass and energy flux; due allowance must be made for the mass change in the energy balance equation (also mentioned by Hoffman and Pan (98)). This method was used to study the losses of liquids from porous matrices.

Reasonably accurate (agreement with literature values better than 5%) heats of evaporation of some solvents (water, ether, dioxane) and NH_4NO_3 solutions were obtained by TG methods (99). This technique also was used (100) to study the volatilization rates of some organic compounds that are of interest as environmental contaminants (naphthalene, hexachlorobenzene, 4-chlorobiphenyl, n-decane). Evaporation rates were influenced by the rates of heating (5, 10, and 25 K min^{-1}). A good representation of behavior was provided by the evaporation model, described in detail, provided that the surface area of the substance was known. It was assumed that equilibrium was established between condensed phase and vapor and that there was convective transport and diffusion to the container outlet. It was concluded that TG methods provide a useful method for studies of the evaporation of organic compounds.

TG–DTA measurements were used (101) to determine solubilities in inorganic salt–water systems. The rates of evaporation of water through a small hole in a capsule containing the solution were measured. The rate of solute escape changes when the solution curve crosses the solubility curve. The solubilities of NaCl, KCl, and Na_2SO_4 in water were determined to 1% accuracy, rapidly and using only small sample quantities.

5.4 INTRODUCTION TO THE THEORY OF SOLID-STATE REACTION KINETICS

Factors controlling the rates of chemical changes involving crystalline reactants differ in several essential respects from those operating during homogeneous reactions. Nevertheless, the kinetic theory applied in the analysis and mechanistic interpretation of rate data for crystolysis reactions (102) is comprehensively based upon that developed originally to account for rate processes occurring in gases or in solution. It is well established and (probably of greater significance) it is generally accepted that reactions in crystals are satisfactorily represented by the Arrhenius equation. As a consequence, solid-state reactions are often portrayed as proceeding through an activated complex, a transition state of the type formulated by the absolute reaction rate theory. This model was originally developed on the assumption that chemical changes follow energetic interactions (collisions) between randomly moving precursor molecules. This assumption, however, is clearly inapplicable to reactions within crystals, where the precursors to chemical change are effectively immobilized species. These precursors, ions, molecules, or atoms in solids are often regarded as being located at, or in the immediate vicinity of, a contact zone of locally heightened reactivity, the *reaction interface,* located between the solid reactant and the residual product (usually crystalline though some are amorphous). The particles of residual material develop (grow) by deposition of product, formed through chemical change at the reactive (advancing) contact interface. It follows that various aspects of interface reactions differ fundamentally from those occurring in homogeneous processes and, therefore, require appropriate reconsideration of the theory. This includes identifying reasons for the relative increase of reactivity that is often associated with the advancing interface. Continued chemical change in the immediate vicinity of the reactant/product contact zone maintains the progress of the active zone into the reactant solid. Thus, two complementary aspects of behavior that feature in solid-state decompositions require consideration in their kinetic analysis: first, the form of the isothermal kinetic rate equation, involving geometric and spatial parameters and, second, the temperature dependence of reaction rate (possibly) determined by bond redistribution steps and other mechanistic controls.

The summary of the theory of kinetics of solid-state reactions given here is applicable to TG investigations and also equally to rate measurements by other methods of thermal analyses. For more extensive treatments and further background information see References 4,5,10,29,71,76,103.

Note: The term *crystolysis reaction* was proposed for use in describing the thermal decomposition of a solid reactant (31,72). This specific label is of value as a keyword and index entry, while also confirming that the reaction of interest proceeded in the crystalline state. This important aspect of behavior is not always explicitly stated in reports of many thermal investigations, where the phase in which the change occurs is not positively identified.

5.4.1 The Isothermal Kinetic Rate Equation

The occurrence of a chemical change, preferentially, often almost exclusively, within the advancing active zone at, or close to, the reactant/product contact interface, introduces the concept of *reaction geometry* into the kinetic analyses of solid-state reactions. The dispositions of these active reaction zones change systematically as each maintains its advance into unchanged material, and the product formed through its progress accumulates behind the interface. Reasons for the locally enhanced reactivity associated with active interfaces have not been established in detail. There is also the probability that mechanisms of reactions in solids are different from those that follow energetic collisions between molecules in a gas. Interface reactions may be relatively simple but, alternatively, may involve processes of the following, as well as other, types:

1. An essential (perhaps unstable) intermediate may be generated, which reacts within the interface, where it may be immobile and thus maintained at locally high concentration, in contrast with reactions in a gas, where such a participant would become dispersed.
2. There may be a stereochemical relationship adopted, which is favorable for reactions between immobilized reactant species (at a crystal boundary), perhaps extending to groups of several reactant species capable of undergoing cooperative interactions. Such reactions of high molecularity are of a type that would be highly improbable (and, consequently, contributing little to product formation) in a homogeneous phase.
3. A product phase may adsorb and catalyze the breakdown of a reactant component.

Little is known about the detailed steps and their controls for reactions at the interface of reactant/product contact, for which investigative methods are difficult or unavailable.

Rates of homogeneous reactions are directly proportional to reactant concentrations because every equivalent reactant species possesses an identical (constant) probability of undergoing chemical change in unit time. In contrast, during an interface reaction, the probability that a particular reactant species will undergo chemical change (in unit time) is markedly increased on the close approach of an active reaction interface. This discontinuous change in reaction probability is inconsistent with the assumptions underlying the kinetic theory of homogeneous reactions, and interface reactions must therefore be based on a different reaction model. The model found to be most successful, and generally accepted, is formulated through consideration of the variations of the interface geometry that occur as reactions progress. Assumptions in this model are that all areas of an established interface undergo equivalent reactions and that all reaction zones advance at constant rate. It follows, therefore, that the overall reaction rate, at any time, is directly proportional to the area of interface actively contributing to product formation. The rate equations derived by this approach are conveniently referred to as the *kinetic model* to distinguish these expressions from the *reaction mechanism*, which is the sequence of *chemical* steps, including intermediate formation, through which reactant is transformed into product.

The principles of kinetic model derivation (to give rate equations of the form $g(\alpha) = kt$) can conveniently be illustrated by the diagrammatic representation of a *nucleation and growth reaction* (Figure 5.2). This pattern of reaction rates is characteristic of many solid-state decompositions. Reaction is initiated by *nucleation* at a limited number of reactive sites, usually at crystal surfaces and identified as points of enhanced reactivity, even instability, often ascribed to local damage, imperfections, etc. Identification of the reasons for nucleation has been found to be particularly difficult, the number of sites capable of initiating reaction is often relatively very small, and the specific features responsible for enabling reaction to commence are necessarily destroyed by the appearance of the nucleus. Completion of the nucleation process generates a *growth nucleus*, a particle of product embedded in, and maintaining contact with, the reactant phase across the active interface, at which the chemical change occurs preferentially (104). The continued growth of each

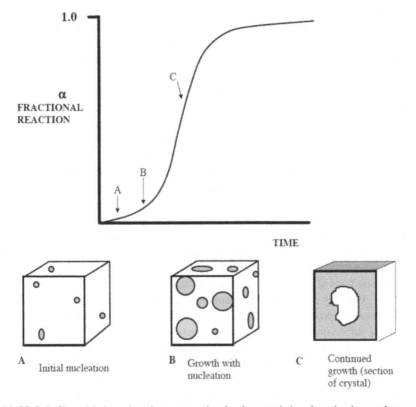

FIGURE 5.2 Sigmoid-shaped α-time curve that is characteristic of nucleation and growth reactions in solids, geometrically represented (product shown shaded): A is initial nucleation, B shows established growth nuclei and continued nucleation, and C shows continued growth as interfaces advance toward the center of reactant particles.

nucleus by facilitated reaction at the outer peripheral contacts with reactant is shown for three-dimensional nuclei in Figure 5.2. Initially, the overall rate of reaction is relatively slow, and the total area of interfaces at these boundaries is small. In the example illustrated here, the effective reaction area increases as the half-hemisphere-shaped aggregates of product grow by equal rates of advance into the crystal in all directions, A to B, as shown. This results in the acceleratory character of the first stages of nucleation and growth processes, illustrated by the (isothermal) α-time plot (Figure 5.2). The formation of additional nuclei may (or may not) continue while areas of unreacted surface remain available. When reaction has advanced to the point of coalescence of all the initially separate nuclei, so that all crystal outer surfaces have reacted, the interface advances inwards towards each reactant particle center (Figure 5.2), and the rate becomes deceleratory during the latter stages. This model accounts for the sigmoid-shaped ('S') yield–time curve, characteristic of many solid-state decompositions (e.g., nickel oxalate, ammonium perchlorate) and dehydrations (alums, copper sulfate pentahydrate). The geometric representation of interface development is supported by microscopic observations through which the individual kinetics of nucleus formation and nucleus growth can be measured (81).

This representation summarizes the conceptual foundation for the theory of reaction kinetics in solids. Essential models used for the formulation of rate equations include those described in the following text; more detailed accounts are given in the references cited earlier.

Several nucleation rate laws have been identified, including *instantaneous* (all nuclei appear immediately at the start of reaction), *linear* (a constant rate of appearance of new nuclei is maintained during the early stages of reaction), *exponential* (the rate of continued nucleation decreases with time), and *power* (the rate of continued nucleation increases with time). These have been distinguished by microscopic observations for various representative reactions (29). It is often assumed that nuclei are randomly distributed across crystal faces.

For a range of reactions, it has been shown, again by microscopic observations, that the rate of (isothermal) interface advance (linear growth of a nucleus with time) is constant, and this is often accepted as having wide applicability. The alternative interface advance rate model arises when the residual phase constitutes a barrier layer to either product escape (outward diffusion) from or the inward access (inward diffusion, perhaps of species from a gaseous reactant) to the reaction zone. In these situations, the rate of interface movement is proportional to $(time)^{1/2}$, decreasing with increase in product layer thickness. A further geometric consideration is the number of dimensions in which interfaces advance, which may be determined by the crystal structure or habit, or both. The example given (Figure 5.2) refers to the three-dimensional growth of nuclei, selected as the most easily envisaged. However, crystal lattices vary, and interfaces in laminar particles often effectively advance in two dimensions, whereas those in needles may advance in a single dimension.

The aforementioned behavior patterns and controls (rates of nucleation followed by growth) may be combined (71) to formulate rate expressions that represent the developing geometry of interface reactions proceeding in solid reactant particles. Through this approach, the following groups of rate equations have been derived, which are widely applied in kinetic studies of crystolysis reactions (see also Table 5.1):

The Avrami–Erofeev equation: Nuclei growing in three (or in two) dimensions, without the inhibiting effect of any diffusion limitation, give sigmoid-shaped curves (Figure 5.2) that are well represented by rate equations of the form

$$[-\ln(1 - \alpha)]^{1/n} = kt \tag{5.4}$$

This is often referred to as the Avrami–Erofeev equation and conventionally labeled A2, A3, or A4, where the number is the same as the exponent, n ($= \beta + \lambda$): β represents the nucleation law, and λ is the number of dimensions in which nuclei grow. When n is unity, this is the first-order equation.

The contracting envelope equation: When nucleation is facile across all surfaces, large numbers of growth nuclei are formed immediately on establishment of reaction conditions. A comprehensive interface is rapidly formed initially and advances inwards thereafter from original particle outer boundaries toward each particle center. This is sometimes referred to as the *contracting cube (or sphere) equation* (depending on the shape) when the particles are equidimensional and $n = 3$ (R3); in two dimensions, $n = 2$ (R2), it is the *contracting square (or area) equation.* Both reactions are deceleratory throughout with the form

$$1 - (1 - \alpha)^{1/n} = kt \qquad (Rn) \tag{5.5}$$

Rate equations involving diffusion control: A characteristic feature of these rate processes is the early or immediate generation of a residual product barrier layer covering all reaction faces. This represents a progressively increasing opposition to volatile product release from (or, in gas–solid reactions, to volatile reactant access to) the reaction interface, which, therefore, advances into the particle at a progressively diminishing rate [proportional to $(time)^{1/2}$]. Such reactions are invariably deceleratory throughout, and the form (the equations D1 to D4 in Table 5.1) depends on the number of dimensions in which the interface advances.

Rate equations based on concentration dependence (reaction order): Under some conditions, the rate characteristics of solid-state processes can be expressed through a concentration-type dependence. For example, if the decomposition of a large number of small crystallites is controlled by an equally probable nucleation step at each particle, then this is a first-order process (A1, F1). These rate equations are also used in nonisothermal kinetic analyses of rate data in the form $g(\alpha) = kt$ and are therefore included in Table 5.1.

TABLE 5.1
Kinetic Models (Rate Equations) Applicable to Reactions of Solids

Reference	Rate Equation $[g(\alpha) = kt]$ (Integral Form)	Name (Description)
A2	$[-\ln(1-\alpha)]^{1/2} = kt$	Avrami–Erofeev equations
A3	$[-\ln(1-\alpha)]^{1/3} = kt$	(nucleation and growth)
A4	$[-\ln(1-\alpha)]^{1/4} = kt$	
B1	$\ln[\alpha/(1-\alpha)] = kt$	Prout–Tompkins equation
R2	$1-(1-\alpha)^{1/2} = kt$	Contracting envelope (2D)
R3	$1-(1-\alpha)^{1/3} = kt$	Contracting envelope (3D)
D1	$\alpha^2 = kt$	Diffusion control (1D)
D2	$(1-\alpha)\ln(1-\alpha) + \alpha = kt$	Diffusion control (2D)
D3	$[1-(1-\alpha)^{1/3}]^2 = kt$	Diffusion control (3D)
D4	$1-(2\alpha/3)-(1-\alpha)^{2/3} = kt$	Alternative (3D)
F0	$\alpha = kt$	Zero order
F1	$-\ln(1-\alpha) = kt$	First order
F2	$(1-\alpha)^{-1}-1 = kt$	Second order
F3	$(1-\alpha)^{-2}-1 = kt$	Third order
Pn	$\alpha^{1/n} = kt$	Power law ($n = 2, 3,$ or 4)
E1	$\ln\alpha = kt$	Exponential law

Note: The integral forms of these kinetic models are shown; differential forms are given by Galwey and Brown (29). These are the most frequently tested rate equations; a limited number of alternatives infrequently appear in the literature. 1D, 2D, and 3D refer to the number of dimensions in which a reaction proceeds.

The most important (isothermal) equations that have been used to represent decompositions and other reactions of solids are listed in Table 5.1. A number of other geometric reaction models occasionally appear; some are mentioned in the references cited. Whereas these geometric controls of reaction rates are central to kinetic modeling in this field, other factors have been shown to influence or control kinetic behavior, including particle sizes and perfection, crystal damage, etc. Such effects are sometimes identified as dependencies of reaction rates on experimental conditions, including the procedural variables.

5.4.1.1 Kinetic Analysis

As with the kinetic analyses of homogeneous rate processes, quantitative comparisons are made between the experimentally measured data for a reaction of interest and the curve shapes of the various rate equations (Table 5.1) to identify the applicable kinetic model. This can be approached in several ways (29,105). One traditional method is to plot graphs of $g(\alpha)$ against time and decide which, from the available expressions (Table 5.1), provides the best (linear) representation, or "fit." There is no general agreement on what criteria constitute a "best" or a "satisfactory" fit. The use of the differential forms of the equations can be expected to provide more sensitive discrimination than the integrated forms between the possibilities that are available (106). An alternative method of comparison is through a scaling factor that provides two or more common points for both experimental data and the theoretical expression, followed by examination of the agreement throughout other parts of the curves. Both these methods enable detection of the α ranges across which agreement is most satisfactory and the magnitudes of deviations elsewhere. This allows consideration of possible reasons for departure from expectation, and local deviations, for example, a minor initial reaction.

Some of the more recent publications use statistical comparisons in which all errors between data and the rate equation are expressed by the magnitude of say, one standard deviation. The "best fit" is then identified as the equation that gives a deviation value nearest to unity or similar criterion. Apparently, this is widely regarded as acceptable practice, but it may not establish whether the deviations arise from random experimental error or as a consequence of systematic variations, perhaps occurring within a limited α range and capable of some other explanation.

It is customary in this subject, to accept the best fit identified by kinetic analysis, as evidence that that reaction proceeds by the geometric model that is assumed in the derivation of the rate expression concerned. Although many analyses of this type appear in the literature, the method is not completely satisfactory because there may be ambiguity in some conclusions. For example, in the interpretation of the magnitude of n in the Avrami–Erofeev equation, the value $n = 3$ could arise either through instantaneous nucleation ($\beta = 0$) followed by three-dimensional nucleus growth ($\lambda = 3$) or through continued nucleation ($\beta = 1$) with growth in two dimensions ($\lambda = 2$), giving $n = 3$ for both models, indistinguishable on kinetic evidence. For this, and other reasons, it is desirable that geometric conclusions from kinetic analysis should, wherever possible, be supported by microscopic observations. Preferably, these should be based on a range of samples partially reacted to various extents within which the sizes and distributions of individual nuclei can be characterized. This is possible only for solids where the product phase can be unambiguously distinguished from the reactant and is more difficult when the reactant is composed of small, irregular, or opaque particles.

5.4.2 The Temperature Dependence of Reaction Rate

The Arrhenius equation is virtually universally applicable to solid state reactions; its use requires no specific justification. The Arrhenius parameters (A and Ea) are often regarded as possessing the same significance as in homogeneous rate processes, for which the theory was first developed. Tests of *linearity* of ln k against T^{-1} plots are rarely made for crystolysis reactions, but the value of this relationship is recognized, as in other reactions (107), for its potential significance in advancing theory. The constants A and E_a are often identified with physical quantities. However, the applicability of the Arrhenius model, founded on the Maxwell–Boltzmann energy distribution function, derived through consideration of collisional energy exchanges, has been questioned by Garn (108–110), who finds it unacceptable to apply this representation to reactions proceeding in a solid. A more recent discussion (102) has pointed out that energy distribution functions for the most energetic electrons and phonons in solids are both similar in form to the requirements of the Arrhenius relation, thus providing a theoretical foundation.

The magnitudes of A and E_a, together with the kinetic model $g(\alpha) = kt$ (the "kinetic triad") constitute a concise, useful, and generally understood, measure of reactivity that can conveniently and succinctly summarize and report experimental observations. Without further interpretation, these may be simply accepted as empirical constants. However, much effort has been directed toward using these parameters to obtain insights into the factors that control reactivity. Early applications of the transition state theory to reactions of solids associated the Arrhenius parameters with the breakdown of activated complexes, distributed at nucleus peripheries, through the Polanyi–Wigner model (29). The barrier to reaction, identified with the magnitude of E_a, was associated with a bond rupture step, and information about the entropy of transition complex formation was inferred from the magnitude of A. However, this representation has found less frequent application in recent work. The probable reason is that there are many anomalous systems, and there has also been a realization that the model has contributed few mechanistic insights into the reactions concerned. We have little information about the detailed interactions between reaction precursors and their controlling factors. In the absence of independent evidence concerning the structure of a possible transition complex and its interactions with neighboring species, it is probably premature to attempt, for reaction of solids, to identify the rate controlling step with the rupture of a particular bond. The interpretation is further complicated by uncertainties in the magnitudes of E_a values measured for reversible reactions, which were among the most thoroughly investigated processes in the early research and to which the Polanyi–Wigner approach was first applied. The nature of the reaction in terms of whether it is isothermal or nonisothermal must also be considered.

> *Isothermal reactions:* A set of isothermal rate constants, measured from exper-
> iments made at intervals across the widest accessible temperature range, are
> frequently used to calculate E_a. The magnitude is often interpreted to iden-
> tify the energy barrier that must be surmounted in the rate-limiting step or
> to provide other mechanistic information. Care must be taken to ensure that
> the data refer to a single-rate process of interest and that rates are not
> influenced by other secondary controls or secondary reactions.
>
> *Nonisothermal reactions:* Numerous kinetic investigations of the thermal
> reactions of solids have used rising temperature techniques, often during a
> linear rate of reactant temperature increase. The kinetic analysis then
> requires the solution of three equations:

$g(\alpha) = kt$ The kinetic model or rate equation (Table 5.1)

$k = \exp(-E/RT)$ The Arrhenius equation

$T = T_o + at$ The rate of reactant temperature rise

A general analytical solution to this combination of equations is not possible, and there is considerable literature devoted to the merits and applicabilities of a wide variety of methods that can extract from the experimental data the magnitude of E_a, usually also the form of the kinetic model $g(\alpha) = kt$, and the value of A. A range of appropriate approximate solutions is available and many have been widely applied. The most usual are associated with the names of their original proposers. An account of this theory is difficult to summarize and is longer than can be conveniently presented here, and the reader is referred to the following detailed surveys (4,5,10,29,103). There are, however, few critical comparisons of the strengths and weaknesses of the alternatives available and diverse methods for kinetic analysis of nonisothermal rate data. Researchers often provide no explanation for a particular choice of the method used for data analysis. A careful and critical comparative analysis of this topic, both theory and applications, would be of great value.

There has been a widespread belief in the literature that the kinetic triad [A, E_a, $g(\alpha) = kt$] can be elucidated through kinetic analysis of data from a *single* nonisothermal experiment. This is an incorrect perception and conclusions calculated from any such restricted data sets must be regarded as unreliable. A useful discussion of the minimum number of experiments required to determine these kinetic parameters has been given by Gotor et al. (111).

Two further problems merit particular comment:

The temperature integral: Incorporation of the exponential temperature dependence into the equations is a matter of difficulty. One method of overcoming the problem in the past has been through the use of an approximation, referred to as the *temperature integral*, which allows for the effect of temperature change on reaction rate. Flynn (112) has pointed out that approximations of this type are no longer necessary because computing facilities now generally available allow temperatures to be calculated to any realistic degree of accuracy regarded as appropriate.

Multiple results: Many articles report results obtained from the analyses of the same set of data by alternative mathematical procedures. In many such articles, the magnitudes of Arrhenius parameters calculated by the methods available give significant variations of the calculated (apparent) magnitudes of A and E_a. These mathematical programs must involve inconsistencies in the calculation methods because the consideration of one reaction, or the use of one set of data, can exhibit only a single temperature coefficient. Furthermore, if it is accepted that the activation energy is a fundamental characteristic of the reaction, then it cannot change with the calculation method. Consequently, the assumptions and procedures used in these kinetic analyses require critical reexamination. One probable explanation for at least some of these variations is that a rate constant can sometimes be defined in two alternative ways, e.g., for the Avrami–Erofeev equation:

$$[-\ln(1 - \alpha)] = k^n t^n \quad \text{or} \quad [-\ln(1 - \alpha)] = K t^n$$

Magnitudes of E_a calculated for the same data from rate constants defined in these alternative expressions will differ by a factor of n. The former definition of k is to be strongly preferred as expressing the rate in units (time)$^{-1}$.

It is desirable to undertake tests of reproducibility in all kinetic measurements to ensure that the reactant sample is representative and to determine the accuracy of all reported data, including the magnitudes of E_a and A. Repeated, nominally identical, and similar experiments do not always give the same results. This is shown in the extensive comparative determinations (113,114) of the kinetic model for the dehydration/isomerization reaction of $K_4[Ni(NO_2)_6] \cdot xH_2O$ and the dehydration of $(NH_4)_2C_2O_4 \cdot H_2O$. Magnitudes of Arrhenius parameters are often reported in the literature to an unrealistic number of significant figures; the use of parallel, repeated experiments usually enables the accuracy of data to be determined. This is to be preferred to the reporting of computer-printed values, without considering their probable significance.

5.4.3 THERMAL ANALYSIS SCIENCE

TG has hitherto found relatively few applications in pharmaceutical sciences other than relatively simple weight loss measurements although the method undoubtedly possesses considerable further potential. In contrast, TG has an extensive literature reporting many most valuable achievements in other branches of chemical sciences. There are currently, however, considerable limitations in the theoretical framework of the thermal sciences yet to be resolved. Some of these have been mentioned earlier, on the premise that awareness of the shortcomings being encountered in the thermal analysis literature generally might help their avoidance in future pharmaceutical developments based on TG and other studies. The following factors may require consideration in undertaking fundamental kinetic measurements in the investigations of reactivity controls and reaction mechanisms:

1. Reaction stoichiometry may require confirmation.
2. Measured reaction rates may be subject to secondary controls.
3. Little is known about interface reaction rate controls and mechanisms.
4. Kinetic parameters calculated for reactions investigated may vary with analytical method.
5. Chemical criteria for classification of thermal rate processes are not available.
6. Thermal analysis results are not usually considered in the widest chemical context.

These comments represent an attempt to focus interest on unresolved weaknesses apparent in the thermal analysis literature. Development of this subject requires careful appraisal of theory and methods available, and awareness of potential problems might facilitate applications and their scientific interpretation in pharmaceutical sciences.

REFERENCES

1. Wilburn, F.W., Sharp, J.H., Tinsley, D.M., and McIntosh, R.M., *J. Therm. Anal.* 37, 2003, 1991.
2. Galwey, A.K., *Thermochim. Acta* 397, 249, 2003.
3. Galwey, A.K., *Thermochim. Acta* 399, 1, 2003.
4. Brown, M.E., *Introduction to Thermal Analysis*, Chapman & Hall, London, 1988.
5. Brown, M.E., *Introduction to Thermal Analysis*, 2nd ed., Kluwer, Dordrecht, 2001.
6. Daniels, T., *Thermal Analysis*, Kogan, London, 1973.
7. Duval, C., in *Wilson and Wilson's Comprehensive Analytical Chemistry*, Vol. VII, Svehla, G., Ed., Elsevier, Amsterdam, 1976, pp. 1–204.
8. Gallagher, P.K., in Brown, M.E., *Handbook of Thermal Analysis and Calorimetry: Principles and Practice*, Vol. 1, Elsevier, Amsterdam, 1998, pp. 225–278.
9. Gallagher, P.K., Series Ed., *Handbook of Thermal Analysis and Calorimetry*, Vol. 1–4, Elsevier, Amsterdam, 1998–2003.
10. Keattch, C.J. and Dollimore, D., *Introduction to Thermogravimetry*, Heyden, London, 1975, pp. 1–164.
11. Oswald, H.R. and Wiedemann, H.G., *J. Therm. Anal.* 12, 147, 1977.
12. Paulik, J. and Paulik, F., in Svehla, G., Ed., *Wilson and Wilson's Comprehensive Analytical Chemistry*, Vol. XIIA, Elsevier, Amsterdam, 1981, pp. 1–277.
13. Sestak, J., Satava, V., and Wendlandt, W.W., *Thermochim. Acta* 7, 333, 1973.
14. Szekely, G., Nebuloni, M., and Zerilli, L.F., *Thermochim. Acta* 196, 511, 1992.
15. Wendlandt, W.W., *Thermochim. Acta*, 26, 19, 1978.
16. Wendlandt, W.W., *Thermal Analysis*, 3rd ed., John Wiley & Sons, New York, 1986, pp. 1–848.
17. Beruto, D. and Searcy, A.W., *J. Chem. Soc., Faraday Trans. I* 70, 2145, 1974.
18. Flanagan, T.B., Simons, J.W., and Fichte, P.M., *J. Chem. Soc., Chem. Commun.* 370, 1971.
19. Bertrand, G., Lallemant, M., and Watelle-Marion, G., *J. Inorg. Nucl. Chem.* 36, 1303, 1974.
20. L'vov, B.V., *Thermochim. Acta*, 315, 145, 1998.
21. Dollimore, D., *Thermochim. Acta*, 203, 7, 1992.
22. Mahgerefteh, H., Khoory, H., and Khodaverdian, A., *Thermochim. Acta* 237, 175, 1994.
23. Okhotnikov, V.B. and Lyakhov, N.Z., *J. Solid State Chem.* 53, 161, 1984.
24. Okhotnikov, V.B., Petrov, S.E., Yakobson, B.I., and Lyakhov, N.Z., *React. Solids* 2, 359, 1987.
25. Galwey, A.K., Laverty, G.M., Baranov, N.A., and Okhotnikov, V.B., *Philos. Trans. R. Soc. Lond.* A347, 139, 1994.
26. Cahn, L. and Schultz, H., *Anal. Chem.* 35, 1729, 1963.
27. Williams, J.R. and Wendlandt, W.W., *Thermochim. Acta* 7, 253, 1973.
28. L'vov, B.V., Novichikhin, A.V., and Dyakov, A.O., *Thermochim. Acta* 315, 169, 1998.
29. Galwey, A.K. and Brown, M.E., *Thermal Decomposition of Ionic Solids*, Elsevier, Amsterdam, 1999.
30. Carr, N.J. and Galwey, A.K., *Proc. R. Soc. Lond.* A404, 101, 1986.
31. Galwey, A.K. and Mohamed, M.A., *J. Chem. Soc., Faraday Trans. I* 81, 2503, 1985.
32. Galwey, A.K., *Thermochim. Acta* 355, 181, 2000.
33. Lallemant, M. and Watelle-Marion, G., *C. R. Acad. Sci. Paris* C264, 2030, 1967.
34. Thomas, J.M. and Renshaw, G.D., *J. Chem. Soc.* A 2749, 1969.
35. Koga, N. and Tanaka, H., *J. Phys.Chem.* 98, 10521, 1994.

36. Galwey, A.K., *J. Pharm. Pharmacol.* 51, 879, 1999.
37. Guarini, G.G.T., *J. Therm. Anal.* 41, 286, 1994.
38. Galwey, A.K., *J. Therm. Anal.* 41, 267, 1994.
39. Herbstein, F.H., Kapon, M., and Weissman, A., *J. Therm. Anal.* 41, 303, 1994.
40. Korsi, R.-M. and Valkonen, J., *Thermochim. Acta* 401, 225, 2003.
41. Valigi, M. and Gazzoli, D., *Thermochim. Acta* 53, 115, 1982.
42. Swallowe, G.M., *Thermochim. Acta* 65, 151, 1983.
43. Gallagher, P.K. and Gyorgy, E.M., *Thermochim. Acta* 31, 380, 1979.
44. Paulik, F. and Paulik, J., *J. Thermal Anal.* 8, 556, 1975.
45. Chang, H., Murugan, R., and Ghule, A., *Thermochim. Acta* 374, 45, 2001.
46. Honda, K., *Sci. Rep. Tohuku Univ.* 4, 97, 1915.
47. El-Houte, S., El-Sayed Ali, M., and Sorensen, O.T., *Thermochim. Acta* 138, 107, 1989.
48. Rouquerol, F. and Rouquerol, J., *Proc. 3rd ICTAC*, Vol. 1, Wiedermann, H.G., Ed., Birkhauser, Bassel-Stuttgart, 1972, p. 373.
49. Rouquerol, J., *Thermochim. Acta* 144, 209, 1989.
50. Criado, J.M., Rouquerol, F., and Rouquerol, J., *Thermochim. Acta* 38, 117, 1980.
51. Ortega, A., Akhouayri, S., Rouquerol, F., and Rouquerol, J., *Thermochim. Acta* 235, 197, 1994.
52. Barrall, E.M. and Gritter, R.J., in *Systematic Materials Analysis*, Vol. IV, Richardson, J.H. and Peterson, R.V., Ed., Academic Press, New York, 1978, pp. 343–405.
53. Mu, J. and Kloos, D., *Thermochim. Acta* 57, 105, 1982.
54. Alves, S.S., *Thermochim. Acta* 157, 249, 1990.
55. Norem, S.D., O'Neill, M.J., and Gray, A.P., *Thermochim. Acta* 1, 29, 1970.
56. Brown, M.E., Bhengu, T.T., and Sanyal, D.K., *Thermochim. Acta* 242, 141, 1994.
57. Garn, P.D. and Alamolhoda, A.A., *Thermochim. Acta* 92, 833, 1985.
58. Kanari, N., Allain, E., and Gaballah, I., *Thermochim. Acta* 335, 79, 1999.
59. Dollimore, D., Gamlen, G.A., and Taylor, T.J., *Thermochim. Acta* 75, 59, 1984.
60. Holdiness, M.R., *Thermochim. Acta* 75, 361, 1984.
61. Mittleman, M., *Thermochim. Acta* 166, 301, 1990.
62. Arii, T., Senda, T., and Fujii, N., *Thermochim. Acta* 267, 209, 1995.
63. Groenewoud, W.M. and de Jong, W., *Thermochim. Acta* 286, 341, 1996.
64. Kaisersberger, E., Ed., Thermal analysis and gas analysis methods, special issue, *Thermochim. Acta* 295, 1–186, 1997.
65. Bonnet, E. and White, R.L., *Thermochim. Acta* 311, 81, 1988.
66. Raemaekers, K.G.H. and Bart, J.C.J., *Thermochim. Acta* 295, 1, 1997.
67. Gupchup, G., Alexander, K., and Dollimore, D., *Thermochim. Acta* 196, 267, 1992.
68. Shopova, N., *Thermochim. Acta* 273, 195, 1996.
69. Brown, M.E., Galwey, A.K., and Li Wan Po, A., *Thermochim. Acta*, 203, 221, 1992.
70. Brown, M.E., Galwey, A.K., and Li Wan Po, A., *Thermochim. Acta*, 220, 131, 1993.
71. Jacobs, P.W.M. and Tompkins, F.C., in *Chemistry of the Solid State*, Garner, W.E., Ed., Butterworth, London, 1955, pp. 184–212.
72. Carr, N.J. and Galwey, A.K., *Thermochim. Acta* 79, 323, 1984.
73. Galwey, A.K., Poppl, L., and Rajam, S., *J. Chem. Soc., Faraday Trans. I* 79, 2143, 1983.
74. Galwey, A.K. and Mohamed, M.A., *Proc. R. Soc. Lond.* A396, 425, 1984.
75. Christy, A.A., Nodland, E., Burnham, A.K., Kvalheim, O.M., and Dahl, B., *Appl. Spectrosc.* 48, 561, 1994.
76. Young, D., *Decompositions of Solids*, Pergamon, Oxford, 1966, pp. 1–209.
77. Guarini, G.G.T. and Dei, L., *J. Chem. Soc., Faraday Trans. I* 79, 1599, 1983.

78. Reading, M., Dollimore, D., Rouquerol, J., and Rouquerol, F., *J. Therm. Anal.*, 29, 775, 1984.
79. Reading, M., Dollimore, D., and Whitehead, R., *J. Therm. Anal.* 37, 2165, 1991.
80. Cooper, J.A. and Garner, W.E., *Proc. R. Soc. Lond.* A174, 487, 1940.
81. Galwey, A.K., Spinicci, R., and Guarini, G.G.T., *Proc. R. Soc. Lond.* A378, 477, 1981.
82. Galwey, A.K. and Guarini, G.G.T., *Proc. R. Soc. Lond.* A441, 313, 1993.
83. Draper, A.L., *Proc. Robert A. Welch Foundation, Conferences on Chemical Research, XIV, Solid State Chemistry*, Houston, TX, 1970, pp. 214–219.
84. Galwey, A.K. and Laverty, G.M., *Proc. R. Soc. Lond.* A440, 77, 1993.
85. Galwey, A.K. and Poppl, L., *Philos. Trans. R. Soc. Lond.* A311, 159, 1984.
86. Bhattamisra, S.D., Laverty, G.M., Baranov, N.A., Okhotnikov, V.B., and Galwey, A.K., *Philos. Trans. R. Soc. Lond.* A 341, 479, 1992.
87. Galwey, A.K., Laverty, G.M., Okhotnikov, V.B. and O'Neill, J., *J. Therm. Anal.* 38, 421, 1992.
88. Acock, G.P., Garner, W.E., Milsted, J., and Willavoys, H.J., *Proc. R. Soc. Lond.* A189, 508, 1947.
89. L'vov, B.V., *Thermochim. Acta* 291, 179, 1997.
90. L'vov, B.V., *Thermochim. Acta* 373, 97, 2001.
91. Price, D.M., Bashr, S., and Derrick, P.R., *Thermochim. Acta* 327, 167, 1999.
92. Youssef, A.M., Alaya, M.N., and Hamada, M.A., *Thermochim. Acta* 235, 91, 1994.
93. Goworek, J., Stefaniak, W., and Dabrowski, A., *Thermochim. Acta* 259, 87, 1995.
94. Staszczuk, P., Jaroniec, M., and Gilpin, R.K., *Thermochim. Acta* 287, 225, 1996.
95. Sigrist, K. and Stach, H., *Thermochim. Acta* 278, 145, 1996.
96. Goodrum, J.W., Geller, D.P., and Lee, S.A., *Thermochim. Acta* 311, 71, 1998.
97. Etzler, F.M. and Conners, J.J., *Thermochim. Acta* 189, 185, 1991.
98. Hoffman, R. and Pan, W.-P., *Thermochim. Acta* 192, 135, 1991.
99. Topor, N.D., Logasheva, A.I., and Tsoy, G.K., *J. Therm. Anal.* 17, 427, 1979.
100. Pichon, C., Risoul, V., Trouve, G., Peters, W.A., Gilot, P., and Prado, G., *Thermochim. Acta* 306, 143, 1997.
101. Endoh, K. and Suga, H., *Thermochim. Acta* 327, 133, 1999.
102. Galwey, A.K. and Brown, M.E., *Proc. R. Soc. Lond.* A 450, 501, 1995.
103. Brown, M.E., Dollimore, D., and Galwey, A.K., *Comprehensive Chemical Kinetics: Reactions in the Solid State*, Vol. 22, Elsevier, Amsterdam, 1980, pp. 1–340.
104. Galwey, A.K. and Laverty, G.M., *Solid State Ionics* 38, 155, 1990.
105. Galwey, A.K. and Brown, M.E., *Thermochim. Acta* 29, 129, 1979.
106. Galwey, A.K. and Brown, M.E., *Thermochim. Acta* 269/270, 1, 1995.
107. Laidler, K.J., *J. Chem. Educ.* 61, 494, 1984.
108. Garn, P.D., *J. Therm. Anal.* 13, 581, 1978.
109. Garn, P.D., *Thermochim. Acta* 135, 71, 1988.
110. Garn, P.D., *Thermochim. Acta* 160, 135, 1990.
111. Gotor, F.J., Criado, J.M., Malek, J., and Koga, N., *J. Phys. Chem. A* 104, 10777, 2000.
112. Flynn J.H., *Thermochim. Acta* 300, 83, 1997.
113. House, J.E. and Ralston, R.P., *Thermochim. Acta* 214, 255, 1993.
114. House, J.E., Muehling, J.K., and Williams, C.C., *Thermochim. Acta* 222, 53, 1993.

6 Thermogravimetric Analysis: Pharmaceutical Applications

Duncan Q.M. Craig and Andrew K. Galwey

CONTENTS

6.1 THE SCOPE OF TGA USAGE WITHIN PHARMACEUTICAL DEVELOPMENT

The preceding chapter has outlined the many possibilities afforded by thermogravimetric analysis (TGA), together with a brief overview of the practicalities associated with such measurements. However, in relating the technique to current pharmaceutical practice, it can be reasonably argued that, at present, TGA is almost certainly underexploited, with the method used largely as a means of estimating water content. This is clearly an important application from a drug design and delivery viewpoint but does not reflect the full range of possibilities afforded by the approach, particularly in terms of understanding the nature of the binding of the volatilized component to the substrate or assessing the kinetics of the weight loss process, from which information on longer-term stability may potentially be obtained. One of the contributing factors to the poor crossover between the basic chemistry and pharmaceutical fields in this respect may be the difference in emphasis placed on the technique within the two scientific cultures. The majority of work using TGA outside the drug-related arena is concerned with the examination of thermal decomposition reactions, and a wealth of literature exists in which the chemical pathway, loss profile, and kinetics have been interrelated. However, thermal decomposition induced by temperature ramping (with the important exception of dehydration) is usually not a major issue within the pharmaceutical sciences owing to the materials in question being processed and stored at temperatures well below those at which such reactions are discernible by TGA. Similarly, extrapolation of thermally induced decomposition rate data to normal storage temperatures to predict long-term stability is unreliable because of the strong possibility of the mechanisms involved being different.

There are nevertheless highly convincing arguments for the encouragement of more sophisticated use of TGA within the pharmaceutical sciences. In the first instance, there is a relatively small but highly interesting body of work concerned with detailed analyses of dehydrations of drugs and excipients. As discussed later, this work has shown the potential of using TGA to obtain a more detailed analysis of the kinetics of dehydration and the nature of the drug or excipient–water interaction, particularly when used in conjunction with complementary techniques such as x-ray diffraction (XRD) and differential scanning calorimetry (DSC). Second, there is considerable interest in optimizing the removal of residual organic solvents from dosage forms such as polylactide microspheres. There is a very important potential role for TGA in this respect, as appropriate kinetic analysis of the loss process may allow the design of manufacturing protocols whereby the solvent is removed more effectively. Finally, the issue of volatility or sublimation of components in dosage forms is widely recognized but poorly understood, with problems including the loss of the drug itself, preservatives, or plasticizers. This is again an area whereby a more sophisticated understanding of the kinetics of loss, obtained through relatively simple TGA experiments, could prove to be of great value to the formulator.

In the following discussions, representative examples of the use of TGA to characterize drugs, excipients, conventional dosage forms, and modified release dosage forms are outlined. The selection of suitable illustrative studies is made difficult by the very widespread use of the technique within the pharmaceutical field, although with varying levels of detail. Consequently, a summary of every study would be not only extremely lengthy but would also serve little useful purpose, as the generic principles underlying the use of the approach are identical in the majority of such studies. This is not a criticism of the authors concerned but merely a reflection of the "add-on" status of TGA within the field at present. The present examples have been selected either because they are illustrative of typical usage of the technique, or because they offer a different facet of the use of TGA methods that may be of interest to readers working on similar systems.

6.2 CHARACTERIZATION OF DRUGS AND EXCIPIENTS

6.2.1 Basic Characterization of Drugs and Excipients

The most basic pharmaceutical use of TGA is as a means of measuring water contents, and examples of this abound in the literature. In particular, TGA provides a simple and rapid method for weight loss determination that is usually much more convenient to perform than, for example, Karl Fischer analysis. However, as mentioned in Chapter 5, great care must be taken with regard to assumptions concerning the identity of the material being lost. It is very usual for workers to ascribe any weight loss in the region of 80 to 100°C to water, whereas in many cases, the cause may be decomposition of the substrate material or volatilization of other residual solvents.

A pertinent example of the use of TGA to determine total water content is that of Naini et al. (1), who used the technique to measure the water contents of spray-dried and crystalline lactose, trehalose, sucrose, and mannitol as a function of the storage relative humidity, using the technique to obtain a profile of the absorption or desorption behavior of these important materials. In addition, it should be noted that, when studying amorphous materials, it is essential to know the water content of the material when measuring the glass transition temperature. It follows that TGA provides a potentially highly useful tool in this respect, particularly when run in conjunction with DSC. An example of this is provided by Passerini et al. (2) who demonstrated that, even after drying using normal protocols, polylactide micro-spheres may contain significant amounts of water depending on the chemical composition of the spheres (Figure 6.1a), which, in turn, may influence the glass transition temperature of the dried product (Figure 6.1b). The authors also measured the residual organic solvent contents of the spheres and were thus able to establish strong circumstantial evidence for the lost material being almost entirely water. It is perhaps worth mentioning that Karl Fischer was not performed on these systems owing to the quantity of material required for accurate analysis being prohibitive, a further advantage of the TGA approach.

6.2.2 Decomposition and Vaporization Studies

As stated previously, the most common use of TGA within the organic and inorganic fields, namely the study of decomposition, is not of great relevance to the pharmaceutical field owing to drugs and excipients seldom being heated to the decomposition temperatures, the important exception being dehydration reactions. Nevertheless, there are a limited number of studies that have addressed the decomposition issue. For example, Beyers et al. (3) have studied the decomposition profile of furosemide, a high-ceiling diuretic, to establish the activation energy of the decomposition process. The authors measured the TGA responses of the drug over a range of heating rates (Figure 6.2a), from which the activation energy E_a was calculated at four different weight loss fractions (Figure 6.2b) via

$$E_a = 4.35 \cdot \frac{\Delta \log \beta}{\Delta 1/T} \qquad (6.1)$$

where β is the heating rate and T the temperature; the activation energy obtained was 47.7 kcal/mol (199.6 kJ/mol). These authors also performed a detailed compositional analysis of the degradation products using mass spectroscopy, 1H and ^{13}C-NMR. Clearly, it is necessary to exercise great caution in extrapolating the activation energy measured between 220–240°C to that corresponding to long-term degradation

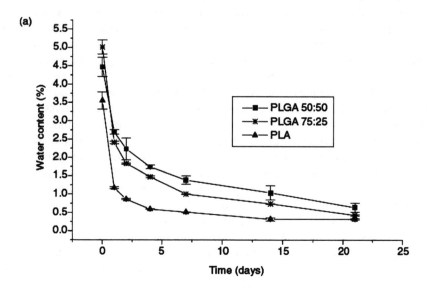

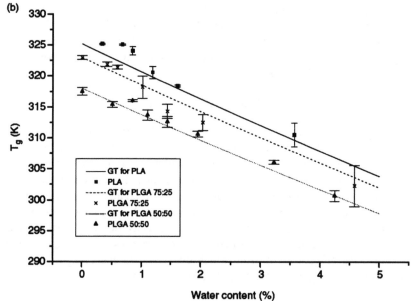

FIGURE 6.1 (a) The relationship between drying time at 35°C and residual water content for PLA, PLGA 75:25, and PLGA 50:50 microspheres. (Reproduced from Passerini, N. and Craig, D.Q.M., *J. Controlled Release*, 73, 111, 2001. With permission.) (b) The effect of water on the glass transition of PLA, PLGA 75:25, and PLGA 50:50 microspheres. The lines represent Gordon–Taylor fitting for the experimental data. (Reproduced from Passerini, N. and Craig, D.Q.M., *J. Controlled Release*, 73, 111, 2001. With permission.)

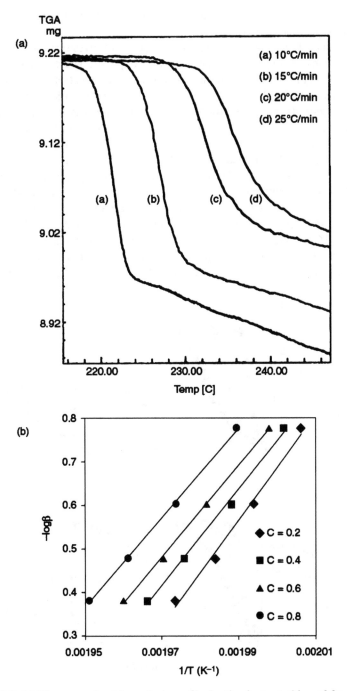

FIGURE 6.2 (a) Thermogravimetric analysis profile for the decomposition of frusemide at different heating rates, (b) corresponding relationship between heating rate (β) and weight loss fraction (C) as a function of temperature (T) according to Equation 6.1. (Reproduced from Beyers, H. et al., *Drug Dev. Ind. Pharm.*, 26, 1077, 2000. With permission.)

on storage. Nevertheless, this is a useful example of a thermal decomposition study applied to a pharmaceutical material.

The group of Dollimore has conducted a series of investigations, involving the decompositions of drugs and pharmaceutical excipients. These include studies by Huang et al. (4) into the isothermal and nonisothermal kinetics of captopril degradation and a study by Burnham et al. (5) into the degradation kinetics of procainamide hydrochloride, noting a two-stage endothermic process at temperatures above the melting point. The degradation of excipients has also been studied by the group, including work on the degradation of starches (6) and ibuprofen–starch mixtures (7). Other such studies include that of Leung and Grant (8), who examined the stability of two model dipeptides, aspartame and asphartylphenylamine. They noted that both peptides underwent solid-state intramolecular aminolysis producing a solid product and gaseous material on heating, with aspartame yielding methanol at 167–180°C and asphartylphenylamine yielding water at 186–202°C. They also used isothermal TGA studies to monitor the rate of the cyclization reaction, using the gas loss as a marker for the overall process (Figure 6.3). The authors then used the Prout–Tompkins equation (see Table 5.1 in Chapter 5) to calculate the activation

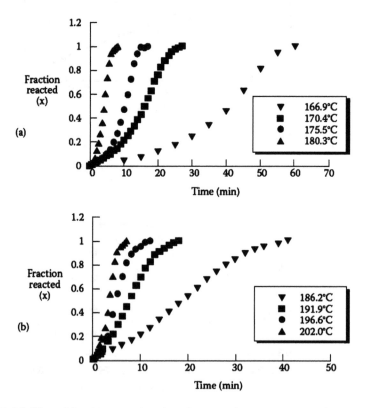

FIGURE 6.3 Plots of fraction reacted against time at various temperatures (isothermal reactions), calculated from the weight loss for (a) aspartame and (b) asphartylphenylamine. (Reproduced from Leung, S. and Grant, D.J., *J. Pharm. Sci.*, 86, 64–71, 1997. With permission.)

energies for the two reactions (268 kJ/mol and 242 kJ/mol for aspartame and asphartylphenylamine, respectively).

An aspect of behavior that has received attention in the thermal analysis literature but, with some notable exceptions, has not been widely explored in the pharmaceutical arena is the measurement of vaporization of drugs and excipients. Such information is of potential significance, particularly for preformulation studies in which the derivation of the activation energy and enthalpy of vaporization or sublimation may be a highly useful means of characterizing thermodynamic parameters associated with drugs such as solubility parameters. In addition, knowledge of these parameters may facilitate anticipation of issues such as drug or excipient loss during processing or storage. Chatterjee et al. (9) describe the use of TGA to determine the vaporization and sublimation characteristics of new drug compounds as a means of deriving the activation energy and enthalpy for these processes and to identify and anticipate possible formulation issues that may arise. This chapter contains a discussion of the different approaches that may be used and also a brief description of some of these.

The most frequently used equations that relate the vapor pressure of a material to temperature are as follows. In moderate temperature ranges, the Antoine equation provides a reasonable empirical prediction, given by

$$\ln P = a - \frac{b}{T+c} \tag{6.2}$$

where P is the vapor pressure, T is the absolute temperature, and a, b, and c are constants that may be obtained from the reference source of Stephenson and Malamowski (10); clearly, the necessity to obtain or derive these parameters is a limitation of the approach. Other methods include use of the Langmuir equation (11) given by

$$\frac{dm}{dt} = P\alpha \sqrt{\frac{M}{2\pi RT}} \tag{6.3}$$

where dm/dt is the rate of mass loss per unit area, P is the vapor pressure, α is the vaporization constant, M is the molecular weight of the evaporating vapor, T is the absolute temperature, and R is the gas constant. The original premise was that the vaporization constant α is unity if the surrounding medium is a vacuum; hence, one must assume a constant but nonunity value for nonvacuum conditions such as would be found in a typical TGA experiment. For such studies, Equation 6.3 may be modified to

$$P = \frac{\sqrt{2\pi R}}{\alpha} \cdot \sqrt{\frac{T}{M}} \cdot \frac{dm}{dt} = kv \tag{6.4}$$

where $k = \sqrt{2\pi R} / \alpha$ and $v = (\sqrt{T/M}) \cdot (dm/dt)$. By plotting the vapor pressure P against v, then the vaporization constant may be obtained from the slope k. However, a limitation to the approach is the nonconstant value for α with respect to temperature; hence, corrections with concomitant assumptions are required to account for this variation.

The most widely used relationship for evaluating vapor pressures is the Clausius–Clapeyron equation, given by

$$\log \frac{p_2}{p_1} = \frac{\Delta H_{vap}}{2.303R} \cdot \frac{T_2 - T_1}{T_2 T_1} \tag{6.5}$$

where p_2 and p_1 are the vapor pressures at temperatures T_2 and T_1, and ΔH_{vap} is the enthalpy of vaporization. Chatterjee et al. (9) have suggested a modification for this equation that takes account of the change in heat capacity of the material as a function of temperature.

A number of studies have explored ways in which partial vapor pressures may be obtained using TGA data, thereby allowing both prediction of vapor pressure under a range of circumstances and calculation of the constants associated with the approaches described previously. In particular, Price and Hawkins (12) have argued that the rate of mass loss for vaporization and sublimation within a TGA should be a zero-order process, and hence should be constant for any given temperature, subject to the important condition that the available surface area also remains constant. This means that the value of v from Equation 6.4 should be easily calculated from the TGA data. If one performs this experiment for materials with known vapor pressure and temperature relationships (the authors used discs of acetamide, benzoic acid, benzophenone, and phenanthrene), then the constant k for the given set of TGA experimental conditions may be found. Once this parameter is known, the vapor pressure may be assessed for an unknown material in the same manner.

Related approaches have been used by Phang and Dollimore (13) who examined the vapor pressures of a range of antioxidants, the evaporation of which has been recognized as potentially leading to the rancidity of cosmetic and food products. Similarly, Cheng et al. (14) studied the evaporation of methyl salicylate using simultaneous TGA and differential thermo analysis (DTA) and analyzed the kinetics of vaporization using the Arrhenius, Kissinger, and Ozawa equations, collating these data with the latent heat of vaporization calculated from the Clausius–Clapeyron equation. These approaches therefore suggest that TGA may have further application in the fields of degradation and evaporation or sublimation than has been the case to date.

6.2.3 CHARACTERIZATION OF HYDRATES

The majority of studies on drugs and excipients involving TGA have used the technique as a means of identifying and characterizing pharmaceutical hydrates. The method presents several opportunities in this respect. In the first instance, the presence of a hydrate (as opposed to sorbed water) is usually easily discerned by the sharpness of the instrumental response corresponding to the water loss process. More specifically, loss of water of hydration usually occurs for pharmaceuticals over a range of approximately 5 to 20°C, depending on the experimental conditions used, and is usually within the 80 to 140°C temperature interval, often with stoichiometric ratios of water to substrate. Loss of sorbed water tends to occur over a wider range (typically 20–30°C) and in the temperature range of 60 to 90°C, with the amounts of water usually being <2% w/w for most low-molecular weight drugs, although for more hydrophilic, higher-molecular weight substances, this value may be much higher. The authors have, for example, measured the water content of alginates as being anything up to 13% w/w in the as-received state; this may represent a major source of error when calculating polysaccharide concentrations.

The structure and physicochemical properties of drug hydrates is a substantial subject in its own right and more information can be obtained from the excellent texts by Byrn (15), Morris and Rodriguez-Hornado (16), and Morris (17). In brief, hydrates are crystalline materials in which molecular water is incorporated into the crystal lattice as opposed to being sorbed onto surfaces. There are three classes of hydrates. Class 1 systems (isolated site hydrates) contain individually retained water molecules isolated from each other by the intervening drug molecules. On this basis, the TGA loss profile may be expected to be relatively narrow due to catastrophic breakdown of the structure over a narrow temperature range, leading to sudden and stoichiometric water release. Class 2 systems (channel hydrates) contain water in channels rather than in isolated sites. For these, the TGA response is expected to show loss, occurring over a wider temperature range, commencing at a lower temperature than for most Class 1 systems owing to initial water loss taking place from the ends of the channels rather than requiring structural collapse (although, this may occur as a consequence of the water loss process). Class 3 systems are ion-associated hydrates, involving the presence of a metal ion. This class of material is of considerable significance within the pharmaceutical sciences because of the widespread practice of preparing drugs in salt forms to facilitate dissolution and bioavailability. Such hydrates tend to lose water at higher temperatures than Class 1 or Class 2 systems owing to the relatively strong bonding between the ion and the water molecules. On losing water, such hydrates may retain the same basic conformation without the water molecules, they may change to a new conformation, or they may become amorphous. A knowledge of their dehydration behavior is highly useful not only to assist structural elucidation but also to aid prediction of stability issues and processing effects. TGA has been extensively used as a means of characterizing such systems and is, a routine method of identification and stoichiometric assessment, although again there may be opportunities for more sophisticated measurement that have not yet been fully exploited within the pharmaceutical field.

A large number of investigations are available in which TGA has been used as a component of an array of techniques used to study the water loss characteristics of hydrates. For example, Michel et al. (18) used TGA in conjunction with single-crystal x-ray diffraction to elucidate the structure of a sesquihydrate form of (–)-hyoscine hydrobromide, an anticholinergic drug isolated from the belladonna plant. The study arose from earlier work (19) in which the authors attempted to elucidate the structure of the recognized hemihydrate form and, in so doing, found that the drug could also exist as both a sesquihydrate and anhydrate form. An excellent correlation was found between the stoichiometry predicted by x-ray diffraction (XRD) and the water loss measured by TGA, with the total mass loss (about

6% w/w), corresponding well to the predicted 1.5 molecules of water per molecule of free base. In addition, the authors noted that the water loss occurred in two stages at about 90°C and 102°C, corresponding to the loss of one molecule and half a molecule of water and free base molecule, respectively. This study illustrates several points pertinent to the previous discussions. In the first instance, the investigation was (perfectly appropriately) very much led by the XRD studies, with TGA being used as a confirmatory technique, as is often the case. However, the structure of the sesquihydrate was found to be conformationally identical to the hemihydrate in terms of the unit cell dimensions, molecular structure, packing arrangement, and crystal system, thereby emphasizing the usefulness of TGA as a complementary method. Indeed, the authors imply that previous XRD studies into what was thought to be the hemihydrate must now be considered unreliable, outlining an important limitation in accepting XRD data as the sole definitive technique for identifying and characterizing hydrates.

The investigation by Bettinetti et al. (20) into the solid-state properties of the antiplatelet agent picotamide monohydrate provides a further example of such a study. The authors examined the effect of dehydration on the subsequent structure using XRD, IR, DSC, TGA, and hot-stage microscopy (HSM). They noted a biphasic TGA profile for the monohydrate that altered to a single loss process on gentle grinding, this being ascribed to changes in morphological features, resulting in modifications to the water loss pathways. In addition, the anhydrous material was shown to exist in two polymorphs depending on dehydration conditions, with slow conversion to the hydrate noted on storage under ambient conditions.

A further example is provided by Dash and Tyle (21), who studied the solid-state properties of AG337 ($C_{14}H_{12}N_4OS \cdot 2HCl$), a novel antitumor compound, using DSC, TGA, thermomicroscopy, Karl Fischer titrimetry, XRD, fourier transform infrared spectroscopy (FTIR) and scanning electron microscopy (SEM). This material showed a complex DSC response, with three endothermic peaks noted at 112, 213, and 312°C and two exotherms at 261 and 320°C (Figure 6.4). TGA data over the same temperature range indicated three partially overlapping loss processes totaling a weight loss of 9.73% w/w, with Karl Fischer analysis yielding a value of 9.28% w/w water. These figures correspond reasonably well to the value of 9.15% w/w water predicted for a stoichiometric dihydrate of this drug; hence, the authors (quite reasonably) assumed that a dihydrate was present. HSM studies using samples immersed in silicon oil indicated that the liberation of bubbles through the oil, associated with water loss, was not seen sharply but was observed throughout the range of 30–210°C. The DSC data indicated that the first endothermic event (peak at 112°C) occurred over a wide temperature range of 30–160°C. This corresponded to two of the TGA loss processes, resulting in a composite mass loss of about 7.5% w/w, equivalent to the loss of 1.5 molecules of water per molecule of drug. The third loss process (onset 180°C) corresponded to a loss of a further half molecule of water. Taken together, these data indicate that the water is lost in a series of stoichiometric steps rather than as a single process, reflecting the different binding states of the water within the structure. A further point to arise from this study is that the second endothermic peak (213°C) was ascribed to melting of the anhydrate. However, this peak coincided with the third weight loss process; this is clearly

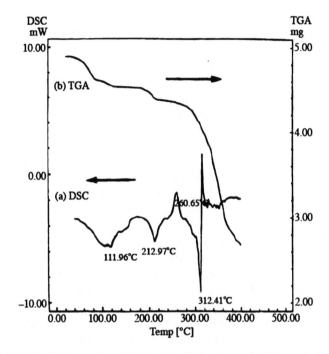

FIGURE 6.4 Differential scanning calorimetry and thermogravimetric analysis data for the model drug AG337. (Reproduced from Dash, A.K. and Tyle, P., *J. Pharm. Sci.*, 85, 1123, 1996. With permission.)

contradictory because a material cannot simultaneously be an anhydrate and also lose water of hydration. The authors ascribed this anomaly to the use of different pans in the DSC and TGA studies that resulted in varying water loss kinetics, indicating the need for care when directly comparing TGA and DSC data. Our own experience is also that the different purge conditions within the two instruments may lead to difficulties in reconciling observations. One solution to this problem is to use apparatus capable of simultaneous DSC and TGA measurement; this approach is not dealt within this chapter, more information can be found in the text by Haines (22) among others. The exotherm seen at 261°C was ascribed to the recrystallization of the anhydrate, possibly into a further polymorph that then underwent melt/decomposition. The authors then mimicked the heating conditions used to produce the different forms identified to isolate the different species: the initial dihydrate, the anhydrous phase formed after initial and secondary dehydration prior to melting (nominally, a metastable form), and the anhydrate formed after recrystallization (nominally, a stable form). They noted that the XRD patterns were identical for the metastable and dihydrate forms, indicating that the dehydration process resulted in the lattice structure being retained after water loss. Melt recrystallization then yielded the stable form that exhibited a distinct XRD profile, indicating molecular rearrangement between the forms.

The point highlighted earlier with regard to the choice of pans, although possibly appearing trivial at first, may be of some considerable importance with regard to the comparison of DSC and TGA data, either when performing separate studies or when using simultaneous DSC and TGA equipments. The current practice within the pharmaceutical sciences is to use open pans for TGA studies under almost all circumstances to allow free escape of volatilized components, whereas nonhermetically crimped pans or pinholed hermetically sealed pans are commonly used for DSC studies to maximize sample-pan contact and to maintain baseline stability. However, these different crimping techniques will almost inevitably render comparison of data between the two techniques uncertain as the ease of removal of gaseous components may significantly differ. Consequently, the temperatures at which the mass loss and the response endotherm are observed may not coincide. Our own experience is that the use of pinholed pans for both sets of measurements allows valid comparisons to be made without compromising the heat flow conditions during DSC studies, although care is still required owing to the differences associated with the purge gas atmosphere flow rates within the two instruments.

A number of useful studies have been performed on similarly complex drug hydrate systems. In particular, the group of Grant have performed a series of studies on metal salt hydrates of neocromil. This drug is used in the treatment of asthma and may be presented both as a range of metal salts and a number of different hydrate forms; hence, rendering this molecule a useful means by which the hydration behavior may be varied and studied while maintaining the same drug substrate. In the first such study (23), the authors prepared the pentahydrate, the heptahydrate, and the decahydrate of the bivalent metal salt, nedocromil magnesium. They used DSC, TGA, Karl Fischer titrimetry, powder XRD, HSM, SEM, FTIR, and ^{13}C solid-state nuclear magnetic resonance (NMR) in conjunction, thereby providing a very wide and complementary array of analytical approaches. The mass loss profiles of the three hydrates showed significant differences over and above the absolute mass loss values in that the heptahydrate and decahydrate showed no initial level TGA baseline when heated from ambient conditions, implying that water loss commenced before or immediately on heating, with implications for the storage stability of these forms (Figure 6.5a–c). In addition, it was noted from the derivative curve that the mass loss appeared to be biphasic. In contrast, the pentahydrate showed two distinct loss processes at 75–130°C and 200–230°C, corresponding to the loss of four and one molecules of water, respectively. Subsequent spectroscopic analysis indicated that the structure of the decahydrate was similar to that of the heptahydrate for drug conformation, indicating that the three additional water molecules in the former are loosely bound and do not significantly contribute to the crystalline structure. Moreover, XRD studies suggested that these two hydrates are less crystalline than the pentahydrate. In addition, the pentahydrate showed strong characteristic reflections at 10.7 and 33.9°C, which were absent from the other two forms, further highlighting the distinct nature of this hydrate. The authors also cited previous spectroscopic work (24) that suggested that four of the five water molecules in the pentahydrate exist as a tetramer whereas the fifth exists as a stable monomer. Taken with the TGA data, it is reasonable to suggest that the two loss processes correspond to these two different water-binding states, with the tetramer showing greater temperature instability than the monomer. These studies have indicated that TGA data may be extremely useful for yielding insights into the state of water binding in complex

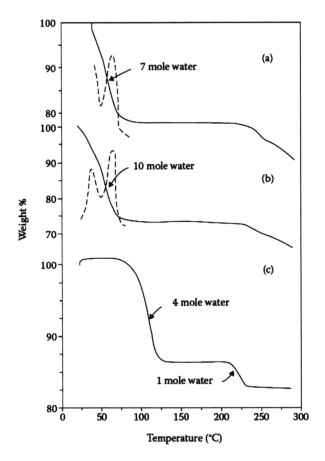

FIGURE 6.5 TGA curves (full line) and derivative TGA curves (dotted line) of the nedocromil magnesium hydrates: (a) heptahydrate, (b) decahydrate, and (c) pentahydrate. (Reproduced from Zhu, H. et al., *J. Pharm. Sci.*, 85, 1026, 1996. With permission.)

hydrates, particularly when used in conjunction with spectroscopic techniques that may provide specific structural information.

The second paper in this series (25) concerned the zinc salt, which was prepared in the octahydrate, heptahydrate, and two pentahydrate forms, the last being prepared by dehydration of either the heptahydrate (modification A) or the octahydrate (modification B). An interesting aspect of this study, in the context of the present discussion, is the use of different pan types for the TGA and DSC studies. Figure 6.6 and Figure 6.7 show the responses of the four samples in both open and crimped pans. Figure 6.6 indicates the steps corresponding to the loss of 4, 2, 2, and 3 water molecules (Table 6.1) in the loss processes for the octahydrate, heptahydrate, pentahydrate A, and pentahydrate B, with a remarkably close correlation noted between the derivative TGA peaks and the corresponding DSC responses. It may be seen that the octahydrate does not show a simple stoichiometric water loss profile (unlike the heptahydrate) whereas the two pentahydrate forms show different loss profiles.

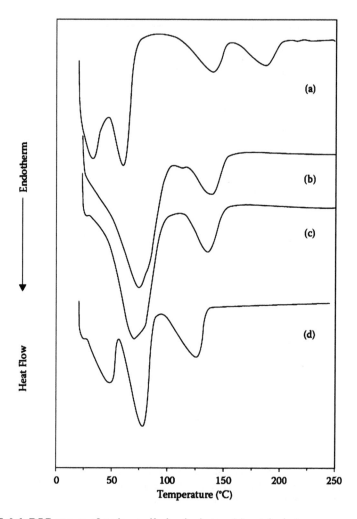

FIGURE 6.6 DSC curves of nedocromil zinc hydrates: (a) octahydrate open pan, (b) heptahydrate open pan, (c) pentahydrate modification A open pan, and (d) pentahydrate modification B open pan.   *(Continued)*

Use of crimped pans resulted in marked variations in the response profiles. The authors ascribed this difference, at least for the heptahydrate, to conversion to the pentahydrate under conditions of high humidity generated within the pan as a result of the impeded escape of the water vapor compared to the open pans. Indeed, they suggest that the use of crimped (or by extrapolation, pinholed) and open pans may be a very useful means of assessing the effect of water vapor on the transformations between different hydrate types. They were also able to relate the structural properties of the hydrates, obtained from spectroscopic studies, to the water loss profile. In particular, they suggested that the octahydrate is partially amorphous and contains loosely bound water, leading to a high dependence of the water loss profile on the local environment. In contrast, the high temperature loss of one molecule of water

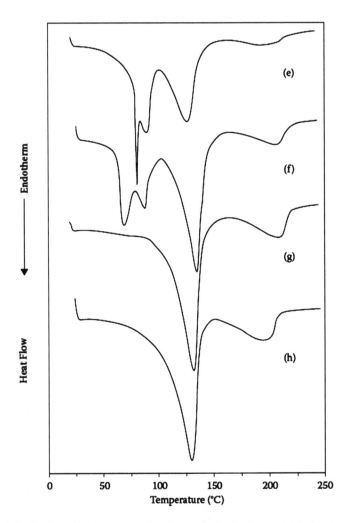

FIGURE 6.6 *Continued.* DSC curves of nedocromil zinc hydrates: (e) octahydrate crimped pan, (f) heptahydrate crimped pan, (g) pentahydrate modification A crimped pan, and (h) pentahydrate modification B crimped pan. (Reproduced from Zhu, H. et al., *J. Pharm. Sci.*, 86, 418, 1997. With permission.)

from the heptahydrate and from the two pentahydrates was ascribed to direct linkage of this water molecule to the zinc ion and to two carboxylate oxygen atoms, demonstrating the strength of water bonding that may occur in pharmaceutical salts.

The authors also studied the behavior of the calcium salt (26). Again, this material showed distinct characteristics in that a pentahydrate and 8/3 (2.66 stoichiometic ratio) hydrate were formed. Whereas TGA in crimped and open pans were performed as before, the emphasis of this paper was on the spectroscopic characterization of the hydrates, particularly given the fractional stoichiometry outlined previously. TGA studies indicated three water loss stages, whereas single-crystal XRD indicated three binding states for the water. This series of papers has therefore

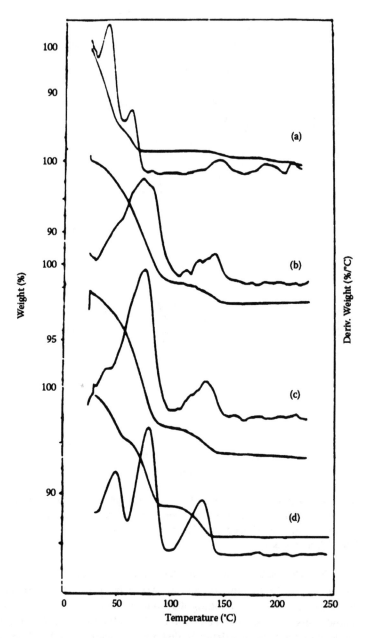

FIGURE 6.7 TGA and derivative TGA curves of nedocromil zinc hydrates: (a) octahydrate open pan, (b) heptahydrate open pan, (c) pentahydrate modification A open pan, and (d) pentahydrate modification B open pan.   *(Continued)*

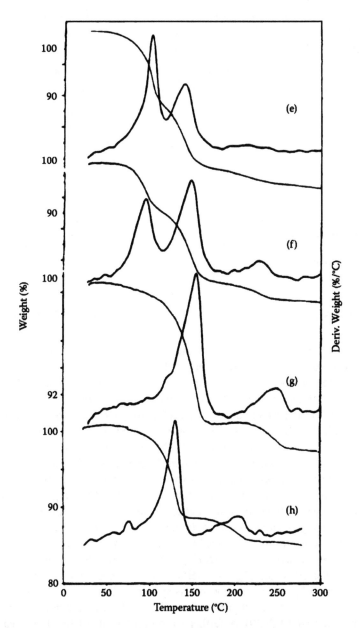

FIGURE 6.7 *Continued.* TGA and derivative TGA curves of nedocromil zinc hydrates: (e) octahydrate crimped pan, (f) heptahydrate crimped pan, (g) pentahydrate modification A crimped pan, and (h) pentahydrate modification B crimped pan. (Reproduced from Zhu, H. et al., *J. Pharm. Sci.*, 86, 418, 1997. With permission.)

TABLE 6.1
Peak Maximum Temperatures (°C) from DSC and Derivative TGA (DTG) and Water Loss (moles water per mole drug) from TGA Data for Nedocromil Zinc Hydrates

Hydrate	Open Pan			Crimped Pan		
	DSC	DTG	Moles H_2O	DSC	DTG	Moles H_2O
Octahydrate	32	31	5.5	80	98	3.3
	61	63	2.2	90	135	4.3
	138	143	0.3	127		
	185	186	0.2	197	215	0.6
Heptahydrate				68		
				88	89	1.9
	75	74	6.0	134	145	3.9
	140	140	1.0	207	222	1.0
Pentahydrate (A)	72	75	4.1	131	146	3.9
	136	132	0.9	209	239	0.9
Pentahydrate (B)	49	50	1.9			
	79	81	2.0	130	131	3.8
	127	130	0.9	194	205	0.9

Source: Reproduced from Zhu, H. et al., *J. Pharm. Sci.*, 86, 418, 1997. With permission.

provided a very valuable link between the crystal structure of a series of pharmaceutical hydrates and their dehydration profiles.

It was stated earlier that the vast majority of pharmaceutical solvates are hydrates. There are a few studies, however, in which other solvates of drugs have been studied. Ghosh et al. (27) examined a range of dialkylhydroxypyridones (iron chelators with possible application for the treatment of anemias) and compared their structures to their corresponding formic acid solvates. TGA was able to monitor the loss of formic acid, providing complementary information for spectroscopic studies that in turn were able to provide a molecular-level explanation for the desolvation profiles.

In addition to characterizing the water loss processes of drugs, TGA has also been extensively used to study equivalent processes associated with excipients. One material that has generated considerable interest is α,α-trehalose, a natural disaccharide that is believed to be responsible for the ability of many plants and animals to survive severe dehydration or freezing. As a raw material, trehalose is usually presented as the dihydrate (T_h). In addition, the material may be obtained in the anhydrous form (T_β) and as an amorphous system. It has also been suggested that a further polymorph of the anhydrous form (T_α), with a melting point of about 403 K, may be generated (28,29). A further form (T_γ) has also been described that is now thought to comprise a mixture of the anhydrous and dihydrate materials (30). In addition to (or perhaps parallel to) the controversy concerning the number of forms and the temperature-dependent interconversion of those forms is also highly complex. Several reports have indicated that this is, at least partially, owing to the sensitivity of this material to the experimental conditions used for DSC or TGA

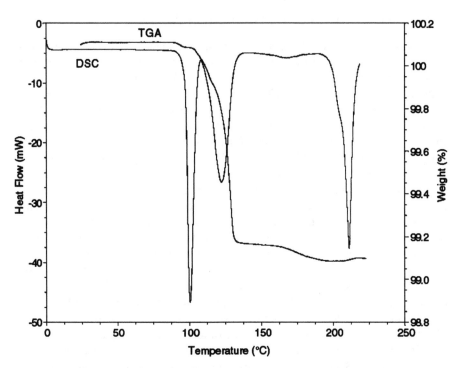

FIGURE 6.8 DSC and TGA response for α,α-trehalose dihydrate in pinholed pans (10°C/min). (Reproduced from McGarvey, O.S. et al., *J. Phys. Chem. B*, 107, 6614, 2003. With permission.)

experiments. For example, Taylor and York (31) have indicated that the dehydration profile of trehalose dihydrate is highly dependent on particle size, with smaller particles tending to form the amorphous material and larger ones the anhydrate. McGarvey et al. (32) have attempted to examine the effects of experimental variables systematically, looking particularly at the thermal behavior of the amorphous and dihydrate forms in relation to the pan type used. They noted that on performing DSC and TGA studies on the dihydrate in pinholed pans, two endothermic peaks were seen followed by a baseline discontinuity (Figure 6.8). The corresponding TGA studies indicated that the water loss corresponded largely (although not completely) to the second rather than first endothermic peak. On isolating the material formed immediately after the first endotherm, it was noted that the T_y form was found, indicating that the water loss process was biphasic. The initial loss process (possibly, from the particle surfaces) was regarded as resulting in a partially dehydrated system, whereby the anhydrous layer acted as a barrier to further water loss. On heating the sample, this barrier was overcome, and the majority of the water loss process then took place, resulting in the appearance of two distinct endothermic peaks. In a later study, Horvat et al. (33) used TGA to study the effect of a model drug (paracetamol) on the dehydration process, noting that the presence of the drug lowered the temperature of recrystallization of amorphous trehalose to the anhydrate form.

6.2.4 KINETIC ANALYSIS OF THE DEHYDRATION PROCESS

In addition to the use of TGA as a complementary technique in the characterization of the structures and properties of hydrates, a number of scholars have attempted to quantify the kinetic parameters of dehydration. As mentioned previously, such approaches have also been used for vaporization and decomposition reactions. In addition, it must be stressed that the limitations outlined in Chapter 5 (reversibility, self-cooling, etc.) must be considered when attempting to use measured rate data for quantitative purposes or as a basis for chemical and mechanistic interpretation. Chapter 5 may be consulted for a detailed discussion of the possibilities, limitations, and principal approaches associated with kinetic measurements using TGA.

An example of the kinetic approach is provided by Suzuki et al. (34) in a study of the isothermal dehydration of theophylline monohydrate, as indicated in Figure 6.9a. The authors modeled the data according to the Avrami–Erofeev equation (A2; see Chapter 5, Table 5.1) and found a reasonable fit (Figure 6.9b) from which the rate constants obtained were utilized to plot the temperature dependence of the reaction rate (Figure 6.9c), which in turn yielded the value of *Ea*. It was concluded that the reaction occurs via random nucleation followed by two-dimensional growth of nuclei. These authors highlight a discrepancy with the earlier work of Shefter et al. (35), who concluded that this dehydration fitted zero-order kinetics. Suzuki et al. (34) ascribed these different results to the different sample preparation techniques used for the two studies, with their work involving the use of sieved, nontriturated material whereas Shefter et al. (35) used ground crystals. Agbada and York (36) extended these studies, particularly to identify the possible role of particle size in determining the kinetics. It was shown that although larger particles fitted the Avrami–Erofeev equation, finer fractions were described by the two-dimensional phase boundary equation (R2; Chapter 5, Table 5.1).

The issue of particle size was also addressed by Sekiguchi et al. (37), who investigated the dehydration of sulfaguanidine monohydrate, with particular emphasis on the kinetics of dehydrations of intact and ground crystals. They noted that the dehydration of intact crystals was represented by the phase boundary equation (R2; Chapter 5, Table 5.1), although the Avrami–Erofeev equation (A2; Chapter 5, Table 5.1) was also applicable up to 80°C. When the ground crystals were dehydrated under dry conditions at atmospheric pressure, the best linearity was found with the Avrami–Erofeev equation (A2; Chapter 5, Table 5.1). However, under reduced pressure the contracting phase boundary equation R3 (Chapter 5, Table 5.1) gave the best fit. The authors also studied dehydrated material that had been allowed to resorb water to reform the hydrate, finding that this material fitted equations R2 or R3 (Chapter 5, Table 5.1). These studies highlight the influences of experimental conditions when attempting to obtain kinetic parameters, as highlighted in Chapter 5.

The isothermal dehydrations of more complex drug hydrates have also been studied. Zhu and Grant (38) examined the dehydration of nedocromil magnesium pentahydrate, which shows the two distinct dehydration processes discussed previously (23). They found that the lower-temperature dehydration process (about 60–90°C) fitted the Avrami–Erofeev equation (A2) whereas the higher-temperature process (about 180–200°C) fitted the Prout–Tompkins and the Avrami–Erofeev

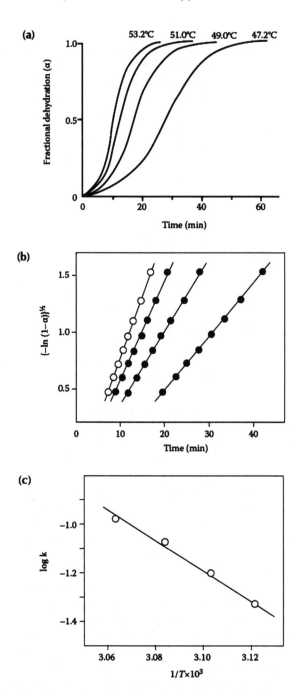

FIGURE 6.9 (a) Isothermal α–time curves for the dehydration of theophylline monohydrate, (b) corresponding fitting to the Avrami–Erofeev equation, and (c) temperature dependence of the calculated rate constant. (Reproduced from Suzuki, E. et al., *Chem. Pharm. Bull.*, 37, 493, 1989. With permission.)

equations (A3; Chapter 5, Table 5.1) equally satisfactorily. They also noted a higher activation energy for the higher-temperature process (120.64 kJ/mol compared to 69.69 kJ/mol for the lower-temperature dehydration process). This was discussed in the context of their previous suggestion that the lower-temperature process represents the loss of tetramer water whereas the higher-temperature loss corresponds to isolated site water that is more tightly bound within the crystal lattice. The authors also studied the effect of a range of variables including sample size, particle size, and water vapor. An increase in rate was found for the smaller particles, ascribed to their higher surface area. Similarly, a faster rate was found for the studies using a smaller sample mass. Chapter 5 contains more comments regarding the effect of these variables on kinetic measurements.

6.3 CHARACTERIZATION OF CONVENTIONAL DOSAGE FORMS

A limited number of studies have examined the weight loss profiles of finished conventional dosage forms, the obvious difficulty being the sample size and form that can be conveniently accommodated into most conventional instruments. Examples of such studies include that of Pyramides et al. (39), who used a range of thermoanalytical techniques to study atenolol tablets along with the corresponding constituent excipients. They found that the finished tablet showed the following mass loss processes: one from ambient temperature to 100°C, a large loss between 200 and 380°C, and a smaller extended temperature loss above 400°C. By comparing this composite profile with those of microcrystalline cellulose, sodium starch glycolate, and magnesium stearate, the authors suggested that it is possible to at least qualitatively identify certain components within finished tablets. This postulate is clearly dependent on the assumption that the presence of one component does not influence the decomposition of any other. This possibility was addressed by "spiking" the tablets with one of the components to ascertain whether the responses were indeed additive. TGA may also be used to study particulate dosage forms. Najafabadi et al. (40) used TGA to characterize solvent levels in disodium chromoglycate spray-dried particles for inhalation, cosprayed with L-leucine as a deaggregation agent.

A further area in which TGA could arguably be used to a greater extent is the study of water distribution in aqueous creams. These dosage forms are composed of oil droplets dispersed in an aqueous medium, interspersed with liquid crystalline lamellae that are largely responsible for the rheological properties of the material. Water may be present in several forms within such a system, including "free" water as an aqueous phase (meaning, in this context, that the water was not associated with any particular structure), bound within the lamellae, or bound within precipitated crystals of cetyl or cetostearyl alcohol, which is used as a cosurfactant. For such systems, there is a potential role for TGA as a means of characterizing the water distribution within and between the phases. Junginger (41,42) suggested that two weight loss peaks may be distinguished, corresponding to the loss of bulk water and interlamellar water following disruption of the lamellar structure on heating, with subsequent higher-temperature processes, corresponding to loss of water of hydration from long chain alcohols. Although these (reasonable) interpretations require further verification, the possibilities afforded by using the technique as a quality control tool are also apparent. This is exemplified in the study by Peramal et al. (43), whereby the TGA profiles of a range of commercial creams were comparatively studied (Figure 6.10), with marked differences noted between the batches despite all four complying with the required total water content. This is evidence that, whereas the chemical compositions of these systems are almost

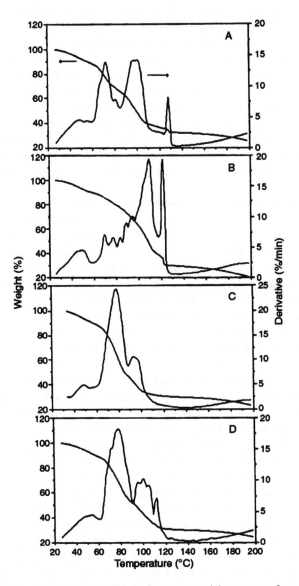

FIGURE 6.10 Thermogravimetric profiles for four commercial creams, performed at 10°C/min. (Reproduced from Peramal, V.L. et al., *Int. J. Pharm.*, 155, 91, 1997. With permission.)

certainly very similar, the physical structure may be less so. The rheological properties of the creams were also examined, although no clear relationship between the TGA profile and the viscosity was apparent.

6.4 CHARACTERIZATION OF CONTROLLED-RELEASE DOSAGE FORMS

In addition to the characterization of conventional drugs, excipients, and finished dosage forms, TGA may also be used to characterize controlled-release systems. For example, a number of studies have examined hydrogels, including an investigation by Yamini et al. (44) of biodegradable gels based on N-vinyl-2-pyrrolidone and polyethylene glycol diacrylate (PAC). The study included spectroscopic measurements to determine the molecular structure, equilibrium swelling, and *in vitro* release. In addition, the authors examined the TGA profiles of the two polymers, showing a biphasic loss that they attributed to bound water between 50 and 100°C and degradation between 390 and 475°C. The relatively high temperature of the latter process indicated reasonable thermal stability for these polymers, with concomitant implications for the range of processing conditions that could be applied to the materials.

Investigations have also been conducted into drug complexes. For example, Ambrogi et al. (45) studied the incorporation of ibuprofen among the lamellae of hydrotalcites having the general formula $[M(II)_{1-x}M(III)_x(OH)_2]^{x+}[A_{x/n}{}^{n-}]^x mS$ where M(II) is a divalent cation (usually Mg), M(III) is a trivalent cation (usually Al), A^{n-} is an exchangeable inorganic or organic anion, and m is the proportionate number of moles of solvent S (usually water) intercalated per mole of compound. The authors argued that by interchelating ibuprofen into a hydrotalcite system, a controlled release system could be generated that would overcome the problems of short half-life and the gastrointestinal and central nervous system side effects of the drug. The latter is associated with the high initial plasma levels that arise from conventional ibuprofen dosage forms. The authors used TGA to determine the stoichiometry of the complex by ascribing the total weight loss between room temperature and 600°C to water and ibuprofen decomposition, on the basis of observation that decomposition of the base talcite does not occur until about 1000°C. By combining this information with H and C elemental analysis, it was possible to determine the stoichiometry of the complex as $Mg_{0.67}Al_{0.33}(OH)_2 \cdot IBU_{0.33} \cdot 0.5H_2O$, where IBU represents ibuprofen. Similarly, Winters et al. (46) used TGA to study gliclazide–β-cyclodextrin complexes, both individually and as a solid-state complex. It was possible to estimate the extent of entrapment by observation of the change in weight loss for the β-cyclodextrin over the temperature range of 85 to 105°C that is ascribed to interchelated water. More specifically, it is known that water within the cavity of β-cyclodextrin stabilizes the internal ring structure; indeed, on heating cyclodextrin to about 350°C, the ring structure collapses and the system decomposes. By examination of the water loss profile before and after complexation with the drug, the authors were able to postulate that the drug was replacing the entrapped water and was similarly stabilizing the structure. This conclusion was on the basis of observed

similarity of the decomposition profile before and after complexation. Other investigations into oral dosage forms include that of Weuts et al. (47), whereby TGA is
used to measure residual solvent in solid dispersions prepared using polyacrylic acid.

Thermogravimetry has also been used to characterize pharmaceutical micro- and
nanocapsules. For example, Chow et al. (48) used TGA to study sustained release
cefaclor polyvinylpyrrolidone microspheres prepared by a solvent evaporation
method. The drug is incorporated as the monohydrate, characterized by a weight
loss of 0.044% between 40 and 80°C (corresponding to one mole of water) followed
by decomposition at about 185°C. By comparing the weight loss profiles of the
encapsulated and unencapsulated drug, the authors were able to ascertain that the
drug was still in the monohydrate form in the finished product, a nontrivial determination by many other techniques. Similarly, Chauvet et al. (49) used TGA to
determine the composition of nanoparticles based on poly(alkylcyanoacrylates) by
using the degradation profile of the components (dextrans, pluonic F68, and
poly(butyl-2-cyanoacrylate)) as fingerprints, thereby enabling the compositions of
different formulations to be estimated. In particular, the authors showed that, by
using different molecular weight dextrans as suspending agents, both the composition and size distribution of the nanoparticles could be manipulated.

6.5   CONCLUSIONS

This chapter has attempted to highlight some of the possibilities afforded by TGA
for the characterization of pharmaceuticals and is intended as a companion text to
Chapter 5, which outlined the fundamental principles involved. In brief, the method
may be used to assess the quantity of solvent present, with the rider that great care
is required in drawing conclusions with regard to the identity of the materials lost.
This is in itself of great value, not only as a rapid alternative to methods such as
Karl Fischer titrimetry but also as a means of assessing solvent levels in very scarce
or expensive samples. The method may also be used as a complementary technique
to DSC when assessing glassy materials, particularly, in terms of measuring plasticizer levels such as water. Given that many pharmaceuticals are to some extent
hygroscopic, such measurements are essential for meaningful T_g assessment. The
most common use of TGA, however, is in the characterization of hydrates, whereby
the stoichiometry and binding characteristics of water within the solid may be easily
assessed. The level of sophistication with which these measurements are made may
vary greatly according to the needs of the operator. However, without wishing to
make invidious comments, those who wish to see the full extent of the possibilities
of the approach for the investigation of pharmaceutical hydrates could read the works
of the group of Grant, some of which are listed in the references.

The kinetic approach has been used in the study of pharmaceuticals to some
extent, although there remains some uncertainty with regard to the reliability and
predictive power of the data obtained from such modeling. Chapter 5 contains more
details in this respect. Finally, the technique may be used to study both conventional
and controlled release dosage forms, as a means of assessing residual solvent levels
and as a means of understanding solvent binding. Overall, TGA remains a simple,

inexpensive, and conceptually accessible means of characterizing pharmaceutical samples. However, in common with DSC, the apparent ease with which the experiments may be performed tends to mask the sophistication and complexity of the analysis that is required to take full advantage of the possibilities that TGA affords.

REFERENCES

1. Naini, V., Byron, P.R., and Phillips, E.M., *Drug Dev. Ind. Pharm.*, 24, 895, 1998.
2. Passerini, N. and Craig, D.Q.M., *J. Controlled Release*, 73, 111, 2001.
3. Beyers, H., Malan, S.F., van der Watt, J.G., and de Villiers, M.M., *Drug Dev. Ind. Pharm.*, 26, 1077, 2000.
4. Huang, Y., Cheng, Y., Alexander, K., and Dollimore, D., *Thermochim. Acta*, 367–368, 43, 2001.
5. Burnham, L., Dollimore, D., and Alexander, K., *Thermochim. Acta*, 357–358, 15, 2000.
6. Aggarwal, P. and Dollimore, D., *Thermochim. Acta*, 357–358, 57, 2000.
7. Lerdkanchanaporn, S. and Dollimore, D., *Thermochim. Acta*, 357–358, 71–78, 2000.
8. Leung, S. and Grant, D.J., *J. Pharm. Sci.*, 86, 64–71, 1997.
9. Chatterjee, K., Hazra, A., Dollimore, D., and Alexander, K.S., *J. Pharm. Sci.*, 91, 1156, 2002.
10. Stephenson, R.M. and Malamowski, S., *Handbook of the Thermodynamics of Organic Compounds*, Elsevier, New York, 1987, p. 263.
11. Langmuir, I., *Phys. Rev.*, 2, 329, 1913.
12. Price, D.M. and Hawkins, M., *Thermochim. Acta*, 315, 19, 1998.
13. Phang, P. and Dollimore, D., *Thermochim. Acta*, 367–368, 263, 2001.
14. Cheng, Y., Huang, Y., Alexander, K., and Dollimore, D., *Thermochim. Acta*, 367–368, 23, 2001.
15. Byrn, S.R., *Solid-State Chemistry of Drugs*, Academic Press, New York, 1982.
16. Morris, K.R. and Rodriguez-Hornado, N., Hydrates, in *Encyclopaedia of Pharmaceutical Technology*, Swarbrick J. and Vboylan, J., Eds., Vol. 7, Marcel Dekker, New York, 1993.
17. Morris, K.R., Structural aspects of hydrates and solvates, in *Polymorphism in Pharmaceutical Solids*, Brittain, H.G., Ed., Marcel Dekker, New York, 1999.
18. Michel, A., Drouin, M., and Glaser, R., *J. Pharm. Sci.*, 83, 508–513, 1994.
19. Glaser, R., Charland, J.-P., and Michel, A., *J. Chem. Soc., Perkin Trans.*, 2, 1875, 1989.
20. Bettinetti, G., Mura, P., Sorrenti, M., Faucci, M.T. and Negri, A., *J. Pharm. Sci.*, 88, 1133, 1999.
21. Dash, A.K. and Tyle, P., *J. Pharm. Sci.*, 85, 1123, 1996.
22. Haines, P.J., *Thermal Methods of Analysis — Principles, Applications and Problems*, Blackie Academic and Professional, Glasgow, 1995.
23. Zhu, H., Khankari, R.K., Padden, B.E., Munson, E.J., Gleason, W.B., and Grant, D.J.W., *J. Pharm. Sci.*, 85, 1026, 1996.
24. Ojala, W.H., Khankari, R.K., Grant, D.J.W., and Gleason, W.B., *J. Chem. Cryst.*, 26, 167, 1996.
25. Zhu, H., Padden, B.E., Munson, E.J., and Grant, D.J.W., *J. Pharm. Sci.*, 86, 418, 1997.
26. Zhu, H., Halfen, J.A., Young, V.G., Jr., Padden, B.E., Munson, E.J., Menon, V., and Grant, D.J.W., *J. Pharm. Sci.*, 86, 1439, 1997.

27. Ghosh, S., Ojala, W.H., Gleason, W.B., and Grant, D.J.W., *J. Pharm. Sci.*, 84, 1392, 1995.
28. Sussich, F., Urbani, R., Princivalle, F., and Cesaro, A., *J. Am. Chem. Soc.*, 120, 7893, 1998.
29. Sussich, F., Princivalle, F., and Cesaro, A., *Carbohydr. Res.*, 322, 113, 1999.
30. Sussich, F., Skopec, C., Brady, J., and Cesaro, A., *Carbohydr. Res.*, 334, 165, 2001.
31. Taylor, L.S. and York, P., *J. Pharm. Sci.*, 167, 215, 1998.
32. McGarvey, O.S., Kett, V.L., and Craig, D.Q.M., *J. Phys. Chem. B*, 107, 6614, 2003.
33. Horvat, M., Mestrovic, E., Danilovski, A., and Craig, D.Q.M., *Int. J. Pharm.*, 294, 1, 2005.
34. Suzuki, E., Shimomura, K., and Sekiguchi, K., *Chem. Pharm. Bull.*, 37, 493, 1989.
35. Shefter, E., Fung, H., and Mok, O., *J. Pharm. Sci.*, 62, 791, 1973.
36. Agbada, C.O. and York, P., *Int. J. Pharm.*, 106, 33, 1994.
37. Sekiguchi, K., Shirotani, K.-I., Sakata, O., and Suzuki, E., *Chem. Pharm. Bull.*, 32, 1558, 1984.
38. Zhu, H. and Grant, D.J.W., *Int. J. Pharm.*, 215, 251, 2001.
39. Pyramides, G., Robinson, J.W., and Zito, S.W., *J. Pharm. Biomed. Anal.*, 13, 103, 1995.
40. Najafabadi, A.R., Gilani, K., Barghi, M., and Rafiee-Tehrani, M., *Int. J. Pharm.*, 285, 97, 2004.
41. Junginger, H.E., *Dtsch. Apoth. Ztg.*, 133, 1988, 1983.
42. Junginger, H.E., Akkermans, A.A.M.D., and Heering, W., *J. Soc. Cosmet. Chem.*, 35, 45, 1984.
43. Peramal, V.L., Tamburic, S., and Craig, D.Q.M., *Int. J. Pharm.*, 155, 91, 1997.
44. Yamini, C., Shantha, K.L., and Rao, K.P., *J. Macromol. Sci., Pure Appl. Chem.*, A34, 2461, 1997.
45. Ambrogi, V., Fardella, G., Grandolini, G., and Perioli, L., *Int. J. Pharm.*, 220, 23, 2001.
46. Winters, C.S., York, P., and Timmins, P., *Eur. J. Pharm. Sci.*, 5, 209–214, 1997.
47. Weuts, I., Kempen, D., Verreck, G., Peeters, J., Brewster, M., Blaton, N., and van den Mooter, G., *Eur. J. Pharm. Sci.*, 25, 387, 2005.
48. Chow, A.H.L., Ho, S.S.S., Tong, H.H.Y., and Ma H.H.M., *Int. J. Pharm.*, 172, 113, 1998.
49. Chauvet, A., Masse, J., Egea, M.A., Valero, J., and Garcia, M.L., *Thermochim. Acta*, 220, 151, 1993.

7 Thermal Microscopy

Imre M. Vitez and Ann W. Newman

CONTENTS

7.1 INTRODUCTION

The objective of this chapter is to present to the reader the many facets of thermal microscopy. Information on the following topics will be presented: the evolution of the technique, background theory, the role of thermal microscopy as a complementary technique, general experimental parameters, industrial applications, references, and vendor directories.

7.2 HISTORY OF THERMAL MICROSCOPY

7.2.1 DEFINITION

Thermoptometry is defined as "a family of thermoanalytical techniques in which an optical property of the sample is monitored against time or temperature, while the temperature of the sample, in a specified atmosphere, is programmed. Two established examples are thermomicroscopy (observation under a microscope) and thermoluminescence" (1). Thermomicroscopy is also referred to as *thermal microscopy*, *optical thermal analysis*, *hot-stage microscopy (HSM)*, or *fusion methods*.

7.2.2 EARLY WORK

The microscope was first applied to chemical analysis in 1833 by Raspail (2), who suggested using crystal habits as a means of identifying chemical compounds. His ideas were further developed by Wormley, as well as Behrens, Kley, Emich, and Chamot into a method of analysis (3).

Lehman proposed in 1891 that the crystallization of an organic compound from its own melt was characteristic for that material (4). Lehman also described the mixed fusion assay for the qualitative determination of the phase diagram for compounds. Although almost completely ignored for 50 years, Lehman's work has been shown to have many applications in research and analysis (3).

7.2.3 PROTOTYPE THERMAL MICROSCOPY SYSTEMS

As there were no commercially available instruments during these developmental years, most early HSM studies were conducted on instrumentation that researchers constructed. An early thermal microscope was developed in the 1930s to 1940s for the high-temperature study of ceramics by C. Wooddell and H. Baumann, Jr., of the Carborundum Company, Niagra Falls, NY. The interesting features of this thermal microscopy instrument were that it could utilize various temperature range furnaces, had a specially designed high-temperature-resistant optical system, could capture time-lapse photographic images, and even generate high-temperature cinemicrophotographic films (5).

In the 1930s, Chamot and Mason defined the state of chemical microscopy (6) by detailing their work on the physical methods and chemical analysis involved with chemical microscopy. In the U.K. in 1954, Welch described a simple microscope attachment for the observation of high-temperature phenomena (7). The small, electrically heated thermocouple could hold a microscopic sample for single-crystal growth assays, melting-point determinations, and glass devitrification studies.

7.2.4 ADVANCES BY A. KOFLER, L. KOFLER, AND McCRONE

In Austria, A. Kofler and L. Kofler used their Kofler hot stage to further explore Lehman's work on the characterization of organic compounds. This instrument was developed by them in the 1930s, and a commercial version was introduced in the U.S. in 1940 by Arthur H. Thomas Company and Reichart (8,9). As the Kofler hot stage was designed so that the thermometer was located near the sample, a representative temperature display was achieved (9). The Kofler hot stage was found by researchers to be a very useful tool at the time for HSM work and, in fact, is still being used in many laboratories. The Koflers published a portion of their work (10) in 1954 and as a result greatly increased the quantity and quality of HSM research conducted in laboratories. An English translation is available (11).

In the U.S. in 1957, McCrone presented a review of fusion methods, techniques, equipment, and applications (12). His definition of fusion methods included the methods and procedures useful in research and analysis, which involved heating a compound or mixture of compounds on a microscope slide (12). His text comprises five parts. Chapter I is an introduction discussing the scope and limitations of fusion microscopy, and Chapter II discusses the commercially available equipment at the time. Chapter III details the general techniques for hot stages, cold stages, and hot bars, characterization and identification of organic compounds, purity estimations, purification methods, analysis of mixtures, determination of composition diagrams, and study of polymorphism. Chapter IV discusses the research applications of fusion microscopy, whereas Chapter V presents charts on the identification of organic compounds based on the melting points and the refractive indices of the melt.

7.2.5 EARLY COMMERCIAL INSTRUMENTATION

In 1967, Mettler introduced the FP-1 hot stage. The following year the FP-2 was introduced, which greatly altered the field of HSM as it permitted direct observation of the sample under study and provided the accuracy obtainable using the microscopic melting-point method (13). The FP-2 consisted of a heated chamber for mounting the sample on a microscope slide, an automatic temperature control display system, and a hand control panel. The use of a temperature feedback system provided automatic linear heating control, which eliminated temperature surges and lags inherent in manually adjusted instruments. With a temperature range of −20 to 300°C, and use of preset heating rates, the instrument provided the microscopist with an automated visual thermal microscopy instrument (9). The Mettler hot-stage instrument greatly increased the use of HSM because of its ease of use and precise temperature control (14).

7.2.6 ONSET OF INDUSTRIAL APPLICATIONS

In the 1960s several advances in the development of a HSM system were made. Hock and Arbogast developed a melting-point apparatus for the study of melting points and crystallization rates in polymers (15). Sommer developed a high-speed differential thermal analyzer (DTA) employing a hot-stage microscope for use at elevated temperatures (16) in which the temperature difference between a sample and a reference was measured. Miller and Sommer described an integrated hot-stage microscope–differential thermal analyzer that incorporated design improvements over earlier equipment (17).

Charsely and Kamp described a versatile HSM unit from Stanton Redcroft in 1971. This unit was useful for characterizing inorganic systems as it permitted observations up to 1000°C and collection of reflected-light intensity measurements (18). Charsely and Tolhurst applied the use of this high-temperature HSM instrument to pyrotechnic systems, and expanded the capabilities of the system by adapting it for the collection of high-speed and time-lapse cinephotography (19).

In 1971, Kuhnert-Brandstätter applied thermomicroscopy to an even narrower segment of chemical compounds, namely, organic compounds of medicinal benefit (20). Her research efforts characterized many polymorphic systems in the pharmaceutical industry. Tynan and Von Gutfeld developed a vacuum sample holder for use on an optical microscope. This accessory had the ability to characterize small samples in the 30 to 800 K temperature range (21). Schultze presented a design for a high-temperature microscope–differential thermal analyzer, which explored this role of the combination technique in the high-temperature research field (22).

In 1980, Perron, Bayer, and Wiedemann described a combination of differential thermal analyzer and transmitted-light microscopy based on the Mettler FP-5 hot-stage system (23). A few years later, the Mettler FP800 system was introduced, which coupled thermomicroscopy with differential scanning calorimetry, to expand the applications of thermal microscopy (24).

7.2.7 RECENT DEVELOPMENTS

As the developments in technology continued in the 1990s, more advanced computer-controlled systems were designed. In 1993, Mettler introduced the FP900 hot-stage system, which incorporated use of a personal computer, a photomonitor for light intensity measurements, transparent crucibles in which to place the material under investigation, and explored the opportunities presented by video technology (25). A photovisual differential scanning calorimeter (DSC) system was developed by Shimadzu that combined the attributes of DSC with the visual data from a microscope and camera (26).

From this summary of the evolution of HSM, it may be seen that many of the technological advances were initiated by scientists applying aspects of thermal analysis and microscopy to problems in their respective industries. Industry continues to utilize thermal microscopy, combining it with evolving technologies to solve new problems. One example is in the polymer industry, where earlier work utilized reflected and depolarized light intensity measurements to study the crystallization rate changes of polymers (27,28), whereas recent work has focused on the use of a combination DSC and optical microscope to study the reflected-light properties of polymeric materials (29,30). Additionally, HSM is often utilized as a complementary tool for the characterization of polymorphic materials, not only in the study of inorganic compounds (31) but also in the study of pharmaceuticals (32–35).

7.3 THEORY

There has been a great deal of effort put into the development of thermal microscopy systems. Although it is important to understand the foundations of this technique, it is more important that the analyst comprehend the mechanism of the analytical technique and how its utilization can provide valuable information.

7.3.1 OPTICAL MICROSCOPES

Because the microscope is the cornerstone of thermal microscopy, the proper use and maintenance of an optical microscope requires primary attention. Although some of the interest that surrounds HSM is often associated with the colorful photomicrographs that are collected during recrystallization or fusion method experiments, it is imperative that the microscope be properly aligned and illuminated. Many microscopes are supplied with detailed instructions for setting up Kohler illumination, focusing and centering the lamp filament, centering objectives, etc. Close attention should be paid to these instructions. In addition, microscope vendors can provide information and training to assist the proper setup and adjustment of the microscope. Other resources (11,36) in the industry are available to aid the microscopist. Although setting up a microscope with the proper illumination and configuration may appear to be a difficult task, this work is as necessary as the proper temperature calibration of the hot stage when reproducible, accurate information is to be obtained. To generate a highly comprehensive characterization of a material using HSM, the microscope from which this information has been collected must be correctly aligned and illuminated.

7.3.2 CROSS POLARIZATION AND BIREFRINGENCE

The use of polarizing filters in an optical light microscope provides the opportunity to study unique features of the material under characterization. Light passing through a polarizing filter will become oriented relative to that filter, so that when a second polarizer is positioned above the sample and perpendicular to the first filter, cross polarization is achieved. When a crystalline material is placed under cross-polarized light, some particles will not be visible and are characterized as *isotropic*, whereas crystals that appear to be white or colored are termed *anisotropic*. An anisotropic crystal is characterized as *birefringent* because the two components of polarized light being transmitted through the particle recombine in a fashion such that one component emerges slightly ahead relative to the other component. The components of polarized light initially are traveling at the same velocity; however, once they enter the crystal, one component is retarded due to the unique properties of the crystal under investigation, such as the varying thickness within a crystalline particle and refractive index. Upon emergence from the crystal, these light components will undergo destructive interference (based on the retardation), recombine, and produce various polarization colors. Light passing through the varying thicknesses of a crystal produces a color pattern known as *Newton's series*. The Michel–Levy chart is a representation of the retardation colors as a function of thickness and birefringence. Given any two of these parameters (color, birefringence, and thickness), the third is determined using this chart. Copies of the Michel–Levy chart may be obtained from microscope vendors and in the literature.

7.3.3 HOT-STAGE ACCESSORY

Through the use of a multidisciplinary approach in which a number of techniques, such as differential scanning calorimetry (DSC), thermogravimetric analysis (TGA), and powder x-ray diffraction (XRD) are utilized to characterize the properties of a material, a comprehensive characterization of the material under investigation may be conducted. Where there are deficiencies or lack of corroborating data, additional techniques must be used to provide supporting evidence of a thermal transition or confirmation of a property. HSM is often used as a technique to provide corroborating data on melting points, recrystallizations, volatilizations, solid–solid transformations, decompositions, etc.

The theory involved in the operation of a hot-stage/DSC accessory is presented. The material under investigation is placed into a transparent crucible and then into the hot-stage accessory, which has a DTA/DSC sensor. A second transparent crucible, which is similar to the sample crucible, is positioned in the hot stage for use as a reference. Upon initiation of the heating program, the sample equilibrates for a moment to allow the crucibles to reach the temperature of the furnace. Once the temperature program starts, the furnace temperature begins to rise, following the heating program. There is a temperature lag between the furnace temperature and the crucible temperature, which is due to the thermal resistance of the measuring sensor and the heat capacity of the crucibles. The difference between the temperatures of the reference and sample crucibles is known as the differential thermal analysis (DTA) signal, and because the DTA signal is proportional to the heat flow of the sample, this differential may be converted into the differential scanning calorimetry (DSC) signal if the proportionality factor is known. This proportionality factor is temperature dependent and is known as the *calorimetric sensitivity* (37).

7.3.4 HOT-STAGE CALIBRATION

Now that some principles regarding the operation of the hot stage have been reviewed, one additional topic relative to the general operation of the hot-stage accessory needs to be addressed, namely, calibration. Instrumentation is often calibrated at the time of manufacture and is delivered to the scientist in a ready-to-assemble and use mode. Depending on the nature of the laboratory, a full calibration, or a calibration check of the temperature, should be conducted on a regular basis (e.g., monthly). Materials that have known melting points may be used; however, materials that are traceable to National Institute of Science and Technology (NIST), or known standards, are recommended. Many thermal analysis vendors supply calibration materials for their instruments. Some materials that are often utilized as standards for thermal microscopy work include: benzophenone (melt maximum at 48.1°C), benzoic acid (melt maximum at 122.4°C), and caffeine (melt maximum at 236.4°C), because the melting of these crystalline particles is clearly apparent under magnification. Standards often utilized for conventional thermal analysis work (DSC and DTA) include indium (melt onset at 156.6°C), tin (melt onset at 231.9°C), and lead (melt onset at 327.5°C). These materials do not lend themselves well to visual temperature calibration, as the melt transitions may not be easily seen under magnification.

It should be noted that accurate temperature data may depend on a number of variables including the type of sample, mounting method, heating rate, type of melting point (equilibrium melting point vs. melt during heating) (38), as well as particle size factors.

7.3.5 PARTICLE SIZE EFFECTS ON MELT

In some samples, a difference between the melt-onset temperatures collected from the DSC and the hot-stage accessory may be observed. These differences may be as small as tenths of °C, or as large as 10 to 20°C. In general, the finer particles melt initially, followed by the larger particles.

Several theories may be proposed to explain the difference in observed melting onsets, one of which is impurities present in the material under investigation. Often, impurities or defects in samples are perceived as specks or spots. Although these indeed are considered impurities, homogeneously or inhomogeneously dispersed impurities should also be considered. Based on the range of particle sizes observed, the concentration of these impurities in the various-sized particles will undoubtedly have an effect on melting behavior. It should also be kept in mind that although the material may be considered pure on a bulk level (as determined by a technique such as high-performance liquid chromatography), on the particulate level, this purity may not be homogeneous. Thus, sampling of any material is critical to the representativeness of the material and to the results obtained.

Additionally, the process from which the material was generated may also play an important role in the melt onset. The recrystallization processes, including the solvent and mechanical systems utilized to manufacture the material, may greatly affect the crystallinity of the material. This would directly influence the amorphous content in the presumably high crystalline material through the production of an "impurity," which would lower the melt onset of the material (39). Although there has been recent interest (40) in investigating these phenomena, their impact on the characterization of materials is not yet fully understood.

7.3.6 Photomicroscopy

The next topic to be addressed is photomicroscopy and, more recently, video microscopy. With the changes in photomicrography in the past few years, including the development of high-resolution video cameras and the advent of digital photography, advances in technology will continue to expand photomicrographic capabilities. Many laboratories utilize instant or 35-mm print media to capture their photomicrographs; however, with the development of video cassette recorders (VCRs) and high-capacity optical drives, other methods of image collection have become more common. The selection of an appropriate camera (color, black and white, digital, still, or video) greatly depends on the application. A high-resolution color video camera may be an appropriate selection for a laboratory involved in the collection of thousands of frames during a hot-stage experiment, but other laboratories may be satisfied with an instant-print camera. Many resources (e.g., photographic equipment and instrument vendors, internal scientific, and multimedia departments) should be explored prior to the selection of a camera, so that one is well versed in the latest products that use the state-of-the-art technologies. Once the available resources have been investigated and the choices defined, an appropriate camera for the desired application may be purchased.

7.3.7 Glossary References

The greatest challenge in considering the utilization of a thermal microscopy instrument is that the scientist must be well versed in thermal analysis, microscopy, and photography, as well as computer image capturing and manipulation. Appropriate nomenclature is also an important tool that scientists need to utilize in documenting their work. There are multiple resources that are available to help build a vocabulary base (41–43). As the scientist becomes trained through many hours of experimentation, their expertise will develop and difficult characterizations become challenges rather than ordeals.

7.3.8 Additional References

A number of resources in the literature are available to thermal analysts, microscopists, and pharmaceutical scientists. Those not mentioned yet in this chapter include references on general thermal analysis (44–47), microscopy (48–54), and pharma-ceutical analysis (55–58).

7.4 EQUIPMENT

The optical microscope is a valuable tool in the laboratory and has numerous applications in most industries. Depending on the type of data that is required to solve a particular problem, optical microscopy can provide information on particle size, particle morphology, color, appearance, birefringence, etc. There are many accessories and techniques for optical microscopy that may be employed for the characterization of the physical properties of materials and the identification of unknowns, etc. Utilization of a hot-stage accessory on the microscope for the char-acterization of materials, including pharmaceutical solids (drug substances, excipi-ents, formulations, etc.), can be extremely valuable. As with any instrument, there are many experimental conditions and techniques for the hot-stage microscope that may be used to collect different types of data. Often, various microscope objectives, optical filters, ramp rates, immersion media, sample preparation techniques, micro-chemical tests, fusion methods, etc., can be utilized.

7.4.1 An Example Configuration for a Hot-Stage Photomicroscopy System

One possible configuration for a HSM system is depicted in Figure 7.1, which shows an example layout for the various components of a hot-stage photomicroscopy system. In this example, the material under investigation is placed in the hot-stage accessory, which in turn is placed onto the microscope stage. The hot-stage controller not only monitors the temperature program for the experiment, but transmits the thermal analysis data to a computer for processing and analysis. A color camera is attached to the microscope for observation of the visual data. Depending on the needs of the laboratory, various accessories may be added to create an integrated hot-stage photomicroscopy system. Additionally, several monitors may be positioned for observation of the live video image, a composite image of the video, thermal,

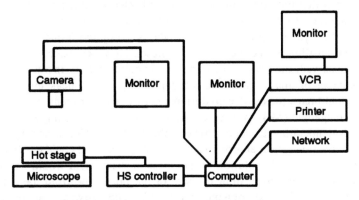

FIGURE 7.1 Example of a configuration for an integrated hot-stage photomicroscopy system.

and temperature stamp, as well as output from a VCR, and lastly, the system may be connected to a local computer network to aid in data storage.

7.4.2 Photography/Video Microscopy Considerations

Because HSM uses images collected from the optical microscope as the basis of the technique, image quality is an important issue. A high-resolution color video camera or a rapid-scanning color digital camera may provide images possessing the required quality. Care should be exercised when the camera is selected, as some scanning digital cameras may require a minute or two to collect a single high-resolution color image. Digital cameras have an additional advantage in that they do not require the use of a frame grabber, because images will be saved digitally. (A frame grabber is a computer component that permits the acquisition of an image, which is then stored digitally.) If the requirement for the laboratory is solely for still images, the selection of a scanning digital camera may be an excellent option to pursue; however, many instant and 35-mm cameras may fulfill the requirements of the laboratory.

For laboratories that routinely conduct dynamic hot-stage experiments to mimic the DSC experimental conditions, the scanning digital camera may limit the analyst to the collection of only a few images during the course of the experiment. In this situation, a high-resolution color video camera that can capture thousands of images during the course of an experiment may provide greater capabilities. A high-resolution color video camera is especially useful when studying crystal forms of chemical substances, in which the material under investigation may exhibit multiple transitions in a small temperature range. The collection of a few images during the course of this experiment may be inadequate to completely represent the numerous phase changes observed in this material during the experiment.

7.4.3 Image Collection and Manipulation

A high frame collection rate is often required for the characterization of materials that exhibit multiple transitions in close proximity to one another. For these samples, a thermal microscopy system that includes a video camera, frame grabber, and image manipulation software may be required. Software packages are available in which individual image frames collected from the camera are saved as individual files. An additional feature that presents some interesting possibilities is the use of a dynamic frame collection rate, whereby the frame collection rate during nontransition temperatures can be preset. However, once thermal transitions are observed, the frame collection rate is automatically increased so that a greater number of images are collected during the area of peak interest. Often, the individual image files collected during a hot-stage experiment may be approximately 1 megabyte (MB) in size, so the images collected during an experiment from room temperature to approximately 300°C at a 10°C/min heating rate and an image collection rate of approximately 1 frame per second could easily generate a 200-MB file. A simple solution to this mass of visual data would be to adjust the frame collection rate to limit the size of the image file from the completed experiment to one that the computer can handle. Alternatively, one could utilize

FIGURE 7.2 Photomicrograph of several single crystals of a pharmaceutical compound. (From DiMarco, J., Gougoutas, J., and Malley, M., Single crystals, Bristol-Myers Squibb Crystallography Group, Lawrenceville, NJ, 1998. With permission.)

an external drive for data handling. Some internal or external optical drives have 4 gigabyte (GB) capability, which should be able to handle these data files easily. Also, some image-capturing software packages compress image files as they are collected and expand them upon viewing. This may prove helpful in data handling. The benefit in having an image manipulation software package that handles the video data from the HSM system is that it provides the microscopist with the ability to view a video immediately following the completion of the actual hot-stage experiment.

A photomicrograph is not only a visual representation of the particles under investigation, but is the singular piece of data that defines that material. To gain an appreciation for the value of a photomicrograph, an image of several single crystals of a pharmaceutical compound (59) is presented in Figure 7.2. This image was collected at 200× magnification with cross polarizers with the first-order red compensator near the onset of a hot-stage experiment conducted to characterize the thermal properties of the material.

7.4.4 DIGITAL IMAGE/VIDEO PRESENTATION

Post-experiment editing of the video files permits individual frames to be chosen for incorporation into reports. It should be noted that most word-processing programs are designed for manipulation of text, and not for the incorporation of large image files. Therefore, post-experiment report generation with the incorporation of several 1-MB images may result in an inefficient use of resources. An option that would negate the need for retrieval of individual frames would be to attach a VCR to the camera output for storage of the video experiment. Although there is a loss of resolution saving to videotape vs. optical disk, the data-handling issue is completely removed. The videotape is portable and may be viewed at project team meetings, conferences, etc. However, it should be noted that the video generated from these experiments may be 15 or 20 min long, with only a moment or two of interesting transitions.

Some digital image manipulation software packages have features that permit rapid editing of image sequences to compress a lengthy hot-stage experiment into a brief video of the thermal properties of the material. This edited video, containing only the images of the interesting transitions, may in turn be saved to videotape to make the video portable. It should be noted that as more advanced video image analysis software packages are developed, their editing and manipulation features will be expanded.

An additional hardware option that should be discussed concerns use of a computer to present the digital videos in the high-resolution format in which they were originally collected. A second computer, configured with an optical drive and video-editing software, would be able to present the same digital video that was collected from the HSM system. An additional monitor would also be required for the presentation. Although this route is more costly, the integrity of the digital video would be preserved as the lower-resolution videotape is not utilized. A third option is the use of digital video disks (DVDs), which possess twice the resolution capability of videotape, together with the ease of use of a VCR.

7.4.5 IMAGE HARDCOPY

The presentation of images in a report to supplement written observations is a great benefit. Whether the images are printed from a black and white laserjet printer, or a color digital printer, images are invaluable to document the thermal transitions. Most high-quality black and white laserjet printers do a fine job in representing the images collected from the HSM system. Recent developments in color ink jet and laser jet printers can produce high-resolution photographic quality prints. A color digital printer that provides high-resolution, photographic-quality images on photographic paper provides the best presentation of the images on paper. At several dollars per printed page, there is a cost factor associated with generation of these reports, in addition to a relatively high cost of the printer. Currently, there are a number of vendors that can provide printers which can generate photographic-quality prints.

7.4.6 COLOR MONITORS

For viewing of the video data from a hot-stage experiment, a high-resolution color monitor connected to the camera output gives the ability to easily view the experiment without having to peer through the microscope ocular. The video (either digital or from videotape) can also be presented on a quality monitor.

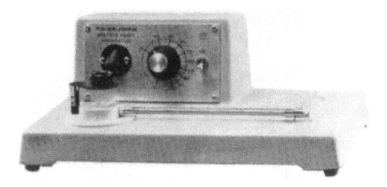

FIGURE 7.3 Fisher-Johns melting-point apparatus. (From Product Literature, Fisher Scientific, Springfield, NJ, 1997. With permission.)

7.4.7 MELTING-POINT EQUIPMENT

Various manufacturers of melting-point systems exist and provide a number of models and accessories for these units. Depending on the application and objective of the experiment, different techniques and equipment are required to achieve the desired result. For applications in which the confirmation of a melt is required without any need for high magnification, a nonmicroscopic melting-point apparatus would suffice. In many units, this assay requires a small quantity of material in a glass capillary tube or between microscope cover slips, which is subsequently heated. Often, a low-magnification lens is provided to view the melt transition. Some units may provide a computer interface such that images during the melt transition may be captured to document the melt. Figure 7.3 (60) and Figure 7.4 (61) present images of Fisher-Johns melting-point and capillary systems, respectively.

7.4.8 HOT-STAGE ACCESSORIES

Several vendors offer various modifications of hot-stage accessories. Depending on the nature of the experiment, different accessories may be utilized. One of the main reasons that the early Kofler hot-stage instruments were so widely used was their ease of use. The Kofler hot-stage comprised an open heated surface onto which the analyst placed a glass slide with the material under characterization. Although the instrument was simple in design and use, it not only provided the opportunity to observe the material on the slide under magnification, but also permitted the microscopist to manipulate the particles with probes and microtools for assays including morphological evaluation, birefringence studies, microchemical tests, and fusion experiments. Although the Kofler hot stage is no longer commercially available, units are still being utilized, indicating there is still a need for a hot-stage accessory that provides for unobstructed access to the material. Figure 7.5 presents an image of the Thomas Model 40 Micro Hot Stage (Kofler type) (62).

Hot-stage accessories are available in a variety of configurations. Although the Kofler hot stage was relatively open to the microscopist, allowing for particle manipulation, many of the commercially available thermal stages (hot, warming, freezing,

FIGURE 7.4 A capillary melting-point apparatus, Electrothermal I9000 Series. (From Thermal Devices Product Brochure, Electrothermal, Inc., Gillette, NJ, 1997. With permission.)

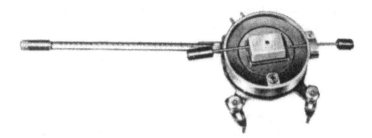

FIGURE 7.5 Thomas model 40 micro hot-stage apparatus (Kofler type). (From 1986–1987 Product Catalog, Thomas Scientific Company, Swedesboro, NJ. With permission.)

etc.), are for all practical purposes closed, meaning that manipulation of the materials on the microscope stage may be obstructed by the housing of the hot-stage accessory. These systems possess this protective housing to ensure uniform temperatures across the sample surface. Although uniform heating is provided, rapid access to the materials under the microscope is diminished. However, the microscopist does possess the ability to observe the material under magnification while a controlled heating or cooling program is applied to the sample. The ability to provide uniform heating is generally more important for combination hot-stage–DSC accessories, which not only provide the visual data necessary for confirmation of thermal transitions, but additionally have the ability to collect DSC data for quantitation of thermal transitions.

FIGURE 7.6 Mettler FP84HT hot-stage accessory. (From Rapid and Automatic Determination of Thermal Values Product Brochure, Mettler-Toledo Company, Hightstown, NJ, 1996. With permission.)

FIGURE 7.7 Instec HCS400 cold stage. (From Product Literature, Instec, Inc., Broomfield, CO, 1997. With permission.)

A microscope, coupled with a hot-stage accessory (either open or closed), can provide an invaluable wealth of information for a physical characterization laboratory. Assays that may be utilized include quench cooling and isothermal experiments for crystal form characterization, use of immersion oils to confirm volatile loss, as well as simple melt determinations. Figure 7.6 to Figure 7.8 present thermal stages for various applications: hot-stage (63), cold-stage (64), and warming-stage (65).

A remote handset is available in some hot-stage units to easily adjust the preprogrammed heating method and to initiate isothermal experiments without having to alter the method in the computer program while the experiment is running. Some hot-stage accessories use glass slides as the media on which the material under investigation may be mounted and then placed into the hot-stage accessory. Other models use glass crucibles to be used with or without cover slips. Depending on the nature of the material and the experiment, the loss of volatiles may obscure observation when

FIGURE 7.8 Physitemp TS-4 diaphot warming stage. (From Custom Stages Product Literature, Physitemp Instruments, Inc., Clifton, NJ, 1994.)

the cover slips are used, as the volatiles may condense on the inside surface of the cover. An alternative would be to not use the cover; however, depending on the corrosive nature of the volatile, the microscope objective is potentially open to damage.

Additionally, a number of other hardware devices may be purchased to add increasing capabilities to the photomicroscopy system, relating to documentation and presentation. Some of the accessories include color cameras, image manipulation software, thermal analysis software, computer hardware for data storage management, color printers, etc. These accessories, used in conjunction with the hot-stage accessory, can develop into a very valuable integrated hot-stage photomicroscopy system that can provide confirmatory information on a wide variety of thermal transitions.

7.4.9 MICROSCOPE COMPONENTS

Different magnification objectives may be fitted onto the optical microscope. Depending on the particle size of the material being evaluated, the appropriate magnification should be utilized. Super long working distance (SLWD) objectives from 10× to 60×, in addition to a 10× ocular, will provide up to 600× magnification for HSM work. In some hot-stage models, the working distance between the top surface of the hot-stage accessory and the lower surface of the objective lens may not accommodate the higher magnification SLWD objectives. Again, various vendors provide different objectives, so several vendors should be evaluated to select equipment that is most appropriate to the goals of the laboratory.

7.4.10 IMAGE/CURVE OVERLAY

Several additional considerations should also be noted. Some laboratories may be interested in developing a hot-stage system that possesses the ability to have an overlay of the temperature and the DSC curve on the image from the microscope.

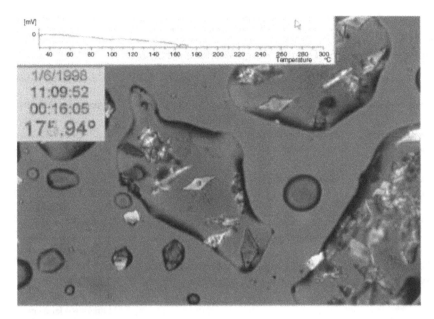

FIGURE 7.9 Example photomicrograph showing overlay of thermal curve and temperature stamp over video image.

Thus, a temperature stamp on each frame can be collected during the experiment. The other benefit in having an overlay of the DSC curve is to corroborate the thermal data that was previously generated on a separate DSC or DTA unit. It should be noted that in some cases, due to the submilligram quantities of sample used for hot-stage studies during which individual crystals or particles were evaluated, the DSC curves may not be equivalent to those generated by conventional DSC or DTA units, which may have used a greater quantity of material. Figure 7.9 presents a photomicrograph showing the DSC curve and time/temperature stamp overlayed over an image of a material nearing the completion of the melt transition. Use of the temperature stamp provides data to support the transition temperature. Figure 7.9 and all subsequent photomicrographs are by I. Vilez.

7.4.11 Vendor Directories

To provide a database of thermal microscopy information, a number of tables containing vendor contacts and products have been generated. It should be noted that these tables are simply a collection of the resources that the authors have collected in the development of their photomicroscopy system and should be viewed neither as an exhaustive list nor an endorsement or recommendation for their use. Because each laboratory has specific goals and requirements, it is suggested that the reader explore all available resources and develop the appropriate photomicroscopy system based on the goals of the laboratory, budgetary limitations, etc.

Information on the following topics may be found in the vendor directory tables: optical microscopy thermal stage vendors (Table 7.1), melting point and nonoptical

TABLE 7.1
Optical Microscopy Thermal Stage Vendor Directory

Vendor	Products	Location	Contact Information
Brooks Industries	Heating, cooling, and warming stages	Lake Villa, Illinois, U.S.A.	Tel/fax: 1-847-356-1045
Creative Devices, Inc.	Heating stages with video character generation	Neshanic Station, New Jersey, U.S.A.	Tel: 1-908-369-4333, fax: 1-908-369-4452
Ernest F. Fullam, Inc.	Custom heating stages	Latham, New York, U.S.A.	Tel: 1-518-785-5533, 1-800-833-4024, fax: 1-518-785-8647
Instec, Inc.	Heating, freezing, and warming stages	Broomfield, Colorado, U.S.A.	Tel: 1-303-404-9347, fax: 1-303-404-9348, http://www.instec.com
Leica Imaging Systems, Inc.	Leitz heating stage	Deerfield, Illinois, U.S.A.	Tel: 1-847-405-0123, 1-800-248-0123, fax: 1-847-405-0147, http://www.leica.com
Linkam Scientific Instruments	Heating, freezing, and warming stages	Waterfield, Tadsworth, Surrey, U.K.	Tel: 44 (0) 1737 36347, fax: 44 (0) 1737 363480
Mettler Inst. Corp.	Heating, freezing, and warming stages	Hightstown, New Jersey, U.S.A.	Tel: 1-609-448-3000, 1-800-METTLER, fax: 1-609-4488-4949, http://www.mt.com
Physitemp Instruments, Inc.	Freezing and warming stages	Clifton, New Jersey, U.S.A.	Tel: 1-732-779-5577, 1-800-452-8510
Shimadzu Scientific Instruments, Inc.	Photovisual DSC	Columbia, Maryland, U.S.A.	Tel: 1-800-447-1227, fax: 1-410-381-1222, http://www.shimadzu.com

microscopy thermal stage vendors (Table 7.2), microscope vendors (Table 7.3), imaging software and hardware vendors (Table 7.4), camera vendors (Table 7.5), and color printer vendors (Table 7.6).

7.5 ROLE AS A COMPLEMENTARY TECHNIQUE

In many industrial laboratories, the analytical chemist is often viewed as a valuable resource for solving manufacturing, research, and development problems. Samples requiring analysis enter the laboratory and results and solutions for problems are generated. To complete this task, the analytical laboratory must possess a wealth of resources, both in instrumentation and in experienced personnel, so that a number of assays may be conducted to generate the data that are required to solve a particular problem. As the techniques in an analytical laboratory are enumerated, each provides data that are unique to that assay. Often, it is only after the initial characterization

TABLE 7.2
Melting-Point and Nonoptical Microscopy Thermal Stage Vendor Directory

Vendor	Products	Location	Contact Information
CIC Photonics, Inc.	Sample compression and hot stage for infrared microscopes	Albuquerque, New Mexico, U.S.A.	Tel: 1-505-343-9500, fax: 1-505-343-9200, http://www.cicp.com
Electrothermal, Inc.	Capillary melting point apparatus, CCTV accessory	Gillette, New Jersey, U.S.A.	Tel: 1-908-647-2900, 1-800-432-8244, fax: 1-908-604-2069
Fisher Scientific	Melting point apparatus	Springfield, New Jersey, U.S.A.	Tel: 1-800-766-7600, http://www.fisher1.com
Gatan, Inc.	Heating and cooling stages for transmission electron microscopes	Warrendale, Pennsylvania, U.S.A.	Tel: 1-412-776-5260, 1-412-776-3360, http://www.gatan.com
Laboratory Devices, Inc.	Capillary melting point apparatus	Holliston, Massachusetts, U.S.A.	Tel: 1-508-429-1716, fax: 1-508-429-6583
Oxford Instruments, Inc.	Heating and freezing stages for electron microscopes	Concord, Massachusetts, U.S.A.	Tel: 1-508-369-9933

of the material under question using several techniques that a solution to the problem can be generated.

The product development process in the pharmaceutical industry is not unlike that of other industries, in which exploratory materials are researched and developed into products, and where problems arise the analytical laboratory is called upon to provide solutions. The characterization of the physical properties of pharmaceutical solids is one of the disciplines utilized early in the drug development process. The characterization of these properties is vital to determining whether the compound under investigation is a candidate for continued development as a drug product. To facilitate the characterization of pharmaceutical solids in laboratories, a conceptual approach for this characterization has been developed (66) that uses decision trees to guide the analyst in the characterization of drug substances.

7.5.1 INTRODUCTION TO MULTIDISCIPLINARY CHARACTERIZATION

The characterization of the physical properties of pharmaceutical compounds under development is often conducted using a variety of techniques including DSC, TGA, XRD, HSM, solid-state nuclear magnetic resonance (NMR), infrared (IR) and Raman spectroscopy, moisture uptake, particle size analysis, scanning electron microscopy (SEM), and micromeritic assays. A typical initial analysis of a pharmaceutical compound under development in a materials characterization group would include DSC, TGA, HSM, and XRD analyses. These four techniques are chosen because the data generated from them, when viewed collectively, comprise a relatively complete initial analysis of the physical properties of the compound. The DSC, TGA, and HSM assays

TABLE 7.3
Microscope Vendor Directory

Vendor	Products	Location	Contact Information
A-Z Microscopes (Meiji)	Optical microscopes	Ontario, Canada	Tel: 1-519-4286, fax: 1-519-672-3378, http://az-microscope.on.ca
Leica Imaging Systems, Inc.	Optical microscopes	Deerfield, Illinois, U.S.A.	Fax: 1-847-405-0147, http://www.leica.com
Lomo Optics, Inc.	Optical microscopes	Gaithersburg, Maryland, U.S.A.	Tel: 1-301-869-0386, fax: 1-301-977-9562, http://www.lomooptics.ru
Nikon, Inc.	Optical microscopes	Melville, New York, U.S.A.	Tel: 1-516-547-8500, fax: 1-516-547-0306, http://www.nikonusa.com
Olympus, Inc.	Optical microscopes	Melville, New York, U.S.A.	Tel: 1-800-455-8236, 1-516-844-5000, fax: 1-516-844-5112, http://www.olympus.com
Swift Technologies, Inc.	Optical microscopes	San Jose, California, U.S.A.	Tel: 1-800-523-4544, fax: 1-408-292-7967
Carl Zeiss, Inc.	Optical microscopes	Thornwood, New York, U.S.A.	Tel: 1-914-747-1800, 1-800-233,2343, fax: 1-914-682-8296, 1-914-681-7446, http://www.zeiss.com

provide information on thermal properties, whereas the XRD provides data on the crystalline or amorphous nature of the material.

It is important to stress that no single technique is adequate to completely characterize the properties of the material; therefore, a multidisciplinary approach to the analysis of these materials should be utilized. Several applications of this approach are documented in the literature (67–69) and indicate that the use of multiple techniques permits a comprehensive characterization of the material.

7.5.2 APPLICATION OF THE MULTIDISCIPLINARY APPROACH

To illustrate the interrelationship of these techniques, a summary of the multidisciplinary approach for the characterization of materials is presented. DSC, TGA, and XRD analyses are initially employed to characterize some of the basic physical properties of the material. DSC analysis of several lots of material produces curves that exhibit endothermic maxima at different temperatures, suggesting the existence of multiple crystal forms. Whereas DSC is utilized to identify the thermal properties of these materials, XRD is the definitive technique utilized to determine the presence of crystal forms. If XRD analysis shows that different powder diffraction patterns exist, corroborating evidence has been generated confirming that the lots of material under investigation comprise different crystal forms. In most cases, DSC and XRD

TABLE 7.4
Imaging Software and Hardware Vendor Directory

Vendor	Products	Location	Contact Information
Advanced Imaging Concepts	Image management packages, digital imaging hardware products	Princeton, New Jersey, U.S.A.	Tel: 1-609-921-3629, fax: 1-609-924-3010
Boeckeler Instruments, Inc.	Video imaging microscopy systems	Tucson, Arizona, U.S.A.	Tel: 1-520-573-7100, 1-800-552-2262, fax: 1-520-573-7101
Buehler Ltd.	Imaging archive and report systems	Lake Bluff, Illinois, U.S.A.	Tel: 1-847-295-6500, 1-800-283-4537, fax: 1-847-295-7979
Configured Systems, Inc.	Systems for and multimedia presentations	East Brunswick, New Jersey, U.S.A.	Tel: 1-732-249-3559, fax: 1-732-220-9406, http://config-sys.com
Digital Processing Systems, Inc.	Professional and desktop video editing equipment	Florence, Kentucky, U.S.A.	Tel: 1-606-371-5533, fax: 1-606-371-3729, http://www.dps.com
Matrox Electronic Systems, Inc.	Video, imaging, and graphics packages	Quebec, Canada	Tel: 1-514-685-2630, http://www.matrox.com
McCrone Accessories	Video microscopy systems and accessories	Westmont, Illinois, U.S.A.	Tel: 1-708-887-7100, 1-800-MAC-8122, fax: 1-708-887, 7764, http://www.mccrone, com
Media Cybernetics, L.P.	Image analysis software products	Silver Spring, Maryland, U.S.A.	Tel: 1-301-495-3305, fax: 1-301-495-5964, http://www.mediacy.com
Universal Image Corp.	Systems for biological video and digital microscopy applications	West Chester, Pennsylvania, U.S.A.	Tel: 1-610-344-9410, fax: 1-610-344-9515, http://image1.com

should be used in conjunction with each other to provide an accurate representation of the physical properties of the material. Additionally, because some materials may consist of mixtures of two or more crystal forms, several techniques should be employed to determine it is a single phase.

DSC, however, is not complete in the characterization of the thermal properties of the material. TGA is required to confirm transitions such as volatilizations and decompositions, and a visual technique, such as capillary melting-point determinations or HSM, is required to confirm transitions such as melts and recrystallizations.

7.5.3 CHARACTERIZATION OF MATERIAL USING OPTICAL MICROSCOPY

Once the material has been characterized by DSC, TGA, and XRD, the next phase of the evaluation of the material may commence. Because the transitions observed

TABLE 7.5
Camera Vendor Directory

Vendor	Products	Location	Contact Information
Dage-MTI, Inc.	Specialized video equipment	Michigan City, Indiana, U.S.A.	Tel: 1-219-872-5514, fax: 1-219-872-5559, http://www.dagemti.com
Leaf Systems, Inc.	Digital printing and video-editing products	East Rutherford, New Jersey, U.S.A.	Tel: 1-201-507-9000, fax: 1-201-507-8448, http://www.scitex.com
Optronics Engineering	High resolution true color cameras	Goleta, California, U.S.A.	Tel: 1-805-968-3568, fax: 1-805-968-0933, http://www.optronics.com
Polaroid Corp.	Instant cameras, digital products	Cambridge, Massachusetts, U.S.A.	Tel: 1-972-401-4014, fax: 1-972-401-4025, http://www.polaroid.com
Photometrics, Ltd.	Scientific imaging products	Tucson, Arizona, U.S.A.	Tel: 1-520-889-9933, fax: 1-520-573-1944, http://www.photomet.com
Princeton Instruments, Inc.	High performance camera systems	Trenton, New Jersey, U.S.A.	Tel: 1-609-587-9797, fax: 1-609-587-1970, http://www.prinst.com
Sony Electronics, Inc.	Audio and video products	Montvale, New Jersey, U.S.A.	Tel: 1-800-35-7669, 1-941-768-7669, fax: 1-941-768-7790, http://www.sony.com

in the DSC and TGA curves may not be clearly understood, additional analysis of the material utilizing optical microscopy, followed by use of HSM, may be employed. A well-trained optical microscopist can generate a great deal of data using a small quantity of sample. The information collected during this visual analysis is extremely valuable in confirming the physical properties of materials.

Because the specifics of an optical and HSM analysis of a material may vary from laboratory to laboratory, an optical characterization example is presented. This may be perceived as a two-step characterization: the first at room temperature, and the second as a function of temperature. The room temperature analysis of the material may be conducted as the material is being prepared for HSM analysis. Once the submilligram quantity of material is loaded into the sample cell and the sample cell is placed into the hot-stage accessory, the hot-stage instrument may then be positioned on the microscope stage, permitting room temperature characterization to commence. Because many microscopes often have variable light conditions, polarizing filters, and objectives of several magnifications, the particles can be analyzed under these different conditions to provide general particle properties. Changes in birefringence can be noted upon observation of the particles under cross-polarizing filters, using a first-order (Red I) compensator, and rotating the microscope stage. Information on particle size, particle uniformity, and homogeneity may be collected.

TABLE 7.6
Color Printer Vendor Directory

Vendor	Products	Location	Contact Information
Alps Electric (USA)	Thermal wax and dye-sublimation printers	San Jose, California, U.S.A.	Tel: 1-800-825-2577, 1-408-432-6000, fax: 1-408-432-6030, http://www.alpsusa.com
Eastman Kodak Co.	Professional and consumer imaging products	Rochester, New York, U.S.A.	Tel: 1-800-235-6325, http://www.kodak.com
Epson, Inc.	Printers, scanners, and cameras	Torrance, California, U.S.A.	Tel: 1-310-782-0770, 1-800-GO-EPSON, http://www.epson.com
Fargo Electronics, Inc.	Sublimation and thermal ink-jet color printers	Eden Prairie, Minnesota, U.S.A.	Tel: 1-612-941-9470, fax: 1-612-941-7836, http://www.fargo.com
Fuji Photo Film (USA), Inc.	Digital imaging equipment	Chicago, Illinois, U.S.A.	Tel: 1-630-773-7200, fax: 1-630-773-7999, http://www.ffei.co.uk
Tektronix, Inc.	Color printers and video networking products	Wilsonville, Oregon, U.S.A.	Tel: 1-800-835-6100, ext. 111, 1-503-682-7377, fax: 1-503-682-2980, htttp://www.tek.com

7.5.4 CHARACTERIZATION OF MATERIAL USING HSM

Once this initial characterization has been completed, continuation of the microscopic analysis using the hot-stage accessory may proceed. As an initial analysis, the ramp rate utilized for the DSC experiment should also be used for the hot-stage analysis. Use of a consistent ramp rate permits direct comparison of the data previously collected by DSC and TGA. If transitions are observed in the thermal data up to 300°C, the hot-stage experiment should also be run to that temperature. Ultimately, the assay should be conducted to generate confirmatory data on all transitions of interest. If available, the color camera should be utilized so that images may be collected as documentation of the transitions observed. Once the experiment is completed, the analyst may be able to compare the DSC, TGA, XRD, optical, and HSM data and develop a comprehensive characterization of the material.

In addition to the hot-stage analysis of the material under a simple heating rate, other hot-stage experiments can also be conducted to gather information. An isothermal experiment could be conducted whereby once the material is molten, the temperature can be held isothermally to facilitate recrystallization. Similarly, if the hot-stage controller has the option to cool the stage, this feature may be employed to cool the material under controlled conditions to initiate recrystallization. If this feature is not available, the heating profile may simply be aborted and the hot stage permitted to cool to room temperature.

Other hot-stage assays that may be conducted include volatilization studies, in which confirmation of volatilization may be achieved by placing a drop of oil on the material under investigation and running the experiment up to the desired temperature. A transparent oil such as mineral oil may be used; however, the analyst should first ensure that the material is insoluble in the oil. As the temperature during the hot-stage experiment reaches the volatilization point (as determined from the onset of the weight loss in the TGA curve), the analyst may observe the evolving gases from volatilization or decomposition forming bubbles in the oil as they evolve from the solid material surrounded by the oil.

Additionally, fusion methods can be conducted for further characterization of these crystal forms. Examples of the application of fusion assays to organic and pharmaceutical materials are available (70–73).

7.5.5 CHARACTERIZATION OF VOLATILES

Follow-up characterization of the volatiles initially analyzed by TGA could also be conducted. Confirmation of the volatiles may be accomplished using one of several techniques: thermogravimetric-infrared (TG-IR) spectroscopy, gas chromatography (GC), or thermogravimetry-mass spectrometry (TG-MS). These techniques may be able to qualitatively and quantitatively determine the content and identification of the solvents present in the material. Additionally, Karl Fisher titrimetric assays may be utilized to quantitate the water content in the material.

7.5.6 ADDITIONAL CHARACTERIZATION

Additional assays may also be conducted on this material to more comprehensively characterize its thermal properties. The following may be considered: variable temperature XRD (VT-XRD) studies may be conducted to observe changes in the powder diffraction patterns of the material as a function of temperature. Selection of appropriate temperatures depends on several factors: the DSC transition temperatures, corroborating TGA and hot-stage evidence, and the accuracy of the heating controller for the XRD unit. It should also be noted that although XRD data may be collected at several different temperatures to simulate a variable temperature experiment, the follow-up VT-XRD assay is essentially a series of isothermal experiments, and the DSC is a dynamic assay. It is therefore reasonable to expect some variation in transition temperatures.

Quench-cooling DSC studies can also be conducted to generate physical property information. In this method, the material is heated to a desired temperature, removed from the DSC cell and quench-cooled by placing the DSC pans onto the laboratory bench or a chilled surface, possibly forcing the material to convert to a nonstandard state. Upon cooling, the samples would be reanalyzed to determine if any nonstandard states were generated.

NMR, IR, and Raman spectroscopy may also be used to identify and characterize the physical properties of materials.

When samples are assayed using this multidisciplinary approach, HSM, as with any other technique, should be viewed as a tool in an analysis regimen to aid in the characterization of materials, and to provide a valuable problem-solving capability for the laboratory. HSM can play a unique role in the comprehensive characterization of materials by providing confirmatory visual and thermal analysis information of the properties of the material.

7.6 PHARMACEUTICAL APPLICATIONS

The importance of a visual technique, such as HSM, lies in its ability to provide information for the confirmation of transitions that are observed using other techniques. As stated earlier, the characterization of the physical properties of many materials, including pharmaceuticals, requires a multidisciplinary approach in which a number of techniques are utilized to confirm suspect transitions. To illustrate the need for this characterization philosophy, several examples, along with numerous photomicrographs showing the benefit that a visual technique such as HSM can provide, are presented.

7.6.1 Melt Responses

Compound 1 is a drug substance under development. The DSC curve of this material, shown in Figure 7.10, exhibits an endotherm whose maximum is at 55.1°C, and the TGA curve shows a negligible weight loss at 75°C. One may assume that the endotherm may be attributable to the melt of the material, because no weight loss is associated with the endotherm. However, because there is no data to support this assumption, incorrect conclusions regarding the nature of the material may be drawn.

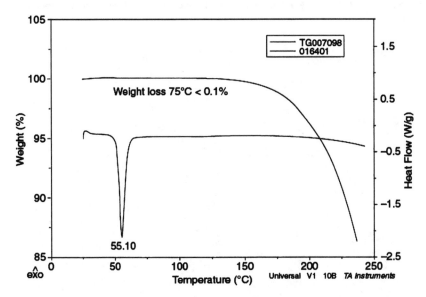

FIGURE 7.10 Compound 1 overlay of the DSC curve (bottom) exhibiting endothermic maximum at 55.1°C, and the TGA curve (top) exhibiting a <0.1% weight loss at 75°C.

HSM analysis of this material, as shown in Figure 7.11 and Figure 7.12, collected at 200× magnification using crossed polarizers with the Red I (first-order) compensator, indicates that the material melts between 45 and 52°C, indicating that the endotherm observed near 55°C is due to the melting of the material. This example shows the ability of a visual thermal analysis technique to assist in confirming melt transitions observed in DSC data.

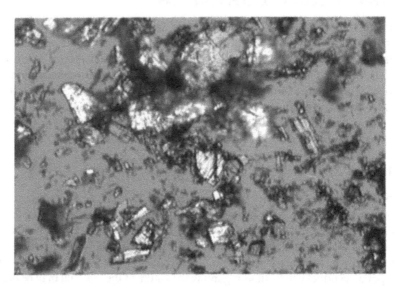

FIGURE 7.11 Photomicrograph of compound 1 collected at 25°C showing the material in a solid state.

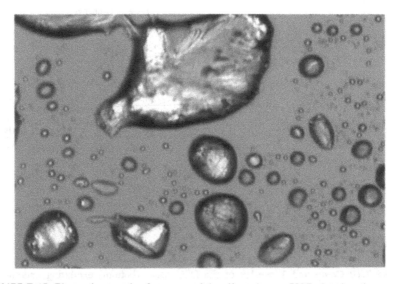

FIGURE 7.12 Photomicrograph of compound 1 collected near 50°C showing the material in a semimolten state.

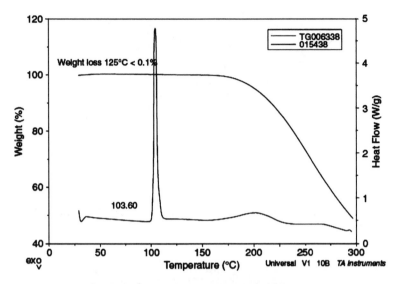

FIGURE 7.13 Compound 2 overlay of DSC curve (bottom) exhibiting endothermic transition at 103.6°C, and TGA curve (top) exhibiting a <0.1% weight loss at 125°C.

7.6.2 RECRYSTALLIZATION UPON COOLING

Although the hot stage is an excellent technique for evaluating materials exhibiting melt transitions, it can also be utilized in the characterization of a number of other systems, such as the investigation of recrystallization phenomena during the cooling profile following a melt; recrystallization studies from one crystal form to another during the initial heating program; systems in which the recrystallization of a different particle morphology is observed following a melt with decomposition transition; and for the observation of solid-state transformations. Examples of these systems will be presented with accompanying photomicrographs, thermal, and XRD data.

Compound 2 is a drug substance generated during manufacturing scaleup. This material was analyzed using XRD, DSC, TGA, and HSM. Figure 7.13 presents the thermal data that shows the DSC curve exhibiting an endothermic transition whose maximum is at 103.6°C, and the TGA curve having a weight loss of < 0.1% at 125°C.

Photomicrographs collected during the HSM experiment presented in Figure 7.14 to Figure 7.17 (collected at 400× using crossed polarizers) show the thermal behavior of this compound as the material was heated from room temperature past the melting point. The hot-stage data was used to confirm that the endothermic transition observed at 103.6°C in the DSC curve was a melt. Following completion of the melt, the sample began to cool as the heating program for the hot-stage accessory had been aborted. Upon continued cooling, the material began to recrystallize and the photomicrographs from this cooling cycle are presented in Figure 7.18 to Figure 7.21. Subsequent reheating of this recrystallized material indicated a melt near 100°C. Use of the image-recording portion of the hot-stage photomicroscopy system permitted collection of images representing recrystallization of this material upon cooling.

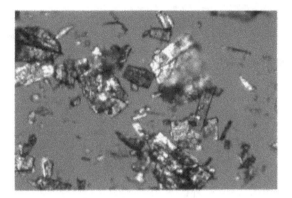

FIGURE 7.14 Photomicrograph of compound 2 collected near 25°C showing the material in a solid state.

FIGURE 7.15 Photomicrograph of compound 2 collected near 90°C showing the melt onset.

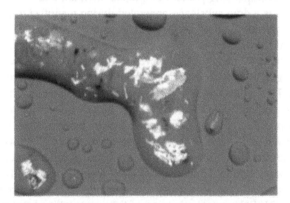

FIGURE 7.16 Photomicrograph of compound 2 collected near 102°C showing the material near completion of the melt.

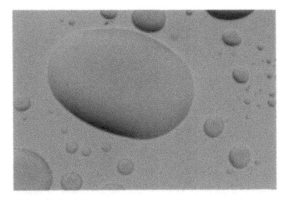

FIGURE 7.17 Photomicrograph of compound 2 collected near 104°C showing the material in a completely molten state.

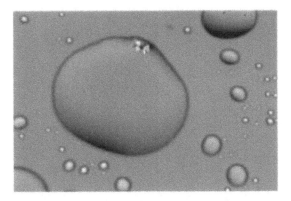

FIGURE 7.18 Photomicrograph of compound 2 collected during cooling at onset of recrystallization.

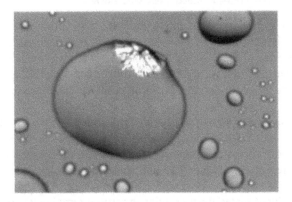

FIGURE 7.19 Photomicrograph of compound 2 collected during cooling as recrystallization continues.

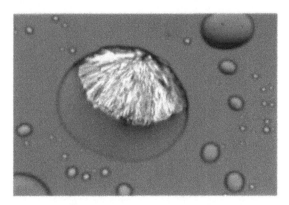

FIGURE 7.20 Photomicrograph of compound 2 collected during cooling as the material nears completion of the recrystallization.

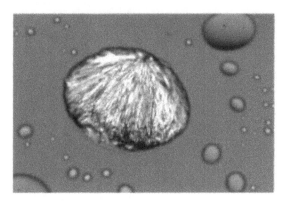

FIGURE 7.21 Photomicrograph of compound 2 collected at completion of recrystallization.

7.6.3 Recrystallization during Heating Profile

Compound 3 is a material that exhibits multiple crystal form changes at elevated temperatures. Thermal data from single crystals of this material are presented in Figure 7.22 (overlay of the DSC and TGA curves) overlay of the DSC curve, which exhibits endothermic maxima at 48.8, 86.4, 170.1, and 190.4°C, exothermic maxima at 119.4 and 173.1°C, and the TGA curve, which exhibits a 6.9% weight loss at 100°C.

This compound was known to have multiple crystal forms based on the DSC, hot-stage, XRD, and VT-XRD data, collected from previous samples analyzed over several months of process development. An overlay of the VT-XRD data collected on this lot is presented in Figure 7.23 and shows the powder patterns collected at initial, 45, 65, 75, and 160°C. The powder pattern at the top of the figure was collected from a standard lot of material representing the crystal form known to exist at 175°C. The data indicates that as the temperature is raised, the crystal form changes. Although the VT-XRD data clearly indicated that a change in crystal form was occurring as the temperature was increasing, confirmation of the crystal form changes

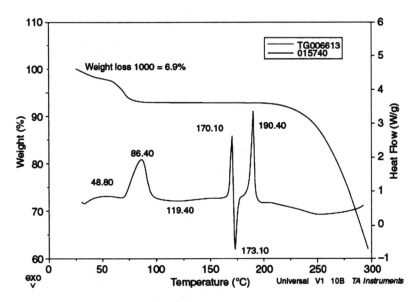

FIGURE 7.22 Compound 3 overlay of DSC curve (bottom) exhibiting endothermic maxima at 48.8°C, 86.4°C, 170.1°C, and 190.4°C, and exothermic maxima at 119.4°C and 173.1°C; and TGA curve (top) exhibiting a 6.9% weight loss at 100°C.

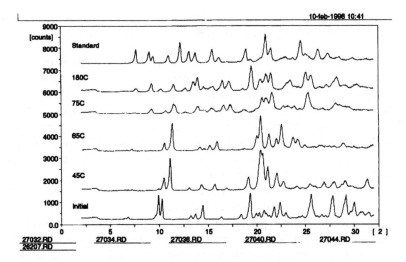

FIGURE 7.23 Overlay of compound 3 VT-XRD powder diffraction patterns at initial (bottom), 45°C, 65°C, 75°C, and 160°C; and standard (top) indicating the different crystal forms as a function of temperature.

FIGURE 7.24 Photomicrograph of compound 3 collected at ambient temperature.

FIGURE 7.25 Photomicrograph of compound 3 collected near 45°C (at the onset of volatilization) showing the fracturing of transparent particles owing to volatile loss.

was required. Additionally, because the DSC curve exhibited multiple transitions, a more comprehensive understanding of the thermal transitions was required.

An HSM analysis of the compound 3 material would provide visual photomicrographs that would confirm whether the suspected transitions are based on the DSC, TGA, and VT-XRD data. Representative images from this hot-stage experiment (collected at 100× using crossed polarizers and the Red I compensator) are shown in Figures 7.24 to 7.31. Figures 7.24 to 7.26 are based on data collected at room temperature, near 45, and near 65°C, respectively, and show the changes in the single crystals as the volatiles evolve (through ~100°C). The appearance of striations on the faces of some of the particles in Figure 7.25, and the change from a transparent to an opaque appearance, as shown in Figure 7.26, illustrate these differences. These images not only corroborate the weight loss observed in the TGA curve and the two volatilization endotherms observed near 49 and 86°C in the DSC curve, but also confirm the changes observed in the XRD patterns from room temperature to 75°C. The solid-state transformation near 119°C was not clearly observed during this hot-stage experiment. Figure 7.27 shows the onset of melt with

FIGURE 7.26 Photomicrograph of compound 3 collected near 65°C showing the recrystallized opaque particles following volatilization.

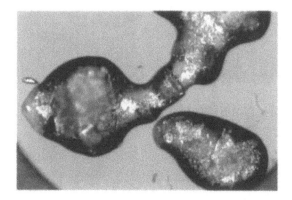

FIGURE 7.27 Photomicrograph of compound 3 collected near 160°C showing the onset of melt.

the nearly simultaneous recrystallization onset near 160°C. An XRD pattern of the material just prior to melting is shown at 160°C in the overlay of XRD patterns. Figure 7.28 and Figure 7.29 show the progression of the recrystallization to approximately 175°C, whereas Figure 7.30 and Figure 7.31 show the melt progression of the recrystallized material, ultimately producing molten pools of material.

7.6.4 Recrystallization into a Different Particle Morphology

The next example, compound 4, is a precipitate formed during a drug substance solution solubility experiment. This material exhibited physical properties that were not similar to the drug substance. Thermal analysis of this material generated the data presented in Figure 7.32, which shows the material exhibiting endothermic transitions at 68.1 and 267.7°C, and an exothermic transition at 220.5°C, and a 3.9% weight loss measured at 150°C was observed in the TGA curve. This material

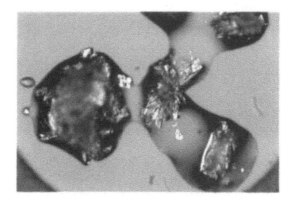

FIGURE 7.28 Photomicrograph of compound 3 collected near 170°C at the recrystallization onset.

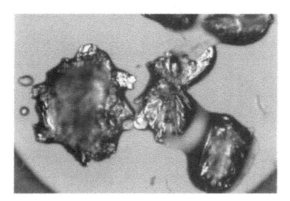

FIGURE 7.29 Photomicrograph of compound 3 collected near 175°C showing the material in a recrystallized state.

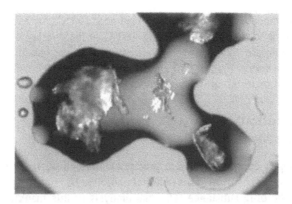

FIGURE 7.30 Photomicrograph of compound 3 collected near 180°C as the recrystallized material is shown melting.

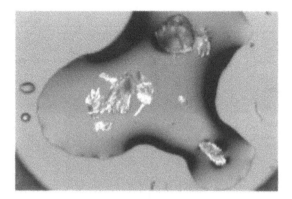

FIGURE 7.31 Photomicrograph of compound 3 collected near 190°C showing the material completing the melt.

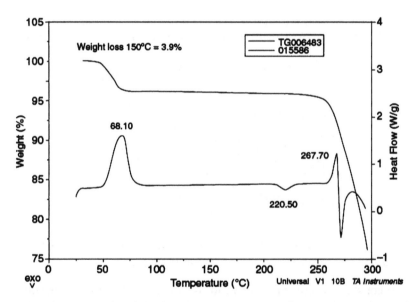

FIGURE 7.32 Compound 4 overlay of DSC curve (bottom) exhibiting endothermic maxima at 68.1°C and 267.7°C, and exothermic maximum at 220.5°C; and TG curve (top) exhibiting a 3.9% weight loss at 150°C.

produced the x-ray powder pattern (bottom) presented in Figure 7.33. Variable-temperature x-ray powder diffraction analysis was subsequently conducted on this material and is also presented in Figure 7.33.

Because no additional physical characterization information was available on this material, HSM was employed to assist in identifying the transitions observed in the DSC curve. Figure 7.34 to Figure 7.36 are representative photomicrographs (collected at 400× with slightly uncrossed polarizers) from the hot-stage experiments that support the data collected from DSC, TGA, and XRD assays.

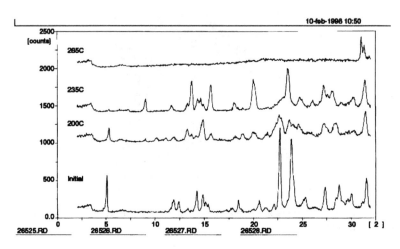

FIGURE 7.33 Compound 4 overlay of VT-XRD powder diffraction patterns at initial (bottom), 200°C, 235°C, and 265°C (top) showing the recrystallization between 200°C and 235°C.

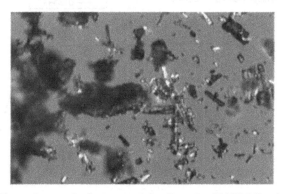

FIGURE 7.34 Photomicrograph of compound 4 collected at 110°C following the solid-state transformation due to the loss of volatiles in columnar morphology.

The hot-stage experiment was able to confirm that following volatilization near 68°C (as observed in the TGA curve), the material exhibited a solid–solid transformation as shown by the XRD powder pattern at 200°C. Between 215 and 230°C, the material recrystallized from columnar and acicular aggregates (Figure 7.34) into spherulite particles (Figure 7.35). The different XRD powder patterns observed at 235°C also confirms this crystal form change. The material then melted with decomposition, as shown by Figure 7.36, and the flat powder pattern at 265°C (top) in Figure 7.33. Again, utilization of the hot-stage system was able to provide corroborating information on the physical properties of this precipitate.

7.6.5 SOLID-STATE TRANSFORMATIONS

Compound 5 provides an example of a material exhibiting several solid-state transformations. Figure 7.37 presents an overlay of the DSC and TGA data for compound 5.

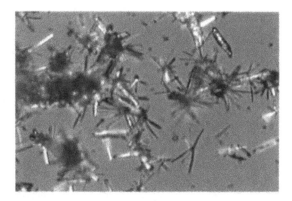

FIGURE 7.35 Photomicrograph of compound 4 collected at 230°C showing the recrystallized material in spherulite morphology.

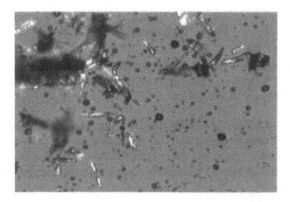

FIGURE 7.36 Photomicrograph of compound 4 collected at 260°C showing the melt and decomposition onset of the recrystallized material.

An endothermic transition was observed at 134.6°C, followed by an endothermic–exothermic combination beginning near 181.6°C, and an additional endothermic transition at 263.5°C. The TGA curve exhibited a 7.8% weight loss measured at 150°C, which correlated with the 134.6°C endotherm observed in the DSC. A large weight loss near 250°C confirmed the 263.5°C endotherm as a decomposition. The logical conclusion that may be drawn from this preliminary data set would be that the material possibly melted and volatilized at 134.6°C, possibly melted and recrystallized at 181.6°C, and then melted with decomposition at 263.5°C. However, no visual confirmation of these transitions was available.

Figure 7.38 to Figure 7.40 present photomicrographs that were collected at 30, 135, and 190°C (at 200× using cross polarizers with the Red I compensator). These hot-stage images, along with the DSC and TGA data, confirm the solid-state transformations. As compound 5 was heated during the hot-stage experiment, it lost volatiles and exhibited a solid-state transformation near 134.6°C (as shown in Figure 7.39), because the particles did not melt but became fractured. The material

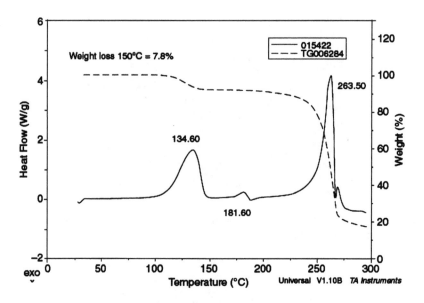

FIGURE 7.37 Compound 5 overlay of DSC curve (bottom) exhibiting endotherm maxima at 134.6 and 263.5°C, and an endotherm–exotherm combination near 181.6°C, and TGA curve (top) exhibiting a 7.8% weight loss at 150°C.

exhibited a second solid-state transformation near 181.6°C (Figure 7.40), as the particles became less birefringent. The material then began to decompose near 190°C. The important point that this example demonstrates is that without visual confirmation of the suspected transitions observed in the thermal data, inaccurate conclusions regarding the physical properties of the material may be drawn; i.e., that no melt occurred.

7.6.6 OTHER PHARMACEUTICAL STUDIES

In addition to the applications of HSM in the study of pharmaceuticals presented, thermal microscopy studies are being conducted at other research facilities. An in-depth survey of the stability behavior of the polymorphs of piracetam was recently conducted (74) by Kuhnert-Brandstätter. A variety of techniques including DSC, IR, and true density, as well as stability testing under humidity and light conditions, were utilized for the characterization of these crystal forms. It is important to note that although this work discusses the crystal forms of a pharmaceutical compound, a large part focuses on the determination of the most stable form and the thermodynamic system that exists between the different forms (enantiotropic or monotropic). Burger and Ramberger's rules for the determination of the most stable crystal form (75) are applied, and discussion of the relative stabilities of the modifications are discussed.

Thermal analysis and infrared spectroscopy are applied to a number of pharmaceutical polymorphic systems (76) in another paper from Kuhnert-Brandstätter. In addition to the characterization of these systems by these two techniques, discussion is provided on the preparation of the modification from either various solvent systems or

FIGURE 7.38 Photomicrograph of compound 5 collected at 30°C showing the material in a solid state.

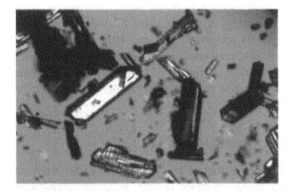

FIGURE 7.39 Photomicrograph of compound 5 collected at 135°C showing the fracturing of the solid material following a volatilization and solid-state transformation.

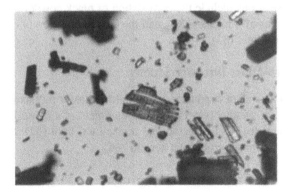

FIGURE 7.40 Photomicrograph of compound 5 collected at 190°C showing the solid material following a second solid-state transformation.

heating and cooling techniques. Details of their characterization of both stable and metastable forms are also provided.

Additional work being conducted in pharmaceutical groups includes chemical microscopical characterization of dilituric acid and its complexes (77), characterization of the physicochemical properties of pharmaceuticals (78), and solid-state investigations incorporating hot-stage microspectroscopy (79,80).

7.6.7 APPLICATION CONCLUSIONS

These application examples, in conjunction with the DSC, TGA, XRD, and hot-stage photomicrographs, hopefully illustrate the theme that no single analytical technique can completely characterize a material. The utilization of multiple techniques is required to conduct a comprehensive characterization of the physical properties of materials, because each technique provides a particular piece of information that is necessary for the characterization of unknowns, process materials, compounds under development, etc.

Additional techniques such as FT-IR microspectroscopy, IR and Raman spectroscopy, NMR, and energy dispersive x-ray, in conjunction with scanning electron microscopy, inductively coupled plasma, etc., may also be utilized to provide additional pieces of information toward the comprehensive analysis of materials and the identification of unknowns. Although HSM may not be a technique that all laboratories require, it is clear that the technique can provide valuable information for the visual confirmation of thermal transitions.

7.7 ADDITIONAL APPLICATIONS

In addition to the applications of HSM detailed earlier in this chapter, there is a great deal of work being conducted in other industrial laboratories using the many variations of thermal microscopy.

7.7.1 PETROCHEMICAL STUDIES

Other HSM applications include the combustion of black powder (81), low-temperature phase changes of n-alkanes (82), and the characterization of crude oils treated with rheology modifiers (83).

7.7.2 REFLECTED-LIGHT INTENSITY STUDIES

Significant work has been conducted using light intensity measurements: studying the birefringence of photopolymers (84), development of a controlled-rate thermo-microscopy system for both reflected and transmitted light (85), and the study of inorganic materials for a pyrotechnic system (86).

7.7.3 MODULATED TEMPERATURE DSC STUDIES

Recent developments combining modulated temperature differential scanning calorimetry with atomic force microscopy have developed a technique called *scanning thermal probe microscopy* (87) that permits the collection of localized atomic thermal microscopy data. An application has been conducted for the study of phase changes in polymer blends (88). A related application in which crystals of an explosive were heated and studied using atomic force microscopy has also been conducted to show defects on the molecular and nanometer scale (89).

7.7.4 MICROSCOPY STUDIES

A nondestructive thermoacoustic material signature (TAMS) imaging method in scanning acoustic microscopy (SAM) was recently utilized to study the matrix and tempers signatures of ancient ceramics (90).

The advances in imaging technology have also been applied to cryomicroscopy, in which a simultaneous DSC and optical video instrument has been developed for the characterization of ice crystallization (91).

The hot stage has not only been applied to optical and atomic force microscopes, but also to scanning electron microscopes. Hot-stage accessories are available on environmental SEMs that can collect ESEM images at elevated temperatures. Applications to the electronics industry include fluxless soldering, intermetallic growth studies, and copper thick-film sintering studies (92–94).

7.7.5 SPECTROSCOPIC STUDIES

Additionally, simultaneous hot-stage DSC/FTIR studies have been conducted to investigate the different cure characteristics of polymeric materials (95). Using this simultaneous technique, information of the thermal, spectroscopic, and visual properties of the material were collected.

7.8 CONCLUSIONS

The application of new technologies to an established analytical technique is an interesting and challenging endeavor. Although thermal microscopy is a technique that has been used for decades for the characterization of materials, new technological developments have essentially reinvented this technique. High-resolution cameras, digital image manipulation and storage, and updated hot-stage accessories have transformed this mature technique into a new research tool with applications in many industries.

The authors have attempted to present a "state-of-the-technique" perspective for HSM in which a wide variety of information regarding this technique has been collected. Although there are applications to industries that were not addressed, it was our intent to present to the reader as much of the available information on the technique that was available. It is our hope that this chapter on thermal microscopy enables other scientists to apply HSM to applications in their respective industries and solve new research, development, and manufacturing problems.

ACKNOWLEDGMENTS

The authors would like to acknowledge the following persons for their direct and indirect contributions to the development of this chapter:

Neal Barlow, Scott Berman, Dave Bugay, Martha Davidovich, John DiMarco, Robert Dubner, Paul Eggermann, Paul Findlay, Jack Gougoutas, Chris Kiesnowski, Berry Kline, Mary Malley, Walter McCrone, Alice McKee, Ken Morris, Ronald Mueller, William, Juliana and Olivia Newman, Sarah Nicholson, Tom Raglione, Chris Rodriguez, Charlee Sevenski, John Steel, Dave Stoney, Arne Swensen, Sandy, Kaitlyn, Zachary and Samuel Vitez, and Glen Young.

REFERENCES

1. Hill, J., Ed., *For Better Thermal Analysis and Calorimetry*, 3rd ed., ICTA, Victoria, Australia, 1991, p. 11.
2. Raspail, F.V., Nouveau Systéme de Chemie Organique Fondù sur des Méthodes Nouvelles d'Observation, Paris, 1833.
3. McCrone, W.C., *Fusion Methods in Chemical Microscopy*, Interscience Publishers, New York, 1957, pp. 1–5, available from McCrone Research Associates, Chicago, IL.
4. Lehman, O., Die Kristallanalyse, Leipzig: Wilhelm Engelmann, 1891.
5. Baumann, H.N., *Bull. Am. Ceram. Soc.*, **27**, 267, 1948.
6. Mason, C.W., *Handbook of Chemical Microscopy*, Vol. I, 4th ed., John Wiley & Sons, New York, 1983.
7. Welch, J.H., *J. Sci. Inst.*, **31**, 458, 1954.
8. Wendlandt, W., *Thermal Analysis*, John Wiley & Sons, New York, 1986, p. 584.
9. Smith, R.V., *American Laboratory*, September 1969, pp. 15–22.
10. Kofler, A. and Kofler, L., Thermo-Mikro-Methoden zur Kennzeichnung Organischer Stoffe und Stoffgemische, Innsbruck: Wagner, 1954.
11. McCrone Research Associates, 2820 S. Michigan Avenue, Chicago, IL.
12. McCrone, W.C., *Fusion Methods in Chemical Microscopy*, Interscience Publishers, New York, 1957, available from McCrone Research Associates, Chicago, IL.
13. Kolb, A.K., Lee, C.L., and Trail, R.M., *Anal. Chem.*, **39**, 1206, 1967.
14. McCrone, W.C., McCrone, L.B., and Delly, J.G., *Polarized Light Microscopy*, McCrone Research Institute, Chicago, IL, 1984, p. 198.
15. Hock, C.W., Arbogast, J.F., *Anal. Chem.*, **32**, 462, 1960.
16. Sommer, G., *Instrum. Technol.* (Southern Africa), **2**, 7, 1965.
17. Miller, R.P. and Sommer, G., *J. Sci. Inst.*, **43**, 293, 1966.
18. Charsely, E.L. and Kamp, A.C.F., A versatile hot stage microscope unit in *Thermal Analysis*, Vol. 1, Wiedemann H.G., Ed., Birkhauser Verlag, 1971, pp. 499–513.
19. Charsely, E.L. and Tolhurst, D., *Microscope*, **23**, 227, 1975.
20. Kuhnert-Brandstätter, M., *Thermomicroscopy in the Analysis of Pharmaceuticals*, Pergamon, Oxford, 1971.
21. Tynan, E.E. and Von Gutfeld, R.T., *Rev. Sci. Instrum.*, **46**, 569, 1975.
22. Schultze, D., *Thermochim. Acta*, **29**, 233, 1979.
23. Perron, W., Bayer, G., and Wiedemann, H.G., in *Thermal Analysis*, Vol. I., Wiedemann, H., Ed., Birkhauser Verlag, 1980, pp. 279–291.
24. Wiedemann, H.G., *Thermochim. Acta*, **85**, 271, 1985.
25. Wiedemann, H.G., *J. Therm. Anal.*, **40**, 1031, 1993.
26. Thermal Analysis Instrument Product Brochure, Shimadzu Scientific Instrument, Inc., Columbia, MD, 1997.
27. Magill, J.H., *Polymer*, **2**, 221, 1961.
28. Barrall, E.M. and Johnson, J.F., *Thermochim. Acta*, **5**, 41, 1972.

29. Haines, P.J. and Skinner, G.A., *Thermochim. Acta*, **134**, 201, 1988.
30. Richardson, P.H., Richards, R.W., Blundell, D.J., MacDonald, W.A., and Mills, P., *Polymer*, **36**, 3059, 1995.
31. Hess, S. and Eysel, W., *J. Therm. Anal.*, **35**, 627, 1989.
32. Kuhnert-Brandstätter, M., *Mikrochim. Acta [Wein]*, **3**, 247, 1990.
33. Burger, A., Henck, J., and Dunser, M., *Mikrochim. Acta*, **122**, 247, 1996.
34. Zhu, H., Padden, B.E., Munson, E.J., and Grant, D.J.W., *J. Pharm. Sci.*, **86**, 418, 1997.
35. Vitez, I.M., Newman, A.W., Davidovich, M., and Kiesnowski, C., *Proceedings from the 25th NATAS Conference*, McLean, VA, 1997, pp. 563–570.
36. New York Microscopical Society, 15 West 77th Street, New York, New York, 10024, U.S.A., Tel/fax: 718-595-1892.
37. Mettler FP900 Thermosystem Operating Instructions, Mettler Instrument Corp., Hightstown, NJ, pp. 5-3–5-7.
38. McCrone, W., personal communication, 1997.
39. Stoney, D., personal communication, 1997.
40. Sadr-Lahijany, M.R., Ray, P., and Stanley, H.E., *Phys. Rev. Lett.*, **79**, 3206, 1997.
41. Inoue, S. and Spring, K.R., *Video Microscopy*, 2nd ed., Plenum Press, New York, 1997, pp. 635–703.
42. Glossary of Microscopical Terms and Definitions, New York Microscopical Society, 1989, available from McCrone Research Associates, Chicago, IL.
43. Lynch, T.F., *American Laboratory*, November 1994, pp. 26–32.
44. Wunderlich, B., *Thermal Analysis*, Academic Press, San Diego, CA, 1990.
45. Haines, P.J., Ed., *Thermal Methods of Analysis*, Blackie Academic and Professional, London, 1995.
46. Speyer, R.F., *Thermal Analysis of Materials*, Marcel Dekker, New York, 1994.
47. Charsely, E.L. and Warrington, S.B., Eds., *Thermal Analysis-Techniques and Applications*, The Royal Society of Chemistry, Cambridge, 1992.
48. Cheronis, N., Ed., *Microchemical Techniques*, Interscience Publishers, New York, 1962.
49. Abramowitz, M., *Optics: A Primer*, Olympus America, Inc., Lake Success, New York, 1994.
50. Abramowitz, M., *Basics and Beyond Series: Microscope*, Vol. 1, Olympus America, Inc., Lake Success, New York, 1985.
51. Abramowitz, M., *Basics and Beyond Series: Contrast Methods in Microscopy*, Vol. 2, Olympus America, Inc., Lake Success, New York, 1985–1990.
52. Abramowitz, M., *Basics and Beyond Series: Reflected Light Microscopy: An Overview*, Vol. 3, Olympus America, Inc., Lake Success, New York, 1990.
53. Abramowitz, M., *Basics and Beyond Series: Fluorescence Microscopy*, Vol. 4, Olympus America, Inc., Lake Success, New York, 1985.
54. Kapitza, H.G., *Microscopy from the Beginning*, Carl Zeiss Company, 1994.
55. Schirmer, R.E., *Modern Methods of Pharmaceutical Analysis*, CRC Press, Boca Raton, FL, 1991.
56. Brittain, H.G., Ed., *Physical Characterization of Pharmaceutical Solids*, Marcel Dekker, New York, 1995.
57. Byrn, S.R., *Solid State Chemistry of Drugs*, Academic Press, New York, 1982.
58. Ford, J.L., Ed., *Thermochim. Acta*, **248**, 1995.
59. DiMarco, J., Gougoutas, J., and Malley, M., Single crystals, Bristol-Myers Squibb Crystallography Group, Lawrenceville, NJ, 1998.
60. Product Literature, Fisher Scientific, Springfield, NJ.
61. Thermal Devices Product Brochure, Electrothermal, Inc., Gillette, NJ.

62. 1986–1987 Product Catalog, Thomas Scientific Company, Swedesboro, NJ.
63. Rapid and Automatic Determination of Thermal Values Product Brochure, Mettler-Toledo Company, Hightstown, NJ.
64. Product Literature, Instec, Inc., Broomfield, CO.
65. Custom Stages Product Literature, Physitemp Instruments, Inc., Clifton, NJ, 1994.
66. Byrn, S., Pfeiffer, R., Ganey, M., Hoiberg, C., and Poochikian, G., *Pharm. Res.*, **12**, 945, 1995.
67. Reffner, J.A. and Ferrillo, R.G., *J. Therm. Anal.*, **34**, 19, 1988.
68. Barnes, A.F., Hardy, M.J., and Lever, T.J., *J. Therm. Anal.*, **40**, 499, 1993.
69. Brittain, H.G., Bogdanowich, S., Bugay, D.E., DeVincentis, J., Lewen, G., and Newman, A.W., *Pharm. Res.*, **8**, 963, 1991.
70. Haleblian, J. and McCrone, W., *J. Pharm. Sci.*, **58**, 911, 1969.
71. Goetz-Luthy, N., *J. Chem. Educ.*, **26**, 159, 1949.
72. Arceneaux, C.J., *Anal. Chem.*, **23**, 906, 1951.
73. Gilpin, V., *Anal. Chem.*, **23**, 365, 1951.
74. Kuhnert-Brandstätter, M., Burger, A., and Venklee, R., *Sci. Pharm.*, **62**, 307, 1994.
75. Burger, A. and Ramberger, R., *Mikrochim. Acta [Wein]*, **2**, 259, 1979.
76. Kuhnert-Brandstätter, M. and Sollinger, H., *Mikrochim. Acta [Wein]*, **3**, 247, 1990.
77. Brittain, H., *J. Pharm. Biomed. Anal.*, **15**, 1143, 1997.
78. Griesser, U., Burger, A., and Mereiter, K., *J. Pharm. Sci.*, **86**, 352, 1997.
79. Bartolomei, M., Ramusino, M.C., and Ghetti, P., *J. Pharm. Biomed. Anal.*, **15**, 1813, 1997.
80. Tsai, S., Keo, S., and Lin, S., *J. Pharm. Sci.*, **82**, 1250, 1993.
81. Hussain, G. and Rees, G.J., *Propellants Explos. Pyrotech.* **16**, 227, 1991.
82. Riga, A., Collins, R., Patterson, G., and Kahn, M., *Proceedings from the 25th NATAS Conference*, 1997, pp. 58–63.
83. Kok, M., Letoffe, J.M., Claudy, P., Martin, D., Garcin, M., and Volle, J.L., *J. Therm. Anal.*, **49**, 727, 1997.
84. Schultz, J.W., Pogue, R.T., Chartoff, R.P., and Ullett, J.S., *J. Therm. Anal.*, **49**,155, 1997.
85. Charsely, E.L., Stewart, C., Barnes, P.A., and Parkes, G.M.B., *Program Book for the 11th International Congress on Thermal Analysis and Calorimetry*, 1996, p. 198.
86. Brammer, A.J., Charsely, E.L., Griffiths, T.T., Rooney, J.J., and Warrington, S.B., *Program Book for the 11th International Congress on Thermal Analysis and Calorimetry*, 1996, p. 289.
87. Reading, M., Hourston, D., Pollock, H., Hamish, A., and Song, M., *Program Book for the 11th International Congress on Thermal Analysis and Calorimetry*, 1996, p. 200.
88. Leckenby, J.N., Mammiche, A., Pollock, H.M., Song, M., Reading, M., and Hourston, D., *Program Book for the 11th International Congress on Thermal Analysis and Calorimetry*, 1996, p. 47.
89. Sharma, J., Coffey, C.S., and Tompa, A.S., *Proceedings from the 25th NATAS Conference*, 1997, pp. 228–234.
90. Golan, G., Shoval, S., Pitt, C.W., and Beck, P., *Program Book for the 11th International Congress on Thermal Analysis and Calorimetry*, 1997, p. 322.
91. Hey, J.M., Mehl, P.M., and MacFarlane, D.R., *J. Therm. Anal.*, **49**, 991, 1997.
92. Koopman, N., *Micro Res. Technol.*, **25**, 493, 1993.
93. Kirchner, K.W., Lucey, G.K., and Geis, J., *Micro Res. Technol.*, **25**, 503, 1993.
94. Link, L.F., Gerristead, W.R., Jr., and Tamhankar, S., *Micro Res. Technol.*, **35**, 518, 1993.
95. Johnson, D.J., Compton, D.A.C., and Canale, P.L., *Thermochim. Acta*, **195**, 5, 1992.

8 Principles and Pharmaceutical Applications of Isothermal Microcalorimetry

Graham Buckton

CONTENTS

8.1   INTRODUCTION

Isothermal microcalorimetry is an exquisitely sensitive, noninvasive analytical technique. It is universally applicable in that almost every process involves a heat change. Thus, conceptually almost every process can be studied by isothermal microcalorimetry. In the following sections the use of this technique will be considered for biological, physical, and chemical processes.

8.2   INSTRUMENT DESIGN

Isothermal instruments are usually of two basic design types: adiabatic and heat conduction calorimeters.

8.2.1   HEAT CONDUCTION CALORIMETERS

Most instruments (CSC, Lindon, VT; Setaram, France; Thermometric, Sweden) work on the heat conduction principle. Here, the heat change in the reaction (endothermic or exothermic) causes a heat flow from or to a heat sink. The heat flow is measured by thermopiles that provide the signal. The flow of heat is to remove the temperature difference between the sample and the heat sink. Heat conduction calorimeters have a space for a sample, and as long as good thermal contact is made between whatever vessel the sample is in and the heat sink, any type of reaction vessel can be used. This gives options for batch, flow (gas of liquid flow), perfusion, and titration vessels. The design of the experiment is limited only by the space available and the imagination and skill of the researcher. As such it would be wrong to talk about heat conduction isothermal microcalorimeters as if they were just a single type of instrument. In reality they offer a whole host of experimental setups on a wide range of possible sample types. Many of these approaches will be reviewed later. A schematic representation of a heat flow microcalorimeter is shown in Figure 8.1.

The output of the calorimeter is a function of the enthalpy change of the process, the concentration reacting, the rate constant, and the order of reaction. The rate equations for a zero-order and a first-order process, and the calorimetric equivalent are shown in Table 8.1. The calorimetric outputs for a zero-order and a first-order process are shown in Figure 8.2. It can be seen that the first-order process has an exponential decay as the concentration remaining falls. The zero-order process is, however, a line parallel to the baseline. The magnitude of the displacement from the baseline depends on the amount reacting, the rate, and the enthalpy change. It is clear that a response that is just at the detection limit of the instrument will be a line just off baseline, and as such it may become rather difficult to distinguish between first-order and zero-order processes.

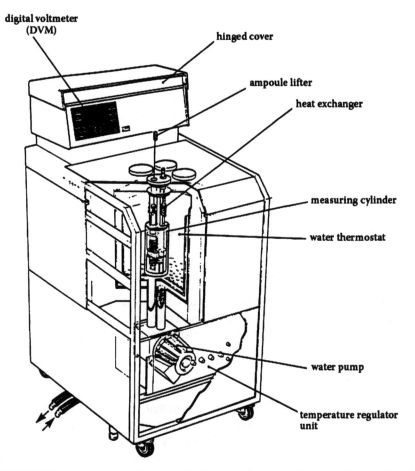

FIGURE 8.1 Schematic representation of a heat conduction isothermal microcalorimeter. (Reproduced from Thermometric Ltd. With permission.)

TABLE 8.1
Standard Rate Equations and Their
Calorimetric Form

Order	Rate Equation	Calorimetric Form
Zero	$dC/dt = -k_0$	$dq/dt = ?\ H(-k_0)C_0$
First	$dC/dt = -k_1 C_t$	$dq/dt = ?\ H(-k_1)C_t$

Note: dC/dt = rate of change of concentration with time (t), k_0 = zero-order rate constant, k_1 = first-order rate constant; ΔH = the enthalpy change; C_0 = concentration available to react at time zero; and C_t = concentration available to react at time t.

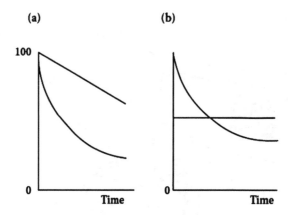

FIGURE 8.2 Schematic representation of a first-order and zero-order process as detected by an isothermal microcalorimeter.

8.2.2 ADIABATIC AND ISOPERIBOL CALORIMETERS

The most commonly used instrument of this type is the isoperibol solution and immersion calorimeter. Here, the heat change caused by a process is retained in the vessel, which is often a vacuum flask, and the temperature change inside the vessel is monitored.

8.2.3 A POTENTIAL PROBLEM

In the discussion that follows it will be shown that isothermal microcalorimeters can be used to detect many different types of processes. The combination of high sensitivity with the ability to detect almost any process makes the instrument a valuable tool. The potential problem, however, is the lack of specificity in many experiments. Consequently, it is quite possible to measure a process that is not the one that was the intended subject of the study. Even in the best experiments, it is extremely likely that one will measure more than just the one process that is the subject of the study. This "Catch-22" aspect of microcalorimetry must always be at the forefront of the researcher's mind, i.e., the ability to measure all things is an advantage, but the disadvantage is that it becomes hard to prove that the measured response was in fact due to the process that was the intended subject of the study.

The questions that should be kept in mind are: What are all the processes being measured? Which other techniques can be used to check what is happening?

8.3 BIOLOGICAL MICROCALORIMETRY

The origins of the isothermal microcalorimetry technique lie in studies on biological systems, ranging from whole animals through microorganisms to enzyme–substrate interactions. It is clear that certain biological experiments are well suited for study by microcalorimetry as they bring their own built-in specificity. This would be true of enzyme–substrate interactions, for example. However, even for specific interactions, other processes, such as dilution effects when the fluids are mixed, will still contribute to the measured response.

It has been shown that the actions of drugs on microorganisms can be studied using isothermal microcalorimetry (1,2), and that the data can be used to obtain structure–activity relationships for the drug substances. Studies on human cells indicate the prospects that isothermal microcalorimetry may be of value in diagnosis, or to give indications of prognosis, in non-Hodgkin's lymphoma patients (3). This is because the instrument is able to detect the metabolic activity of normal and cancerous cells.

Work in the biological and microbiological fields continues. A complexity in such systems is in dealing with the substantial biological variability that often distorts responses from a highly sensitive instrument. No further review of biological microcalorimetry will be given here, but the reader may wish to review three special issues of *Thermochimica Acta* that cover this field (4–6).

8.4 CHEMICAL REACTIONS

8.4.1 A Short Literature Review

The fact that isothermal microcalorimeters may well be able to detect the presence of slow reactions has caused considerable excitement in the pharmaceutical industry. There is a serious need for a rapid method by which the stability of drugs, and the compatibility of drug-excipient mixtures, can be assessed or estimated. To what extent can isothermal microcalorimetry meet that need?

There are only a few reports in the literature on the use of isothermal microcalorimetry for the stability testing of pharmaceuticals (although in other fields, such as explosives, there are many more references testifying to a good degree of success). These include the degradation of β-lactam antibiotics (7), the oxidation of lovastatin (8), the hydrolysis of meclofenoxate hydrochloride (9), and the oxidation of d,l-α-tocopherol (10).

Hansen et al. (8) noted that small containers must be nonreactive. They stated that water transfer through a seal at the rate of 10 μg/d equates to a measurable signal of 0.3? μW. They also noted that oxidation of freshly polished stainless steel and even the absorption and loss of volatiles from fingerprints or labels on the outside of the reaction vessel would be enough to give misleading results (when searching for the true heat flow response for a slow chemical degradation reaction). They argue that the best results (i.e., least chance of the cell making a contribution) are from perfectly clean flame-sealed all-glass containers.

Angberg et al. (10) studied the hydrolysis of acetyl saliclyic acid solutions, for which it was shown that elevated temperatures were needed to follow this rapid hydrolysis process. Thus, the isothermal microcalorimeter was neither more accurate, nor quicker or easier than using a conventional analytical approach, such as titration or chromatography, for this hydrolysis reaction.

Even from this limited data set, it can be seen that some processes appear to have been suitable for study by isothermal microcalorimetry (e.g., oxidation of lovastatin), whereas others (e.g., hydrolysis of aspirin) do not. It is worth considering why this is the case.

8.4.2 Theoretical Detection Limits

Assuming that just one reaction is proceeding in the calorimeter, it can undergo an enthalpy change ranging from close to zero (for example, some ester hydrolysis reactions) to several hundred kJ mol^{-1} (for example, oxidation reactions). It is also possible for rates of reaction to be extremely fast or very slow, and for huge variations in the quantity available for reaction to influence the measured response. Considering all the factors that could alter the measured response for a single reaction, there are wide limits of detection that can be considered possible.

8.4.2.1 The Best-Case Scenario

Wadso (11) has noted that for a reaction with an enthalpy change of 500 kJ mol^{-1}, the smallest detectable reaction would be 0.02% degradation per year. This is calculated on the basis that the molecular mass is 100 (any increase in the molecular mass would result in the detection limit decreasing by the same factor because of the lowering of the number of moles available to react), that 3 g of sample is used and that all of the material is available to react, that the detection limit for the calorimeter is 0.1 µW, and that the process is first order. These assumptions are reasonable as, for example, oxidation reactions can often have high enthalpy changes. It is clear that on this basis the isothermal microcalorimeter would easily be able to detect reactions, and should be able to predict the shelf life parameters for a product.

8.4.2.2 The Worst-Case Scenario

If the assumptions used earlier remain identical, except for the fact that the enthalpy change is as low as 10 kJ mol^{-1}, it would be possible to detect reactions that proceed at 1% degradation per year. An enthalpy change as low as this would be in keeping with values reported for ester hydrolysis reactions. At 1% per year, the instrument is now at the limit of what would be called a shelf life prediction. If, however, the concentration available for reaction was not all of the 3 g of material in the calorimeter cell, then the detection sensitivity would be lowered. For solution-state reactions, or for degradation of amorphous materials in the presence of absorbed moisture, it is probable that all, or most of the sample may well be available to react. However, for surface reactions on crystals, or for powder–powder compatibility studies (in which reactions will be at points of contact between two powders), there is a high probability that the amount reacting will be much lower than estimated earlier, and as such the detection sensitivity will be much lower. Say, for the sake of argument, that the surface of a powder amounted to 1% of its total mass; then a surface reaction (especially one that stopped after the surface had reacted) could have no more than 1% of the material reacting, and as such the detection limit for a hydrolysis reaction could reasonably rise to only be able to detect reactions that proceed at a rate of 100% degradation per year. If, however, the reaction had an enthalpy of 500 kJ mol^{-1}, a reaction at a surface that was 1% of the mass would give a lower detection limit compared to a reaction that degrades at 2% per year. As such, it is possible to imagine scenarios in which this approach could be successful and, equally, those for which it is liable to fail.

The greatest chance of success will be if the entire sample reacts. If only the surface of a crystal reacts, there is a clear advantage in having the largest possible surface area available. This would lead to the prospect of micronizing or milling samples, the disadvantage of which is that the milling process could well change the physical state, for example, making the material partially amorphous. As the reaction rate in an amorphous region may be different from that at the surface of a crystal, data from freshly milled powders could result in different observations from those seen with pure crystalline materials. For reactions that occur at point contacts between two powders, such as tests for excipient compatibility, it is clear that the concentration term will be greatly affected by the particles' sizes, shapes, and packing geometry. As such, the detected signal may be large for two materials that have a large surface contact (e.g., magnesium stearate adhering to a drug surface), but may be low if both materials have a large particle size. This change in measured response does not necessarily correlate with the degradation rate in a product where powder contact may change; for example, magnesium stearate added after granulation may have a more limited contact than when mixed with a fine drug powder, whereas the interaction between drug and diluent may increase as compaction can bring about much greater contact.

8.2.3 Multiple-Stage Reactions

Solid-state reactions, especially those between two solids (e.g., excipient compatibility) can be expected to have complex reaction mechanisms (they are not usually simple first-order processes). It is probable that the reactions will be mediated through the atmospheric conditions, by the presence of humidity, for example. Therefore, it would be essential for humidity to be controlled. It is also possible that rates and mechanisms could change at certain critical RH values when condensation between point contacts of powders, or deliquescence of one component, may begin. It is therefore advantageous to understand the major variables in the reaction and to control experiments accordingly. At present, there has not been much effort to define the most suitable conditions for screening unknown reactions (i.e., use in the preformulation arena). It becomes clear that use in the quality control area may be easier than in early-stage research applications, as by this stage much will be known about degradation mechanisms and kinetics; thus, experimental conditions can be selected with ease. It is possible that on many occasions a microcalorimetry experiment will be an easier quality control test than certain other complex analytical approaches.

This discussion has been centered on estimates based on one reaction process i.e., a reaction of the form

$$A + B \rightarrow C + D$$

It is quite common for degradation processes to be more complicated than this, and to have several intermediate stages. Thus, at any one time there may be a process such as A reacting with oxygen, to give B + C, while B starts reacting with water to give D + E, etc. There will be occasions when more than two reactions cascade. Each reaction will have its own enthalpy change, rate, reacting concentration, and reaction order. Under different environmental conditions different stages in the process may be rate limiting. Clearly, all processes which are above the detection limit of the calorimeter (see earlier case) will contribute to the response. This can cause complexity as it is possible (although not necessarily probable) for some of the intermediate stages to have endothermic reactions that may mask some exothermic response; in any case it makes it difficult to interpret the calorimetric response. Thus, complex processes in isothermal microcalorimetry experiments may be viewed as a bad thing, if the aim is simply to understand if a combination of materials is stable (a quick preformulation experiment). However, it may be argued that isothermal microcalorimetry is uniquely suited to exploration of complex degradation pathways. Recently, Gaisford (12) has shown that calorimetric data can be curve-fitted to models for different reaction mechanism sequences, which may illuminate the progress of complex reactions (especially in the solution state; see previous discussion on detection limits).

Willson et al. (13) and Angberg et al. (14) have used isothermal microcalorimetry to study the oxidation of a solution of ascorbic acid; however, they reached different conclusions. Willson et al. (13) claim that a significant difference exists because they use an ampoule in the calorimeter that is almost full of ascorbic acid solution, whereas Angberg et al. (14) have a considerable head space above their ascorbic acid solution. Willson et al. (13) state that the measured response for ascorbic acid oxidation reported by Angberg et al. (14) was in fact related to the physical process of oxygen diffusion into the solution from the head space. This demonstrates the incredibly complex nature of isothermal microcalorimetry experiments, and shows that minor experimental changes can result in different measured responses and, thus, quite different conclusions. As has been emphasized in each section dealing with isothermal microcalorimetry, good experimental design is fundamental to success, and it may be difficult to distinguish between the response you wish to measure and the other responses that occur in the cell (of which one may initially have no knowledge).

Willson et al. (13) note that most analytical techniques (e.g., high-performance liquid chromatography [HPLC]) look specifically for the arrival of a single product of the reaction (or loss of a reactant), and they are therefore not necessarily of value in investigations of intermediates in the reaction process, but that isothermal microcalorimetry follows the entire process (see earlier text). By comparison of calorimetric data and analytical data, it was possible to propose a sequential reaction mechanism for ascorbic acid oxidation that could not be determined simply from analytical data.

It can be concluded that the ubiquitous presence of multiple physical and chemical reaction processes can make it more difficult to interpret isothermal microcalorimetry data for chemical degradation processes than one may imagine. Thus, there is a potential disadvantage for those who wish to use the calorimeter as a quick screening method. A real advantage is that there is a potential for using isothermal microcalorimetry with suitable theoretical models for sequential processes, which may allow mechanisms for reactions to be understood in greater detail than is possible from conventional chemical analysis.

8.2.4 CALCULATION OF RATE CONSTANTS AND THERMODYNAMIC PARAMETERS

A major breakthrough in accessing rate and thermodynamic data from calorimetric experiments has arisen as a consequence of the work of Willson et al. (15). They manipulated a number of reaction equations into calorimetric form and suggested a route by which it is possible to use the isothermal microcalorimetry output to calculate simultaneously the rate constant, the order of reaction, the enthalpy change, and the quantity of material reacting. Examples of the reactions described include a process where A goes to B, which is described by the equation

$$dq/dt = \Delta Hk(A - (q/\Delta H)^m \tag{8.1}$$

where dq/dt is the rate of change of heat with time, or the heat flow, which is the standard output of the calorimeter (i.e., a power–time relationship), ΔH is the enthalpy change, k is the rate constant, A is the amount reacting, q is the heat output in the defined time period, and m is the order of reaction. By entering the calorimetric data into a suitable computer program, the heat flow can be plotted as a function of the heat output (q) and the other terms estimated by a process of iteration. It has subsequently been realized that to get a reliable estimate of the enthalpy change and of the amount of material reacting, it is necessary to know one of these terms in advance. This is possible for a solution-state reaction (where the quantity reacting is known), but not for a solid-state reaction (for which the quantity reacting will depend on point contacts, and may not be the total mass available). Thus, for a solid-state reaction it is only possible to estimate the rate of reaction (unless either the enthalpy or quantity reacting are known); however, being able to determine the rate constant rapidly is valuable.

8.2.5 REALISTIC EXPECTATIONS: AN OPINION

There is no doubt that isothermal microcalorimetry has the potential to be very useful in studies of chemical degradation. The success will be greatest for solution-state reactions for which the reacting concentration is known, but even these can be limited by physical artifacts by, for example, the slow diffusion of oxygen into the liquid from the head space. The use of this technique for solid-state reactions is certainly not impossible, but is an area where great care is needed. The technique may therefore be used readily in a preformulation environment for solution systems, but is perhaps better applied in a QC (quality control) role for well-characterized solid-state processes.

Very serious consideration is needed to select suitable experimental protocols, especially with respect to the quantities of oxygen and humidity that are present (should they be renewable resources? If so, how are they to be added to the cell?), and with regard to powder–powder interactions (how should they be mixed and in what concentrations?).

There is a danger of excessive swings in the perception for these types of studies. If workers believe that isothermal microcalorimetry will be the simple and ultimate answer to stability- and compatibility-testing requirements, a great deal of excitement and expectation is generated. The problem comes when people suddenly realize that the isothermal microcalorimetry experiment is rather harder to set up and interpret than they had thought, and that it may not work in all cases. In such circumstances, the tendency is to give up and revert to the tried and trusted methods. This is probably too extreme a reaction, and it would be better if everyone realized both the great potential and the serious difficulties in the use of this technique. Isothermal microcalorimetry will be well suited to some situations and poorly suited to others. It is worth remembering that every worthwhile technique was harder to use than anticipated when it was first introduced. For example, HPLC is not a trivial technique to master; operators require skill and knowledge, and it took time to perfect its use in routine analysis. Every situation is more complicated than it seems when viewed in suitable depth. This does not justify turning away from the use of isothermal microcalorimetry, but rather it shows that it is necessary to develop the necessary understanding and expertise.

8.3 SOLID-STATE CHARACTERIZATION OF POWDERS

In the opinion of this author, isothermal microcalorimetry can be useful for the studies of biological systems and in chemical stability testing, whereas it is certainly useful in solid-state characterization of materials.

8.3.1 THE AMORPHOUS CONTENT OF "CRYSTALS"

In recent years there has been a growing realization that "crystalline" powders can have varying amounts of disorder in their structures. Amorphous regions in crystals are generally thermodynamically unstable, i.e., they are in a higher energy state than the crystalline form. Within amorphous regions there will be substantial absorption of water vapor, which can cause physical and chemical transitions. This means that the amorphous regions are reactive "hot spots" in which physical changes or chemical degradation can be initiated. The amorphous material is formed either during crystallization or subsequent processing. It is argued that the processing-induced disruption is mostly at the powder surface, and thus a small amount of disorder (by weight of the sample) can be a very substantial amount of the powder surface. If the powder surface is essentially all amorphous, then the interaction between the powder and other phases will be different from that observed when the material was in the crystalline state. Alterations to interfacial interactions can affect many aspects during the production, storage, and use of products. As processing-induced disruption is seldom a deliberate and controlled event, it tends to lead to batch-to-batch variations, and as the disruption is unstable, recovery to the crystalline form will result in a change in properties with time after processing. Overall, there is an urgent need to characterize materials for amorphous content, and to understand how this relates to the functionality of the material in the subsequent processes and in the product for which it is to be used.

8.3.2 Advantages of Microcalorimetry over DSC

Bulk analytical techniques, such as differential scanning calorimetry, powder x-ray diffraction, and IR spectra, will measure the properties of the sample as a whole. The detection limits for amorphous content with such techniques can vary, but will generally have a lower cutoff of 5 to 10% (16). This detection limit is because these techniques measure the entire sample; thus, the amorphous content becomes a small part of the total signal, and consequently it is difficult to detect with confidence. It may be more appropriate to investigate preferentially the properties of the powder surface only, where amorphous material may predominate. A powerful way of investigating surface properties is by use of vapor sorption studies, which coincidentally also preferentially probe amorphous (over crystalline) regions because of absorption behavior. Hancock and Zografi (17) describe water sorption studies as the "preferred means of studying pharmaceutical systems containing low levels [of amorphous material]."

8.3.3 The Amorphous-To-Crystalline Transition

For all materials the amorphous form is thermodynamically unstable. Any unstable system has to have a mechanism by which to transfer to its stable (lowest-energy) state. The activation barrier that needs to be passed to move to the stable state will determine whether the transition is spontaneous. The transition from the amorphous to the crystalline form will depend on the mobility of the molecules. When below the glass transition temperature (T_g), the molecules lack sufficient mobility to allow spontaneous crystallization (at least within the time scale of observation of *in situ* experiments). However, this does not mean that materials do not crystallize below T_g, it means that there is an issue of the duration over which the system is observed and the activation energy needed to allow crystallization to occur. For example, Yoshioka et al. (18) reported crystallization of indomethacin within a few weeks when stored at 30°C below T_g, but stability of in excess of a year when stored at about 50°C below T_g. Hancock et al. (19) calculated that for all significant (i.e., capable of affecting a shelf life) mobility to cease, indomethacin needed to be stored at a minimum temperature of 50°C below T_g.

At the point when T_g drops to the experimental temperature (T), most amorphous (glass) materials have a viscosity of 10^{12} to 10^{14} Pa sec (20). In reality, this viscosity is too high to allow crystallization to occur rapidly; thus, it is more usual to see crystallization when T_g has been reduced below the experimental temperature. The first observation for particulate materials as T_g drops below T will be a collapse in the structure of particulate samples. Shalaev and Franks (20) have described the softening temperature as being the point at which collapse of structures under gravity will occur within a short period of time. The collapse occurs (at a rapid rate) when the viscosity drops to about 10^8 Pa sec, which for carbohydrates is when $T_g/T = 0.9$. Subsequent to the collapse, the material will crystallize. Again, the rate of onset of crystallization depends on the extent to which T_g has been reduced below T.

The difference between T_g and T can be adjusted by either adding a plasticizer to lower T_g, or by heating the sample to raise T. Many amorphous structures absorb water, which then acts as a plasticizer to lower T_g. However, certain hydrophobic materials may require alternative small molecules to absorb and lower T_g (these could include the vapors of ethanol or chloroform, for example) (21). The use of absorbed vapors to lower T_g, and hence cause collapse and crystallization, will be discussed later. Collapse and crystallization caused by increase of T would be typical of what is observed when differential scanning calorimetry is used to study glass systems. Shalaev and Franks (20) have reported DSC data for freeze-dried sucrose showing the effects of thermocycling. Here, the sample of amorphous sucrose was heated and cooled through the temperature cycle 280-343-280-345-280-365-280-370 K. Two transitions are observed: T_1 (the glass transition) is reversible and is seen on each scan, whereas T_2 (collapse) is seen in the first scan to 365 K but is absent on the second scan to such temperatures. Shalaev and Franks (20) note that unlike the T_g, the collapse response in the DSC is not a true transition in the sense of a change in physical state; i.e., the collapse is not necessarily associated with a change in heat capacity of the sample, but more likely a consequence of the change in sample geometry and conductivity.

8.3.4 SEALED AMPOULES AT CONTROLLED RELATIVE HUMIDITY

Perhaps the simplest and most used calorimetric method of studying amorphous material in powders is to seal the powder in a glass ampoule with a small tube containing a saturated salt solution. The saturated salt solution will not make contact with the powder, but will yield a defined relative humidity (RH) in the airtight ampoule at any set temperature (22).

8.3.4.1 How to Interpret the Shape of the Response for the Sealed Ampoule Experiment

A typical response for amorphous lactose equilibrated at 25°C 75% RH (saturated sodium chloride solution) is shown in Figure 8.3. Figure 8.3 can be seen to have two distinct sections; there is an initial small protracted response followed by a large sharp peak. It is possible to delay the onset of the large sharp peak to a certain extent by either increasing the sample mass or by decreasing the RH (to a lower critical point, which is the RH that will allow sufficient absorption to cause crystallization in a measurable time). Aso et al. (22) have shown that the temperature and humidity used in the calorimetric experiment alter the rate of crystallization, and they concluded that this was due to changes in matrix viscosity affecting the molecular motion of the drug under study (nifedipine).

For samples with lower amorphous content then, the large sharp peak is often seen to split into two separate regions. There is considerable debate concerning the significance of the different sections of the response, and if logical decisions are to be made on how to interpret the data, it is important to consider the processes that occur.

The initial slow peak (Figure 8.3) is not crystallization, as it is possible to stop the experiment at the end of this phase and show that the sample is still amorphous

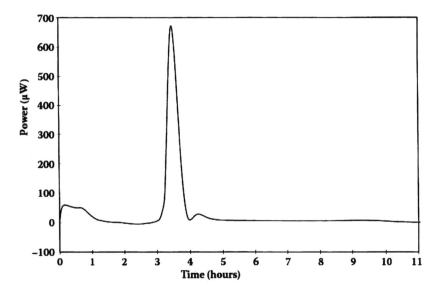

FIGURE 8.3 Typical crystallization response measured for an ampoule experiment in an isothermal microcalorimeter (75% RH; 25°C) for amorphous lactose.

(by x-ray diffraction and DSC). It has been argued (23) that this region represents a vapor-phase wetting of the powder (exotherm) minus a response for the generation of the humid air from the saturated salt solution (endotherm), such that the net response is small. It seems likely that this is too simplistic a description; Buckton and Darcy (24) have shown that collapse occurs during this early stage of the response. Indeed, a sample that had been collapsed (by exposure to 50% RH) did not show a response during this early period. Unfortunately, the process by which collapse was induced will result in retained absorbed water, and as such it cannot be proved as to whether this initial response is due to the collapse process, the water sorption, or (most probably) both. In conclusion, the initial response is a composite of wetting, collapse, and changes in the vapor space, and the saturated salt solution. Thus, an apparently simple thermal event is in fact a complex mixture of numerous different responses.

The large peak in Figure 8.3 includes the response for crystallization. This can be proved by showing that the sample is amorphous before, and crystalline after, this peak. It is noticeable that this peak is extremely sharp; i.e., a rapid rise in heat is followed by a rapid decline. This peak is made up of at least two processes: the crystallization exotherm and the water desorption endotherm; however, it is probable that there will be further contributions from within the salt solution reservoir, and potentially some condensation of the desorbed water in the ampoule.

The fact that the measured crystallization response is often split into two regions causes further confusion. For example, Sebhatu et al. (23) show a schematic version of the calorimetric response for the crystallization of lactose (Figure 8.4) and state that Part II is crystallization and Part III may be mutarotation from β to α-lactose. Therefore, they assess crystallinity by using the area under the curve for Part II,

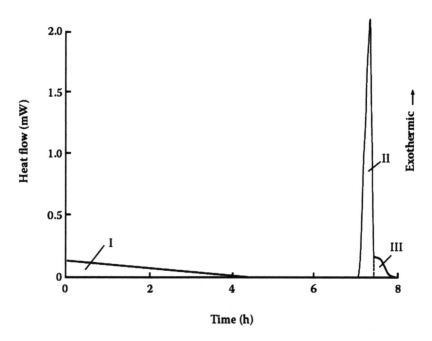

FIGURE 8.4 Schematic microcalorimetric heat flow curve for crystallization of amorphous lactose. (Reproduced from Sebhatu et al., 1994. With permission.)

which equates to about 32 J/g. However, Briggner et al. (25) and subsequent publications use the area under what would be Parts II and III to describe the heat change for the crystallization process. This gives a measured value of about 48 J/g for the crystallization of amorphous lactose. It should be noted that for samples that do not exhibit mutarotation, there is very clear evidence for two distinct peaks during crystallization. This is very obvious during the crystallization of salbutamol sulfate (26), where a distinct endotherm is seen in the first half (before the peak reaches a maximum) of the crystallization exotherm. Given that the contribution of mutarotation to the overall response is not yet proven, it cannot be certain as to which is the best way to treat the lactose data; however, the evidence from the salbutamol example makes it probable that Peaks II and III in Figure 8.4 are both part of the crystallization response.

Irrespective of whether the crystallization response should be taken as Parts II and III or just Part II for lactose (see Figure 8.4, and earlier discussion), it is clear that the crystallization response is reduced in size due to the endotherm for water desorption. Buckton and Darcy (27) have shown that the measured calorimetric crystallization peak, plus the enthalpy of vaporization for an estimate of the amount of water desorbed (i.e., the quantity of water needed to plasticize T_g below T minus the quantity of water that is retained as a hydrate) approximates to the combined endotherms for hydrate loss and melting, which are measured in a DSC for the crystallized lactose. This balance of heats would indicate that the contributions within the saturated salt solution are relatively small during the crystallization response. It

TABLE 8.2
Net Area under the Curve for the Crystallization of Amorphous Lactose at Different Temperatures and Humidities

Temperature	Sodium Chloride	Sodium Nitrite	Magnesium Nitrite
25	48.9	47.9	48.2
35	53.0	47.1	59.4
45	62.1	63.2	66.5

Note: Sodium chloride = 75% RH; sodium nitrite = 65% RH at 25°C, 62% RH at 35°C, and 60% RH at 45°C; and magnesium nitrite = 53%RH at 25°C, 50% RH at 35°C, and 48% RH at 45°C.

Source: Adapted from Buckton and Darcy, 1998.

was concluded by Buckton and Darcy (27) that because the net exotherm in the microcalorimeter was the same at each RH studied at 25°C, then the water desorption must have been similar at each RH. This means that the amount of water sorbed at each RH was the minimum quantity needed to cause rapid crystallization. This would be the case if the supply of water vapor were slow, owing to slow diffusion in the cell and the small surface area of the saturated salt solution. It can be shown that if the supply of water vapor is rapid and plentiful, then the amorphous lactose will equilibrate to a different water load at each RH, prior to crystallization. Bearing this in mind, it can be considered fortuitous that the experimental design caused a balance of kinetics that resulted in the area under the curve at each RH being identical at 25°C for crystallization of both amorphous lactose and amorphous salbutamol sulfate (27). Buckton and Darcy (27) went on to show that at higher temperatures there was a difference in the net area under the curve for different humidities (Table 8.2). This is due to the fact that the amount of water needed to lower Tg below T is less when T is higher, and also that the rate of evaporation and diffusion of the water vapor will increase with T. It is extremely important to realize that the calorimetric response for crystallization contains this substantial balance of exotherms and endotherms. If the data are to be used in a quantitative manner, it is vital that the impact of changes in environmental conditions (temperature and RH) is understood, and that any comparison between data at different temperatures and humidities is undertaken with great care. It should be noted, for example, that the fact that lactose has a constant net area under the crystallization peaks for any RH at 25°C is a consequence of the rate of supply of the water vapor and the viscosity and temperature relationship for lactose (which will determine the point at which crystallization will occur rapidly).

8.3.5 OTHER CALORIMETRIC APPROACHES

8.3.5.1 Gas Flow

This approach uses a calorimeter cell in which the powder is housed, which has been constructed to allow a constant flow of humid air. The humidity can be altered

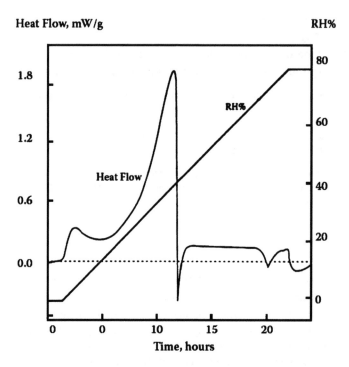

FIGURE 8.5 A micronized drug exposed to an RH ramp. The trace shows vapor wetting the powder, then a very large exotherm (crystallization) at 10 h, followed by a sharp endotherm (water loss). (Reproduced from Briggner, 1993. With permission of Thermometric Ltd.)

in steps or varied continuously as a ramp. The cell is used to study water interaction with a powder, which may be to understand the surface energy of the powder, or to monitor changes in the powder (such as crystallization from the amorphous state, or formation or loss of a hydrate).

In studies of amorphous materials, there will be a clear exotherm as the powder crystallizes and also an endotherm associated with the water loss; these can be seen in Figure 8.5 for the crystallization of amorphous lactose. All water sorption experiments on thermodynamically unstable systems present difficulties in defining an equilibrium. This is because there can be a long lag time between establishing an environmental condition and any change in state of the solid. This lag period can easily be longer than the time for which the sample is allowed to wait at that set condition. Thus, a new environmental condition is established, at that point the sample may change state, giving the impression that this was the key environmental condition that caused the change. However, it could well be that longer storage at the previous condition could have caused the change. This presents a real problem in defining how long to run any experiment. The ramp experiments, in which RH is continually changed at a defined rate, that have been used by Briggner (28) are especially at risk of this failure to reach equilibrium. This is true even if the RH ramp is sufficiently slow to ensure that the extent of water sorption is essentially at equilibrium (as water sorption may be much faster than the physical changes in the solid that result as a consequence of that sorption). Having registered this reservation, it is probable that the onset of crystallization will be seen at about the same point for repeat experiments on the same mass of the same sample. Such ramp experiments have been used by Jakobsen et al. (29) to study the deliquesence of pharmaceutical materials.

A number of studies on the use of these gas flow cells in calorimeters are now being published (e.g., 30–32). These include studies that are designed to characterize the surface (30), and also those that show "critical humidities" at which samples start to deliquesce (29).

8.3.5.2 Solution Calorimetry

Solution calorimetry is undertaken by allowing a sample to equilibrate to temperature by sealing it into a thin-walled glass container. When equilibration has been achieved, the ampoule is broken under the liquid of interest, and the heat change associated with its wetting, dispersion, and solution are measured. In general, the solution process will give rise to the largest heat change, so the experiment is often termed a *heat of solution*, although it is clear that other processes contribute to the response. It is sometimes possible to disperse a powder into a nonsolvent, and thus measure the heat of immersion (also sometimes called the *heat of wetting*). However, it is often hard to find a nonsolvent into which some powders will immerse spontaneously, and if any powder fails to immerse it will make the result invalid.

For most samples, heats of wetting are more easily obtained by vapor sorption experiments using the gas flow cell than by immersion studies. Although the two experiments (vapor sorption and immersion) are different, they both relate to the surface energetics of the powder.

Heats of solution are a valuable method of assessing the lattice energy of a powder; however, the interaction of the powder with the solvent will also be a major contribution to the measured response. This ability to measure a heat that is closely associated with lattice energy has resulted in solution calorimetry being used to assess powder crystallinity. This approach works on the basis that the heat of solution will be different for the crystalline and the amorphous states of a sample. Ward and Shultz (33) showed clear differences in heat of solution between micronized (i.e., partially amorphous) and crystalline salbutamol sulfate samples. These differences were attributed to changes in surface energy; however, the contribution made to the heat of solution by changes in the wetting component of the response is yet to be delineated. It is intuitively more likely that changes in lattice energy were the major contribution to the different measured responses for the amorphous and crystalline materials. However, Hollenbeck et al. (34) have shown that immersion of (insoluble) microcrystalline cellulose can yield valuable information on its physical state, which must be due to changes in wetting. Studies of materials such as microcrystalline cellulose are not possible with the ampoule-style calorimetric experiments, as the vapor does not cause the cellulose to crystallize; thus, only vapor flow experiments and immersion experiments can reasonably be expected to show differences between different celluloses.

Thompson et al. (35) have shown that solution calorimetry can be an effective way of differentiating between drug samples with different degrees of crystallinity.

They also obtained good agreement between heat of solution and thermal activity measured in an isothermal microcalorimeter over the range 0 to 100% crystallinity. Salvetti et al. (36) have demonstrated that different physical forms of carbohydrates can be differentiated by measuring heats of solution. Pikal et al. (1978) used heat of solution measurements to correlate the extent of crystallinity with the chemical stability of antibiotics.

In certain circumstances, solution calorimetry may be more appropriate than vapor sorption approaches as a tool to assess powder crystallinity; for example, if there was any serious fear that the amorphous material may not be accessed by the vapors or that there are no suitable vapors to induce the crystallization response. The potential disadvantage of the solution calorimetry approach is that the responses for both the amorphous and crystalline material are measured, and therefore there is a need for a substantially different heat of solution between the two if small amounts of amorphous material are to be detected. A further difficulty with solution calorimetry is that it may not be easy to find a suitable solvent system that will achieve complete solution in a rapidly enough. Thus, solution calorimetry works as a bulk technique, and measures the response for the entire sample, whereas vapor sorption works by detecting the crystallization response for the amorphous material, with little or no interfering response from the crystalline component. This fundamental difference in approach may mean that on some occasions solution calorimetry will be the preferred option, whereas on others it would be not as good as the vapor sorption approaches.

8.4 TITRATION EXPERIMENTS

Wadso (11) states that "one of the main applications for isothermal microcalorimetry is the investigation of noncovalent binding processes by means of titration techniques." This use of the instrument employs a cell with a liquid (pure liquid, solution, or suspension), which is stirred. It is then possible to inject small quantities of another liquid and to measure the heat flow from the events that take place. One event will be the dilution of the injected material into the fluid in the cell; thus, it is necessary to correct for these dilution effects. Other responses will be due to any interaction between the injected material and the solute or suspended matter in the cell. If it is assumed that the titrated material all interacts with the solute and suspended material, then there will be almost zero free concentration of the titrated sample. This will allow calculations of the enthalpy for the binding process and the stoichiometry.

Pharmaceutical applications of the titration experiment include receptor–enzyme, protein–membrane, and other biomaterial interactions. There are many microcalorimetry publications on binding of drugs to cyclodextrins (e.g., 37,38) however, a real limitation of this approach is that to minimize the heat of dilution and heat of mixing effects, the cyclodextrin and the drug must be dissolved in the same solvent as each other. It would be unusual to have an interest in drug–cyclodextrin complexes in circumstances where both the drug and the cyclodextrin are freely soluble in water.

The injection of a surfactant into a liquid can allow the calculation of the critical micelle concentration and the enthalpy of micellisation (after correction for dilution effects) (e.g., 39). The marginal disadvantage of these experiments is the long duration, as each injection results in a heat decay of about 30 min, making each experiment last about a day. It has been shown (40) that dynamic corrections can be applied to the data that make it possible to make subsequent injections before the first injection has returned to baseline. This approach uses the instrument response time and calculates the decay that would be observed if each injection had been allowed to run to completion. Titration experiments can also be used to study the packing and self-assembly of phospholipids, and the interactions of surfactants with polymers (41,42).

Blackett and Buckton (43,44) have studied surfactant aggregation in model inhalation aerosol systems, and been able to improve our understanding of surfactant aggregation in these nonpolar nonaqueous fluids. Blackett and Buckton (45) showed substantial differences in responses when surfactant was titrated into a chlorofluorocarbon containing salbutamol sulfate, depending on whether the drug had been freshly micronized or allowed to age. The freshly micronized material was partially amorphous (characterized as described in Section 8.5) whereas the aged sample had been exposed to 75% RH to cause it to convert to the crystalline form. The response measured was a composite for dilution effects, adsorption of surfactant to the dispersed drug suspension and, in some circumstances, deaggregation of the drug particles to produce a more uniform suspension. These processes are a mixture of endotherms and exotherms, and the calorimetric data are therefore extremely complex and hard to interpret. The data can simply be used to indicate that changes in formulation, and in the nature of the powder surface have profound effects on how the product will behave. It remains difficult to quantify the amount of surfactant adsorbed to the drug surface in these volatile systems (because of analytical difficulties). As the calorimetric data are indicative of the amount of adsorption, the complex mixtures of exotherms and endotherms make it impossible to quantify the exact amount adsorbed. A further problem with this approach is that it is not possible to use a pressurized container, as the injection into such a system would be too energetic; thus, only model inhalation aerosols can be used. Although for CFC aerosols there were reasonable models that were liquid at room temperature, this is not the case at present for the replacement propellants.

The titration cell for an isothermal microcalorimeter provides an excellent way of following complex interactions for biomaterials, polymers, and surfactants. Thus, this approach will see increasing use in the pharmaceutical sciences in the years to come. As with other calorimetric methods, there will often be parallel processes that will need to be corrected for. Furthermore, the more information that is known about a system from other methods, the easier it will be to understand the microcalorimetry data.

8.5 CONCLUSIONS

Isothermal microcalorimetry is a general term for a wide range of experiments. Some of the approaches have been reviewed here, but others such as perfusion experiments have not been covered. It is possible that perfusion experiments could be of value in the measurement of dissolution of drugs from dosage forms, especially into complex media (44,47).

Of the approaches that have been discussed here, the whole area of biological microcalorimetry has been given only superficial coverage because the subject is too vast to cover here. A candid review of chemical stability testing has been presented to attempt to strike a balance for a technique that offers great promise, but that cannot be expected to deliver in every circumstance.

A considerable amount of detail has been presented on the use of the technique for physical characterization of materials, as in the opinion of the author this is the area that shows greatest promise. It is also an area in which there are very few other techniques matching what can be achieved by isothermal microcalorimetry.

Finally, titration calorimetry is an extremely valuable research tool. It is perhaps too slow and difficult for routine laboratory use, but offers a wealth of information to the research scientist on the properties of complex macromolecules.

Overall, it should be noted that isothermal microcalorimetry is really a bank of related techniques. The fact that almost any system can be studied, in one or more ways, with this technique makes it a valuable tool. However, the fact that concurrent processes and artifacts are also measured with ease makes mistakes probable. Consequently, it is important to remember that it is the easiest thing in the world to use this instrument to obtain data, but it can be very hard to define the exact process to which the data relate, and to provide appropriate interpretation.

REFERENCES

1. Beezer, A.E., Mitchell, J.C., Colegate, R.M. Scally, D.J., Twyman, L.J., and Willson, R.J., *Thermochim. Acta*, 250, 277, 1995.
2. Wadso, I., *Thermochim. Acta*, 267, 45, 1995.
3. Monti, M., Brandt, L., Ikomi-Kumm, J., and Olsson, H., *Eur. J. Haematol.*, 40, 250, 1990.
4. Kemp, R.B., Ed., special issue of *Thermochim. Acta*, 172, 1990
5. Lamprecht, I. and Hemminger, W., Eds., special issue of *Thermochim. Acta*, 193, 1991
6. Kemp, R.B. and Schaarschmidt, B., special issue of *Thermochim. Acta*, 250, 1995
7. Pikal, M.J., Lukes, A.L., Lang, J.E., and Gaines, K., *J. Pharm. Sci.*, 67, 767, 1978.
8. Hansen, L.D., Lewis, E.A., Eatough, D.J., Bergstrom, R.G., and DeGraft-Johnson, D., *Pharm. Res.*, 6, 20, 1989.
9. Otsuka, T., Yoshioka, S., Aso, Y., and Terao, T., *Chem. Pharm. Bull.*, 42, 130, 1994.
10. Angberg, M., Nystrom, C., and Castensson, S., *Acta Pharm. Suec.*, 25, 307, 1988.
11. Wadso, I., *Chem Soc Rev*, 79, 1997.
12. Gaisford, S., Ph.D. thesis, University of Kent, 1997.
13. Willson, R.J., Beezer, A.E., and Mitchell, J.C., *Thermochim. Acta*, 264, 27, 1995.
14. Angberg, M., Nystrom, C., and Castensson, S., *Int. J. Pharm.*, 90, 19, 1993.
15. Willson, R.J., Beezer, A.E., Mitchell, J.C., and Loh, W., *J. Phys. Chem.*, 99, 7108, 1995.
16. Saleki-Gerhardt, A., Ahlneck, C., and Zografi, G., *Int. J. Pharm.*, 101, 237, 1994.
17. Hancock, B. and Zografi, G., *J. Pharm. Sci.*, 86, 1, 1997.
18. Yoshioka, M., Hancock, B., and Zografi, G., *J. Pharm. Sci.*, 83, 1700, 1994.
19. Hancock, B., Shamblin, S.L., and Zografi, G., *Pharm. Res.*, 12, 799, 1995.
20. Shalaev, E.Y. and Franks, F., *J. Chem. Soc. Faraday Trans.*, 91, 1511, 1995.
21. Nyqvist, H., *Int. J. Pharm. Technol. Prod. Manuf.*, 4, 47, 1983.

22. Aso, Y., Yoshioka, S., Otsuka, T., and Kojima, S., *Chem. Pharm. Bull.*, 43, 300, 1995.
23. Sebhatu, T., Angberg, M., and Ahlneck, C., *Int. J. Pharm.*, 104, 135, 1994.
24. Buckton, G. and Darcy, P., *Int. J. Pharm.*, 136, 141, 1996.
25. Briggner, L.-E., Buckton, G., Bystrom, K., and Darcy, P., *Int. J. Pharm.*, 105, 125, 1994.
26. Buckton, G., Darcy, P., Greenleaf, D., and Holbrook, P., *Int. J. Pharm.*, 116, 113, 1995.
27. Buckton, G. and Darcy, P., *Thermochim. Acta*, 316, 29, 1998.
28. Briggner, L.-E., Thermometric Application Note 22022, 1993.
29. Jakobsen, D.F., Frokjaer, S., Larsen, C., Niemann, H., and Buur, A., *Int. J. Pharm.*, 156, 67, 1997.
30. Sheridan, P.L., Buckton, G., and Storey, D.E., *Pharm. Res.* 12, 1025, 1995.
31. Sokoloski, T.D. and Ostovic, J.R., *J. Pharm. Pharmacol.*, 49, 32S, 1997.
32. Puddipeddi, M., Sokoloski, T.D., Duddu, S.P., and Carstensen, J.T., *J. Pharm. Sci.*, 85, 381, 1996.
33. Ward, G.H. and Shultz, R.K., *Pharm. Res.*, 12, 773, 1995.
34. Hollenbeck, R.G., Peck, G.E., and Kildsig, D.O., *J. Pharm. Sci.*, 67, 1599, 1978.
35. Thompson, K.C., Draper, J.P., Kaufman, M.J., and Brenner, G.S., *Pharm. Res.*, 11, 1362, 1994.
36. Salvetti, G., Tognoni, E., Tombari, E., and Johari, G.P., *Thermochim. Acta*, 285, 243, 1996.
37. Tan, X. and Lindenbaum, S., *Int. J. Pharm.*, 74, 127, 1991.
38. Moelands, D., Karnik, N.A., Prankard, R.J., Sloan, K.B., Stone, H.W., and Perrin, J.H., *Int. J. Pharm.*, 86, 263, 1992.
39. Gu, G., Yan, H., Wenhai, C., and Wang, W., *J. Colloid Interface Sci.*, 178, 614, 1996.
40. Bastos, M., Hagg, S., Lonnbro, P., and Wadso, I., *J. Biochem. Biophys. Methods*, 23, 255, 1991.
41. Olofsson, G. and Wang, G., *Pure Appl. Chem.*, 66, 527, 1994.
42. Kevelam, J., van Breemen, J.F.L., Blokzil, W., and Engberts, J.B.F.N., *Langmuir*, 12, 4709, 1996.
43. Blackett, P. and Buckton, G., *J. Mater. Sci. Lett.*, 14, 1182, 1995.
44. Blackett, P. and Buckton, G., *Int. J. Pharm.*, 125, 133, 1995.
45. Blackett, P. and Buckton, G., *Pharm. Res.*, 12, 1689, 1995.
46. Ashby, L., Beezer, A.E., and Buckton, G., *Int. J. Pharm.* 51, 245, 1989.
47. Buckton, G., Beezer, A.E., Chatham, S.M., and Patel, K.K., *Int. J. Pharm.* 56, 151, 1989.

9 High-Sensitivity Differential Scanning Calorimetry

Simon Gaisford

CONTENTS

9.1 INTRODUCTION

In previous chapters, the principles and applications of differential scanning calorimetry (DSC) have been outlined, and it should be clear that the technique is both versatile and extremely sensitive. Using DSC, it is possible to analyze a wide range of systems quickly and cheaply so that thermodynamic parameters may be obtained. These qualities have led to the widespread use of DSC for not only pure research but also for routine thermal analysis. DSC does, however, have some drawbacks. To achieve good thermal contact with a sample, most DSC instruments are equipped with a pair of sample holders into which prepared sample and reference materials are placed. These materials are usually encapsulated in crimped aluminum ampoules, a typical sample mass being 5 to 10 mg. Such a small mass of sample contributes to a good thermal contact between the instrument and the sample and ensures there are no temperature gradients within the ampoule. For dilute samples, compounds with higher molecular weights such as proteins or polymers, or reactions with a small change in reaction enthalpy (ΔH); however, such a small sample mass may preclude the use of DSC, as the magnitude of the heat flow to or from the sample during a reaction may be lower than the detection limit of the instrument. To study such systems it is necessary to use high-sensitivity differential scanning calorimetry (HSDSC), a derivative of DSC. This chapter focuses on understanding the principles and applications of this technique.

9.2 DSC VS. HSDSC

The output from a DSC experiment is a plot of power (measured as the difference in power required to maintain the sample and reference materials at the same temperature) vs. time or temperature. From such data, and dependent on the processes undergone by the sample, it is possible to determine transition onset temperatures (T_o), transition peak temperatures (T_m), glass transition temperatures (T_g), calorimetric enthalpies ($\Delta_{cal}H$, obtained by integration of peak areas), and van't Hoff enthalpies ($\Delta_{vH}H$, obtained by a van't Hoff analysis of peak areas) (see Figure 9.1). The y-axis position of the baseline relative to that of a standard is related to the heat capacity of the sample and, thus, the heat capacity of a sample can also be determined. Depending on the sample to be studied, it is also possible to transform the power signal to apparent molar excess-heat capacity ($C_{p,ex}$), using Equation 9.1, which often gives a more useful thermodynamic insight into the system (especially in the case of biological systems) (1).

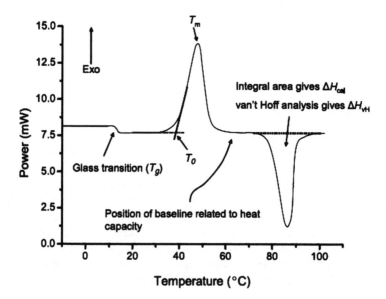

FIGURE 9.1 The parameters that may be derived from a DSC experiment.

$$\frac{dq_p}{dt} \cdot \frac{1}{\sigma \cdot M} = C_{p,xs} \qquad (9.1)$$

Where q_p is the change in power, t is time, σ is the scan rate, and M is the number of moles of material in the sample cell. These parameters give some insight into the thermodynamics of the reaction occurring in the sample cell but, as is always the case with calorimetric techniques, do not give any direct molecular information. For a complete analysis, some insight into the molecular changes occurring in the cell must be gained from supplementary techniques. It is possible to study changes in a thermodynamic property of a series of structurally related molecules and to use those data to build up a quantitative structure–activity relationship (QSAR). QSARs form the basis of many drug development programs.

Although DSC is a highly sensitive technique, typical instruments do not possess the sensitivity of, for example, isothermal microcalorimeters, a constraint imposed by the versatility of the technique. A typical isothermal calorimeter (for example, the thermal activity monitor [TAM] system developed by Thermometric AB, Järfälla, Sweden) is capable of running separate experiments over a range of temperatures and can attain a baseline sensitivity of ±0.1 µW, but it may require 24 h or more to change the temperature of the instrument. A DSC, however, can operate at temperature scan rates of 200°C min^{-1} or higher (2). Because of the instrument-design constraints imposed to attain such versatility, this had led to instruments with a sensitivity, in practice, of ±10 µW. As mentioned previously, typical instruments generally utilize small sample masses of around 5 to 10 mg in crimped aluminum ampoules to ensure a uniform temperature distribution and good thermal contact between the instrument and the sample. For systems that react above 600°C or react with aluminum, copper, platinum, or gold pans may also be employed.

This level of sensitivity has precluded the use of DSC for studying events that occur with a small change in enthalpy (ΔH) in systems where the active component is present in low concentrations or for reactions involving those species with a high molecular weight, because the instrument is incapable of detecting such small changes in heat. Many biologically important molecules, such as proteins and peptides, occur naturally in very low concentrations, have high molecular weights, and are often stabilized in their native state by weak intramolecular forces. These forces may be disrupted with an increase or decrease in temperature, and such disruption can result in conformational or phase changes. Usually, for biological molecules, such changes occur within the temperature range –20 to 130°C, well within the limits for a typical scanning calorimeter.

During the past 20 years, advances in technology have meant that it is now possible to obtain DSC instruments with a much greater level of sensitivity (2–5). Such instruments are generally capable of detecting signals of ±0.5 µW and, to increase both sensitivity and versatility, may utilize larger sample sizes, typically up to 1 ml, than conventional DSC instruments. They are nominally referred to as *high-sensitivity differential scanning calorimeters*, although the terms used by specific manufacturers vary. Among the commercially available instruments are nano-DSCs, ultrasensitive DSCs, high-sensitivity DSCs, medium-sensitivity DSCs, and differential scanning microcalorimeters. The specifications of a range of commercial HSDSC instruments are given in Table 9.1. For the discussion that follows, the generic term HSDSC will be used throughout and refers to all types of commercially available calorimeters. It will be assumed that a typical instrument has a calorimetric sensitivity of $\pm 0.5 \ \mu W$, operates at relatively slow scan rates (0 to 2°C min^{-1}), and holds approximately 0.5 to 1 ml sample.

The increase in sensitivity over standard DSC instruments and the ability to study reactions occurring in solution directly mean that HSDSC may be applied to the study of a range of systems not open to study by standard DSC. Typical examples include the denaturation of proteins, phase changes in lipid bilayers, phase transitions in dilute polymer solutions, and changes in structure of creams and emulsions. Although there are many systems that have been studied using HSDSC, the discussion that follows will concentrate on systems of biological or pharmaceutical relevance.

9.3 INSTRUMENT DESIGN

The use of HSDSC has become more widespread in recent years as the cost of the instruments has fallen to reasonable levels, and there is now a wide range of instruments commercially available (see Table 9.1). In addition to the falling cost of the instruments, advances in computer software have led to more sophisticated methods of data analysis. Indeed, many instruments are supplied as complete systems, aimed at users with no knowledge of thermal analysis, and will analyze data automatically, so that all the operator is required to do is load the sample and print out the results obtained. In addition to advances in instrument design, there is also better availability of highly pure, well-characterized analytes, this being particularly important for experiments using proteins or peptides. As well as being capable of running both heating and cooling experiments, many HSDSC instruments can be programmed to run under isothermal conditions, either as a separate experiment or as part of a scanning experiment, and this versatility increases the number of potential applications of the technique. A discussion of these "step-isothermal" experiments can be found later. Such a quality is particularly important for areas such as pharmaceutical stability testing, in which data need to be acquired over a range of usually elevated temperatures. An experiment can be set up that steps up in temperature and runs isothermally between the temperature steps.

There are two basic designs of HSDSC instrument, as in the case of standard DSC instruments: power compensation and heat flux. The scientific principles behind these designs have been discussed earlier and shall not be repeated here. The difference with HSDSC instruments is that they generally utilize larger sample sizes, allowing a signal of smaller magnitude to be detected, and these samples are held in one of two types of cell: a fixed cell inside the instrument or a removable (batch) cell that can be cleaned and reused. Each system has its own advantages and disadvantages.

9.3.1 FIXED-CELL INSTRUMENTS

With fixed-cell instruments, samples must be loaded directly into the calorimetric chamber, and the need for a metal pan or batch cell is eliminated. Such a system

TABLE 9.1
Specifications and Applications for Some Commercial HSDSC Instruments

Instrument	Micro-DSC III	Micro DSC-VII	CSC Nano-DSC Series III	CSC 4100	B-900	VP-DSC
Manufacturer or supplier	Setaram, 7 rue de l'Oratoire, F-69300, Caluire, France	Setaram, 7 rue de l'Oratoire, F-69300, Caluire, France	Calorimetry Sciences Corp., 515 East 1860 South, P.O. Box 799, Provo, UT	Calorimetry Sciences Corp., 515 East 1860 South, PO Box 799, Provo, UT	Sceres, 16 rue de Chartres, F-91400, Orsay, France	Microcal Inc., 22 Industrial Drive East, Northampton, MA
Type	Heat flux	Heat flux	Power compensation	Heat flux	Power compensation	Power compensation
Scan modes	Heating, cooling, isothermal	Heating, cooling, isothermal	Furnace Heating, cooling (ITC also available)	Heating, cooling, isothermal	Heating, cooling, isothermal	Heating, cooling (ITC also available)
Temp. range	−20 to 120°C	−45 to 120°C	−10 to 130°C (160°C optional)	−40 to 110°C (−40 to 200°C optional)	−190 to 850°C	−10 to 130°C
Scan rates	0.001 to 1.2°C min⁻¹	0.001 to 1.2°C min⁻¹	0 to 2°C min⁻¹	0 to 2°C min⁻¹	0.01 to 15°C min⁻¹	Max. 1.5°C min⁻¹ heating or max. 1°C min⁻¹ cooling
Detection limit	±0.2–2 µW	±1–5 µW	Not stated	Not stated	±0.1 µW	±0.35 µW

Continued

TABLE 9.1 (Continued)
Specifications and Applications for Some Commercial HSDSC Instruments

Instrument	Micro-DSC III	Micro DSC-VII	CSC Nano-DSC Series III	CSC 4100	B-900	VP-DSC
			Sample Holder			
Construction	Hastelloy C	Hastelloy C276	24 K gold	Hastelloy C (tantalum or titanium optional)	Steel, Teflon®, Platinum, glass, Hastelloy, or other	Tantalum
Volume	1 ml	1 ml	0.33 ml	1 ml	1.5 ml	0.5 ml
Type	Batch or flow through cell	Batch or flow through cell	Fixed cell	Batch cell (holds up to three cells)	Batch or flow through cell	Fixed cell
			Uses			
Applications	Biological and pharmaceutical applications, protein denaturation	Foodstuffs and pharmaceuticals, liquid crystal transitions, bacterial growth	Behavior of biopolymers in solution, lipid membrane structure, ligand interactions	Cell metabolism, material stability, high throughput screening	Phase transitions and chemical reactions, pyrolysis, specific heats	Protein denaturation, gel transitions, melting of lipid bilayers

has many advantages in terms of sensitivity, and this was the design used in some of the earliest commercial HSDSC instruments. The most important advantage fixed-cell designs have is that, because the cells are never removed from the calorimeter, each experiment is conducted under conditions in which the thermal contact between the cell and the instrument is identical, increasing the baseline repeatability of the instrument. Because the cells are never removed, there is very little chance of damaging either the cells or the instrument, and the mechanical stability imparted to the cells reduces baseline noise. Usually, to facilitate loading and cleaning of the cells, samples must be liquids. A typical fixed-cell design allows a constant volume of sample to be loaded, minimizing errors that might occur by using a balance to determine the mass of a sample loaded into a removable pan. The drawbacks of the fixed-cell design include the need to use liquid samples and the difficulty that may arise in cleaning the cells after an experiment. A typical fixed-cell, power-compensation instrument is the VP-DSC (Microcal, Inc.) instrument (it should be noted that this instrument is also available with an automatic titration unit attached to the sample cell, the VP-ITC, which is an instrument used widely to study biological systems). The sample and reference cells are fixed within a cylindrical adiabatic shield and are loaded through access tubes. As the temperature of the cell changes during the course of an experiment, the sample is free to expand and contract and, within the measuring space, a constant sample volume is achieved. The temperature difference between the sample and reference cells, ΔT_1, is monitored continuously by a cell feedback (CFB) network during an experiment. If ΔT_1 deviates from zero, because of an exothermic or endothermic process occurring in the sample cell, the CFB network is activated, and heaters under each pan drive ΔT_1 back to zero. The integral of the CFB signal over the course of an experiment is equal to the total heat change of the sample. A similar system is used to maintain the temperature difference between the cells and the adiabatic shield, ΔT_2, at zero, preventing any heat leaks from the cells.

9.3.2 BATCH-CELL INSTRUMENTS

With batch-cell instruments, samples are loaded into external, reusable cells, and these cells are then placed in the measuring chamber of the instrument. Because the cells are removable, they are much easier to load and clean than the fixed cells discussed previously and, hence, can be used to study liquids, solids, or heterogeneous samples. This means that it is possible to investigate directly the behavior of pharmaceutical products, such as creams or powders. Perhaps the biggest advantage of the batch-cell design, however, is that the sample cell can be replaced with different vessels designed to study specific systems. It is possible to obtain vessels designed to study a range of systems, which include circulating liquids, mixing circulating liquids, and mixing of liquids or solids with liquids. There is also the possibility of designing a specific vessel for a specific type of reaction. However, there are some drawbacks to the batch-cell design. The cells must be loaded into the instrument, and the thermal contact between the cells and the instrument may vary between experiments. The cells are loaded to a constant weight rather than a constant volume, and it difficult to load liquid samples without a vapor space (although it is possible to purchase a specially designed cell that enables liquid samples to be loaded without a vapor space. Such a design allows very accurate determination of heat capacities). It is also possible to damage cells though misuse or accident.

Whereas most fixed-cell instruments are power-compensation instruments (because it is possible to place heaters on the base of cells that are not removable), batch-cell instruments are available as either power-compensation or heat-flux designs. One design of a heat-flux, batch-cell instrument is the micro-DSC III (Setaram). The instrument consists of a calorimetric block into which two channels are machined. One channel holds the sample cell, the other holds the reference cell. At the bottom of each channel, between the cell and the block, is a plane-surfaced transducer. The transducers provide a thermal pathway between the cells and the block and are used to maintain the cells at a temperature identical to that of the block. The electrical signal produced by the transducer on the sample side is proportional to the heat evolved or absorbed by the sample. The temperature of the calorimetric block is maintained by a precisely thermostated circulating liquid. The liquid is raised in temperature by a separate heater and is cooled by a supply of circulating water. The precise control of the temperature of the circulating liquid allows scan rates of just $0.001°C$ min^{-1} to be attained and ensures that the calorimetric block is insulated from the surrounding environment.

A more recently developed instrument is the microreaction calorimeter (Thermal Hazard Technology Ltd.). In this instrument, a smaller heat-shield is used around the sample, which permits very rapid changes in temperature at the expense of a slight decrease in sensitivity. The instrument uses a disposable glass ampoule with a rubber septum in the lid that allows provision for titration experiments to be performed. The versatility of the instrument means that it is particularly suited to stability and compatibility testing of pharmaceuticals, and a new six-calorimeter version that uses a common reference chamber is in development.

9.3.3 General Instrument Considerations

Irrespective of whether an instrument employs fixed cells or batch cells, some design considerations are common to all HSDSC instruments. To measure accurately small powers, it is necessary to ensure good baseline stability throughout the course of an experiment. This is usually achieved by maintaining very accurate control of the calorimetric block temperature and ensuring that the properties of the sample and reference cells, such as geometry, cell volume, local environment, thermal conductivity and conduction pathways, and heating rate, are identical.

In standard DSC experiments, scan rates usually employed vary in the range 10 to $20°C$ min^{-1}, although it is observed that the scan rate affects the shape of the thermal trace obtained. In general, slow-heating rates result in high-resolution traces (separating different thermal events that occur over a similar temperature range) with low sensitivity (small peaks). Conversely, fast-heating rates generally lead to an increase in sensitivity at the expense of resolution (6). For high-sensitivity instruments, it follows that a slow scan rate should be employed, because the relative decrease in sensitivity is balanced by the inherent sensitivity of the instrument and the presence of larger sample sizes, allowing high-resolution traces to be obtained. As seen from the data presented in Table 9.1, most commercial HSDSC instruments have a maximum scan rate of around 2°C min^{-1}.

The calorimetric sensitivity of some HSDSC instruments is such that they are just 10-fold less sensitive than a commercial isothermal microcalorimeter, yet they possess the capability to change temperature almost instantly. Indeed, O'Neill et al. (7) recently showed that the performance of two HSDSC instruments (the Micro-DSC III and the microreaction calorimeter) was comparable to an isothermal calorimeter when used to study an International Union of Pure and Applied Chemistry (IUPAC) test and reference reaction. All HSDSC instruments are capable of studying phenomena that occur in dilute solutions and are particularly suitable for use with biological systems, in which the concentrations of species can be kept low enough to avoid molecular interactions. Batch-cell instruments can also be used to study solid or heterogeneous samples. These factors allow a more quantitative understanding of the physicochemical relationships between the thermodynamic parameters obtained from the instrument and the molecular interactions occurring in the sample, and help to explain the growing popularity of the technique over the past decade.

9.3.4 General Experimental Considerations

The data collected from an HSDSC experiment may be affected by the condition of the sample (average particle size, thermal history, etc.) or by the operating conditions of the instrument (reference material, heating rate, etc.). For many systems of interest, samples can be loaded in solution, and there is good thermal contact between the sample and the cell. For samples that are loaded as solids, it is necessary to ensure that the average particle size is as small and uniform as possible, because it has been suggested that the particle size of the sample may affect the thermal trace obtained (8). If the sample melts and subsequently reforms a solid during an experiment, it may be necessary to repeat the heating scan to ensure that only thermal signals from reactions or phase changes are recorded.

As ever, it is good practice to validate the performance of an instrument, using IUPAC recommended test and reference reactions. For DSCs operating in a temperature-scanning mode, the melting of one or more pure compounds is indicated. However, the additional capability of some instruments to run isothermally means that other test reactions (for short-term reactions, the base-catalyzed hydrolysis of methyl paraben [BCHMP], or for long reactions, the imidazole-catalyzed hydrolysis of triacetin [ICHT]) can be employed. Finnin et al. (9) showed the results of a validation exercise, conducted by Thermal Hazard Technology as part of an installation routine, using BCHMP. In the case described, a systematic misreporting of the reported temperature of a calorimeter was identified, caused by an upgrade to the calorimeter's firmware, a discrepancy that might not have been noted using traditional electrical calibration methods. The work highlighted the importance of both manufacturers and end users adopting chemical test reactions into their test and validation routines.

Some HSDSC instruments have the capability of running experiments with more than one sample simultaneously; for example, the CSC 4100 (Calorimetry Sciences

Corp.) has the capability to run three samples at the same time. In such cases, the thermal behavior of samples with different thermal histories may be investigated, or "good" and "bad" batches of the same product may be compared directly.

9.4 PHARMACEUTICAL AND RELATED APPLICATIONS

The most common systems studied using HSDSC are those that comprise components of relatively high molecular weight in dilute solutions because, as has been discussed previously, most other systems can be studied using conventional DSC. Pharmaceutically, typical systems that are studied include lipids and interactions or phase changes of lipid membranes, folding and unfolding of proteins and peptides, interactions between biological molecules, and phase changes or interactions of polymers. In the following sections, some of the published work in these areas will be discussed so that the reader may gain familiarity with the typical uses of an HSDSC instrument.

9.4.1 STUDIES OF DILUTE POLYMER SOLUTIONS

Polymers are used in many pharmaceutical preparations, typically as emulsifiers or thickeners, and may even be present as an active ingredient.* As such they must be nontoxic when taken in relatively large doses and, as many formulations are subjected to a temperature ramp during manufacture, their properties over specific temperature ranges must be well defined. It is also necessary to ensure that any polymers used in a particular formulation do not interact with other excipients present. Typical examples of polymers that may be used pharmaceutically are polyethylene glycols (PEGs) (10) and poly(ethylene oxide)-poly(propylene oxide) (PEO-PPO) block copolymers, of general formula PEO_A-PPO_B-PEO_A (poloxamers, Pluronics [BASF] or Synperonic PE [ICI]) (11).

The study of the aqueous phase properties of the poloxamers has created a large body of literature (12–19), mainly because these polymers are observed to aggregate with an increase in solution temperature or concentration (12–14, 20–22). Aggregation occurs as a result of dehydration of the (hydrophobic) PPO moiety, leading to aggregates with a core of PPO surrounded by a corona of PEO. If the solution temperature or polymer concentration is raised high enough, then the poloxamers cannot be solubilized and precipitate out (resulting in a "cloud point"). In the past, studies of dilute solutions of poloxamers have been conducted using traditional chemical assay techniques, such as surface tension measurements (16,17,23) or small angle neutron scattering (SANS) (18,19,24), because thermal techniques were not sensitive enough to detect the small heats associated with the transitions occurring. HSDSC, however, has proved a sensitive enough technique with which to investigate these changes (12–14,25,26).

Early work showed that HSDSC was sensitive enough to allow the study of the transitions occurring in a 5 mg ml^{-1} sample of poloxamer P237, the data obtained

---

* For example, poloxamer P407 is found as the active ingredient in products such as *Baby Oragel®*, a tooth and gum cleanser, and *Clerz®* and *Pliagel®*, soft contact lens cleansers.

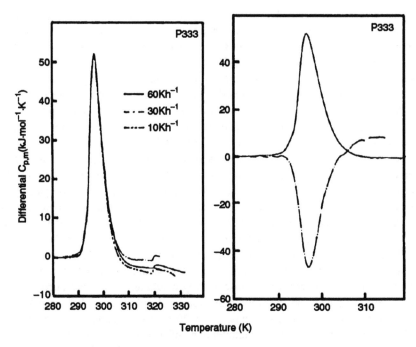

FIGURE 9.2 HSDSC traces for 5 mg ml⁻¹ P333. The data show no dependence on scan rate (left) and are reversible (right). (Reprinted from Beezer et al., *Langmuir* 10:4001, 1995. With permission. Copyright (1995) American Chemical Society.)

being stated as the first calorimetrically recorded example of a phase transition of a synthetic block copolymer in dilute aqueous solution (25). Subsequently, phase transitions of 27 poloxamers, at 5 mg ml⁻¹, were studied using HSDSC, and the thermodynamic parameters derived from the data were tabulated (12). That study contains data for all the poloxamers manufactured by ICI (now Uniqema) and forms the most complete listing published, although at only one concentration. The transition observed for a typical poloxamer phase transition is broad, extending over a range of 10 to 20°C, an observation noted when using any technique to study the process (Figure 9.2). This is usually ascribed to polydispersity within the polymer sample. The HSDSC trace is also reversible, forming a mirror image on the temperature axis (Figure 9.2), and reproducible.

From the HSDSC data, it is possible to derive onset temperatures of aggregation (T_o) and peak temperatures of aggregation (T_m) (12). T_o values define the temperature at which aggregation initiates at a given concentration, and T_m values define the shape of the transition curve. It is possible to relate transition temperatures to polymer composition such that QSARs may be derived. Transition temperatures were observed to decrease approximately linearly with increasing molecular weight, and, in a separate study using UV spectroscopy, with increasing concentration (27) for those polymers containing the same percentage of PEO.

Recent studies have suggested that HSDSC may be similarly used to study the interactions among binary mixtures of poloxamers in dilute aqueous solution (28), because it is not unreasonable to expect more than one particular polymer to be present in a specific formulation. These data may be of importance when considering formulation compositions. The study showed that most binary pairs of poloxamers behaved cooperatively, i.e., the HSDSC curves showed the presence of only one transition, implying that the polymers present formed mixed aggregates. In such cases, the transition temperature recorded always matched with that recorded previously for the polymer with the lower transition temperature. Interestingly, for a few pairs of poloxamers, noncooperative behavior was noted, i.e., the HSDSC curves showed the presence of two transitions (Figure 9.3 and Figure 9.4), implying that the poloxamers present formed homogeneous aggregates. More important, where formulation is concerned, it was found that, of the poloxamers commonly used in pharmaceutical products, some showed cooperative behavior (for example, poloxamers P217 and P237) whereas others demonstrated noncooperative behavior (for example, poloxamers P237 and P407). Such observations may be of importance when these polymers are formulated together. It might be expected, for example, that the properties of a cream containing poloxamers P217 and P237, two polymers that were shown to aggregate cooperatively, might be altered dramatically by heating during manufacture or storage, because heating would lead to the presence of mixed aggregates. Moreover, the transition temperature of poloxamer P407, an extremely widely used excipient, was found to be below 20°C at some concentrations; it seems likely that fluctuations in storage temperature of a product containing P407 could lead to an appreciable change in performance, especially if the polymer is formulated as an active ingredient.

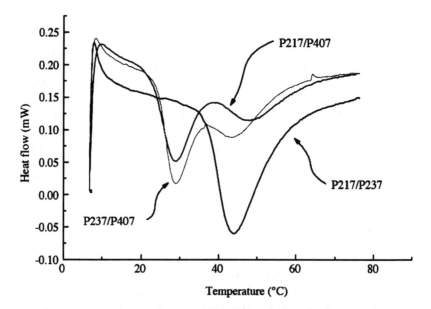

FIGURE 9.3 HSDSC traces for binary mixtures of poloxamers at 20 mg ml⁻¹, showing two transitions for P217/P407 and P237/P407 and one transition for P217/P237. (From Gaisford, S., Beezer, A.E., and Mitchell, J.C., *Langmuir* 13, 2606, 1997. With permission.)

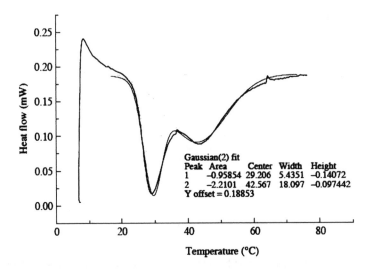

FIGURE 9.4 HSDSC trace for 20 mg ml⁻¹ P237/P407 and the fit of the data to a double Gaussian model. (From Gaisford, S., Beezer, A.E., and Mitchell, J.C., *Langmuir* 13, 2606, 1997. With permission.)

9.4.2 Studies of Biological Molecules

An advantage of HSDSC over other analytical techniques is that it is sensitive enough to allow the direct study of inter- and intramolecular interactions occurring in dilute aqueous solution. This is especially important for biological molecules, because it allows the study of these systems in their native state. Typical classes of biological molecules that have been investigated using HSDSC include proteins, enzymes, lipids, and nucleic acids.

9.4.2.1 Proteins

By using HSDSC, it is possible to obtain information on both the forces that stabilize proteins in their native state and on the nature of the unfolding of the protein as the temperature of the solution is increased. It should be noted that in many cases protein denaturation is irreversible and that some proteins cannot be investigated using calorimetric methods because they precipitate after denaturing. It is also a common feature of most protein denaturation HSDSC traces that there is a difference in heat capacity between the initial and final states of the protein (29), reflecting the fact that the initial and final states of the protein in solution are physically distinct, contributing to the irreversibility mentioned earlier. As in the case of the poloxamers discussed previously, the properties of a formulation containing a protein may therefore be altered significantly by an increase in temperature, because the majority of proteins will not function correctly in their denatured state.

Proteins are commercially important in foodstuffs, along with carbohydrates and fats, and are often used as emulsifiers during food processing. Inducing protein denaturation is also important when cooking food; for example, egg white changing from a liquid to a solid arises from the denaturation of albumin. Because protein

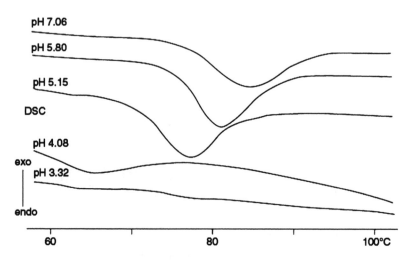

FIGURE 9.5 HSDSC traces of beef blood hemoglobin at various pHs, showing denaturation at high pH and no denaturation at low pH. (Reprinted from Riesen et al., *Thermochim. Acta*, p. 279, 1993. Copyright 1993. With permission from Elsevier Science.)

denaturation can be induced by, among other factors, an increase in temperature, it is possible to characterize different proteins by their thermal response over specific temperature ranges and, thus, to use HSDSC as a quality assurance tool. It has been shown that HSDSC can be used to study proteins that denature with time at room temperature or that denature with changes in pH (30). For instance, the HSDSC trace of beef blood hemoglobin recorded at pH 7.06 shows a clear denaturation endotherm whereas a similar trace recorded at pH 3.32 shows no endothermic signal, indicating that the protein must have denatured completely before the experiment was conducted (Figure 9.5). Using HSDSC as a quality control tool, a protein sample could be removed from storage and placed in the calorimeter and the thermal trace obtained; when compared with a standard, it would indicate whether the material had degraded significantly or not.

Of course, it is possible to use HSDSC to study protein interactions in much greater detail by fitting the data obtained to a suitable model. In simplistic terms, protein denaturation reactions can be classified into two groups: two-state transitions and non-two-state transitions (3). For two-state reactions, there are essentially two macroscopic states, the native and denatured states, and the populations of the intermediate states are not significantly populated at equilibrium. For non-two-state reactions, some or all the intermediate states are populated at equilibrium, and it might seem reasonable to expect that this is the most likely case for large molecules such as proteins. It has been demonstrated, mostly as a result of calorimetric measurements, that the unfolding of some proteins shows only a two-state character, the "all or none" mechanism, although most proteins exhibit more complex unfolding mechanisms (3). Fitting calorimetric data to a two-state model is achieved by application of the van't Hoff analysis; in this analysis, the log of the fractional area under the transition curve is plotted against $1/T$, the slope of the resulting plot being equal to $-\Delta H/R$, where R is the gas constant. Finding that the van't Hoff equality holds for a particular set of data gives confidence that the protein denaturation being studied is of two-state or all or none character. A further check that the study reaction is two-state in character is that the value of T_m or $T_{1/2}$ (the temperature at which half the process has occurred) should be independent of concentration. An increase in the value of T_m with an increase in concentration suggests a decrease in the degree of oligomerization, and a decrease in the value of T_m with an increase in concentration suggests an increase in the degree of oligomerization. Fitting data that do not follow a two-state mechanism requires a suitable model. Once a model has been found, the data may be fitted to obtain the required parameters. Having obtained the values of the parameters, it is possible to deconvolute the data into their component parts. By selectively fitting the data to different models, it is possible to infer the type of mechanism the reaction is following, because incorrect models should not give a good fit. It follows that it is possible to obtain useful information on both the thermodynamics and mechanism of relatively complex protein reactions from a single HSDSC experiment.

An excellent example of the use of HSDSC to study the unfolding of various forms of a protein — in this case, tropomyosin — is provided by Sturtevant et al. (31). Tropomyosin exists in a two-chain α-helical coiled coil conformation and in two forms: αα-tropomyosin and ββ-tropomyosin. The αα form has a single sulf-hydryl group at position 190 whereas the ββ form has an additional sulfhydryl at position 36. If the sulfhydryl groups are oxidized, then disulfide bonds will be made between the two chains: one bond in the αα form and the other in the ββ form (32). Some studies have been conducted using forms of tropomyosin that have had the sulfhydryls blocked to prevent chain cross-linking, and it is usually assumed that this does not affect the thermal unfolding of the protein (31). HSDSC was used to study the oxidized and reduced species of the αα and ββ forms and the sulfhydryl blocked (using both carboxymethylation [CM] or carboxyamidomethylation [CAM]) species of the αα form, and it was shown that for the oxidized and reduced αα forms, the data could be fitted to a three-domain model, whereas the data for all other species was fitted to a four-domain model. As well as giving information on the structural changes occurring in the different species, the HSDSC data also showed clearly that blocking the sulfhydryl groups led to an appreciable change in unfolding behavior. For the sulfhydryl-blocked αα-tropomyosin, it was observed that the blocked forms exhibited considerably different behavior from the non-blocked form. The T_m values for the lower transitions were markedly affected, whereas the T_m values for the upper transitions were similar.

The sensitivity of HSDSC also allows the study of protein interactions with various biological species. Typical experiments include the binding of a protein with a ligand, a new drug compound, for instance, or the behavior of a protein in a membrane. The binding of proteins with ligands may be followed, using isothermal titration calorimetry (ITC) or by observing the effects of ligand binding on the reversible unfolding of the protein (33).

Another application of HSDSC is the study of the thermodynamic effects of protein mutations, an excellent review of which is provided by Sturtevant (34). In such studies, amino acid changes are made in proteins of known structure, and the resultant changes in the thermodynamics of processes, such as unfolding, are determined quantitatively.

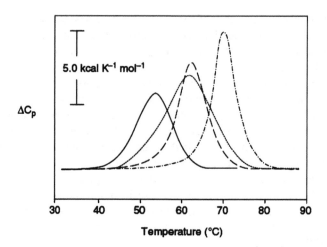

FIGURE 9.6 HSDSC traces for the denaturation of wild type and mutant versions of the protein λ repressor (--Wild type, ······ Ala-46/Ala-48, - - - Cys-88, - · - · - Triple mutant). (Reprinted from Stearman et al., *Biochemistry* 27:7571, 1988. With permission. Copyright [1988] American Chemical Society.)

An example of this type of study is provided by Stearman et al. (36) where the protein λ repressor was investigated. Three mutant proteins were created; in the first mutant, two glycines were replaced with alanines in an attempt to stabilize the α-helix structure (the Ala-46/Ala-48 mutant); in the second mutant, a tyrosine was replaced with a cysteine, allowing a di-sulfide bond to form (the Cys-88 mutant), and the third mutant contained both these sets of mutations (the triple mutant). The thermal unfolding of the wild type and mutant proteins was studied using HSDSC because it is possible to obtain information on stability from changes in thermal behavior; higher transition temperatures suggest the protein is more stable. It was observed that the wild type protein denatured at 53.9°C, the Cys-88 mutant at 62.7°C, the Ala-46/Ala-48 mutant at 62°C, and the triple mutant denatured at 70.3°C (Figure 9.6). The data show that the Cys-88 mutant possessing one extra di-sulfide bond was more stable than the wild type but equally stable as the Ala-46/Ala-48 mutant, which possessed a stabilized α-helix structure. The triple mutant was the most stable variant.

The previous examples demonstrate how HSDSC may be applied to the study of proteins in their native state, in a way that is not possible using conventional DSC. The sensitivity of the technique is such that it is possible to investigate changes in behavior induced not only by radical changes in structure — denaturation, for example — but all the way through to changes induced by minor variations of amino acids. It can be seen that HSDSC may be used to build up a complete picture of the thermal behavior of many proteins, and that the technique may assist in the manu-facture of proteins that possess specific thermal properties.

9.4.2.2 Nucleic Acids

Calorimetry has been used to study the thermal behavior of nucleic acids for many years but, until fairly recently, the lack of suitable techniques for obtaining highly purified nucleic acids limited the samples studied to a small number of molecules (36). HSDSC, in conjunction with new synthetic methodologies, allows the study of many types of nucleic acid behavior, including the thermal characterization of helix to coil transitions (37), B-Z transitions in DNA (38), the melting of a DNA duplex containing an abasic site (39), and the influence of sequence, base modification, and solution conditions on DNA triplex stability (40). For instance, Chaires and Sturtevant (38) showed that HSDSC can be used to study the B-Z transition of the polymer poly(m^{5}dG-dC). The B form of a polynucleotide is right-handed, and the Z form is left-handed, and an increase in temperature is often sufficient to convert a sample from one form to the other. The authors report that previous studies on the transition offered some evidence on the enthalpy of the transition; one suggested a ΔH_{vH} of approximately zero whereas another reported temperature dependent conformational transitions to the Z form under a variety of solution conditions. Using HSDSC, the authors obtained a direct measurement of the enthalpy change for the transition, and complementary techniques (circular dichroism [CD] and spectroscopy) were used to ascertain information on the conformation of the nucleotide with temperature. The calorimetric experiments showed a B-Z transition centered around 38°C and a second transition, not previously observed, around 54°C. Calorimetric data also showed a helix-coil transition at a higher temperature (approximately 120°C). All three transitions were shown to be two-state in character by using the van't Hoff analysis. CD and spectroscopic data suggested that the second transition corresponded to a change to an alternate left-handed conformation of unknown structure. The calorimetric data allowed enthalpy values for the transitions to be determined: $\Delta_{cal}H$ values of 0.61 kcal (mol · bp)$^{-1}$ for the lower transition and 1.05 kcal (mol.bp) for the higher transition, and ΔH_{vH} values of 68 kcal mol^{-1} for the lower transition and 263 kcal mol^{-1} for the higher transition. The ratio $\Delta_{vH}H/\Delta_{cal}H$ gives the length of the cooperative unit: 110 ± 20 bp, in this case. The data suggested that the transition from the B- to the Z form was enthalpically driven and not entropically driven, a conclusion that had been drawn from previous studies.

HSDSC has also been used to study nucleotides that contain abasic sites (39). An abasic site (usually abbreviated to AP) is produced when a heterocyclic base is removed by the selective hydrolysis of an N-glycosidic bond. Abasic sites can form under physiological conditions — as the first step in DNA repair, for example — and are believed to play a significant role in mutagenesis. Using HSDSC, the thermal behavior of a series of DNA duplexes, shown later, was investigated.

Strand 1: CGCATGAGTACGC
Strand 2: GCGTACXCATGCG

The duplexes differed only by the presence or absence of an abasic site at position X. In the absence of an abasic site, X becomes T, and the two strands can associate to form a fully paired Watson–Crick helix (referred to as the *A•T duplex*). Three abasic derivatives were created by placing different moieties in site X, each containing a synthetically modified 2-deoxyribose moiety (modified with either a tetrahydrofuran group, A•F; an ethyl group, A•E; or a propyl group, A•P). HSDSC measurements showed that the A•T duplex indicated a transition centered around

75°C, whereas the three abasic derivatives showed smaller transitions centered 60°C. In other words, the presence of an abasic site can decrease the stability, transition enthalpy, and transition entropy of a duplex structure compared with its Watson–Crick form.

9.4.2.3 Lipids

Using HSDSC, it is possible to investigate the changes in phase transition properties as a function of temperature (thermotropic mesomorphism) and concentration (lyo-tropic mesomorphism) of both synthetic and natural lipids (4). The major problem with investigations of lipids is that the lipid is usually studied in liposome form, whose properties may not wholly reflect the state of the lipid in its naturally occurring membranous form. The ability to understand and manipulate membrane function would lead to many applications in the biotechnological, pharmaceutical, and med-ical fields, although liposomes also have many important applications. It is important to study lipids of high purity because the phase transitions observed are much sharper than those observed for proteins or nucleic acids. Some phospholipids exhibit phase transitions that equate to first-order behavior; i.e., all the molecules are in one state at a certain temperature and are all in a different state at a slightly different temper-ature (4). It is also important to ensure that the presence of impurities is kept to an absolute minimum because even low levels of impurities (up to 1% w/w) can affect the value of T_m.

As a routine analysis tool, HSDSC may be used to determine the absolute purity of lipid samples. For example, HSDSC studies of recrystallized 1,2-dipalmitoyl-sn-glycero-phosphorylcholine (1,2-DPPC) showed that the sample was >99.94 mol% pure (41). HSDSC may also be used to investigate the nature of the interaction between a ligand and a lipid phase. For instance, the phase behavior of DPPC was investigated as a function of n-butanol concentration by Zhang and Rowe (42). Phospholipids are known to undergo a series of transitions with an increase in temperature; a typical series is shown in Figure 9.7. In the L_c phase, the lipid exists in an ordered, condensed crystalline subgel, in which the hydrocarbon chains are all fully extended in the trans configuration. An increase in temperature leads to the conversion to the L_β phase, the subtransition, in which the polar head groups have increased mobility, and there is greater penetration of water into the interfacial region of the bilayer. A further increase in temperature gives rise to the pretransition to the P_β phase, in which the polar head groups rotate, and there is cooperative movement of the hydrocarbon chains prior to melting. High temperatures induce the main transition to the L_α phase and the gel to liquid crystalline transition, in which the

$$L_c \longrightarrow L_\beta \longrightarrow P_\beta \longrightarrow L_\alpha$$
$$\downarrow$$
$$L_\beta I \longrightarrow L_\alpha$$

FIGURE 9.7 Typical transitions that a phospholipid system may undergo upon heating (def-initions of symbols are given in the main text).

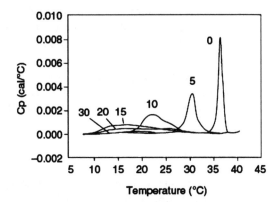

FIGURE 9.8 HSDSC scans of the pretransition observed for DPPC as a function of *n*-butanol concentration (numbers on the plot represent the concentration of *n*-butanol in mg ml^{-1}). (Reprinted from Zhang et al., *Biochemistry* 31:2005, 1992. With permission. Copyright [1992] American Chemical Society.)

hydrocarbon chains "melt" (43). In the presence of a second, smaller lipid or other suitable ligand, the L_β phase can undergo a pretransition to the $L_\beta I$ phase, in which interdigitation of the two lipids has occurred. The $L_\beta I$ phase can then convert directly to the L_α phase with an increase in temperature.

Zhang and Rowe (42) noted that the main transition between the P_β and L_α phases was very narrow (the width at half-height equaled 0.07°C) but was broadened in the presence of *n*-butanol (the width at half-height increasing to 0.37°C) at 30 mg ml^{-1}, a concentration at which the lipid is interdigitated prior to melting (such that the main transition is between the $L_\beta I$ and L_α phases). HSDSC data showed that the transition temperature for the P_β to L_α transition became lower with an increase in *n*-butanol concentration. Above the threshold concentration for interdigitation, the temperatures for the $L_\beta I$ to L_α transition also decreased with increasing *n*-butanol concentration. Figure 9.8 shows the HSDSC traces obtained for the pretransition of DPPC in the presence and absence of *n*-butanol. The transition temperature decreases as the *n*-butanol concentration increases. The shape of the transition changes between 5 and 10 mg ml^{-1} *n*-butanol, and it is suggested that this reflects a change in the type of transition occurring: L_β to P_β at low or zero added *n*-butanol and L_β to $L_\beta I$ at higher *n*-butanol concentrations.

The enthalpy of the transitions was also determined, the $L_\beta I$ to L_α transition having a slightly higher enthalpy than the P_β to L_α transition, reflecting the greater order of the $L_\beta I$ phase compared with the P_β phase. The enthalpy values were observed to differ slightly with different concentrations of *n*-butanol, presumably reflecting a contribution of the enthalpy of binding of *n*-butanol to each phase. At higher temperatures, further phase changes can occur involving the formation of three-dimensional, macromolecular structures. Typical phase changes include the formation of inverted cubic (Q_{II}) and hexagonal (H_{II}) structures. Siegel and Banschbach (44) used HSDSC to investigate higher-temperature L_α to Q_{II} transitions in concentrated solutions of the hydrated form of *N*-methylated dioleoylphosphatidylethanolamine (DOPE-me). High concentration samples of DOPE-me (33% wt%) in TES buffer showed the presence of a small endotherm at 61.2 to 61.6°C, corresponding with the formation of a Q_{II} phase, as well as an endotherm at 66.0°C, corresponding with the conversion to the H_{II} phase, at a scan rate of 9°C h^{-1}. As the scan rate was reduced, the lower temperature endotherm grew in size at the expense of the higher temperature endotherm. At a scan rate of 1.1°C h^{-1} the endotherm at 66°C had disappeared completely, indicating that the entire sample formed the Q_{II} phase.

9.4.3 Industrial Applications

HSDSC instruments that employ batch cells are capable of studying heterogeneous systems and, hence, can be used to directly study industrial products, such as creams, powders, or liquids. Perhaps the two most important areas in which HSDSC may be used industrially are stability testing and excipient compatibility.

Determining the stability of pharmaceutical products is of importance, and the conventional method at present is to use elevated temperature DSC. Samples are studied at elevated temperatures, typically 70 to 90°C, because it would be impractical, in most cases, to wait the several years it may require for the products of any reaction to build up to detectable levels. In this case, data are recorded at higher temperatures, and the Arrhenius equation is employed to calculate the rate at any desired (usually lower) temperature (45). Of course, any extrapolation requires the assumption that the mechanism of the reaction does not change over the temperature range of both the experiment and the extrapolation. As HSDSC is, by definition, more sensitive than DSC, stability studies can be conducted at lower temperatures, reducing the potential errors introduced by application of the Arrhenius equation.

HSDSC may also be used as a rapid screening tool to test the efficacy of pharmaceutical preparations. For instance, recent work has shown that HSDSC may be used to screen batches of a wax-coated drug (46). Cefuroxime axetil (CA) is the synthetic prodrug of cefuroxime, a broad spectrum and second-generation cephalosporin antibiotic, and is formulated as an aqueous suspension for administration to children. The formulation consists of small granules of CA coated with stearic acid (SA), the wax masking the intense bitter taste of the drug. Dissolution tests showed that optimal drug release into solution was obtained in phosphate buffers of high (>7.0) pH, whereas suboptimal release was obtained in low pH phosphate buffers and in water. HSDSC experiments showed the presence of multiple transitions for those samples providing conditions for optimal release and single transitions for samples providing conditions for suboptimal release (Figure 9.9). Studies on the structure of the wax spheres after removal from various buffers, using electron microscopy, revealed that the particles suspended in high pH buffers appeared to break down into flakes, whereas those suspended in low pH buffers did not. The HSDSC experiments suggest the presence of more than one type of substance in the high pH media indicating that, for the spheres to break down, some interaction with the buffer ions must occur. As optimal drug release is observed in those cases in which multiple peaks are observed in the HSDSC, it is

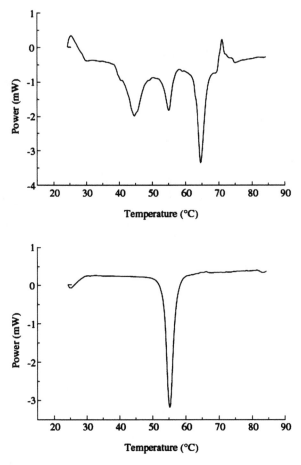

FIGURE 9.9 HSDSC traces obtained for SACA in pH 8.0 phosphate buffer (top), a disso-lution medium that gives excellent dissolution results, and distilled water (bottom), a disso-lution medium that gives very poor dissolution results. (Robson, 1997, unpublished data.)

possible to employ the instrument as a rapid screen for the suitability of different dissolution media.

The ability of some HSDSC instruments to run in both isothermal and scanning mode during one experiment allows rapid screens for excipient compatibility (47). For example, Figure 9.10 shows the HSDSC traces obtained for 1:1 mixtures of acetylsalicylic acid (aspirin) with magnesium stearate and aspirin with lactose. It is known that aspirin and magnesium stearate react, and they are not formulated together. It can be seen that at lower temperatures, there is no reaction in either of the mixtures but at 55°C the aspirin and magnesium stearate mixture gives an endothermic signal. The aspirin and lactose mixture gives rise to no thermal signal over the course of the whole experiment. Such an observation suggests that aspirin is incompatible with magnesium stearate but not with lactose, and lactose would be the most suitable excipient for use in a particular formulation.

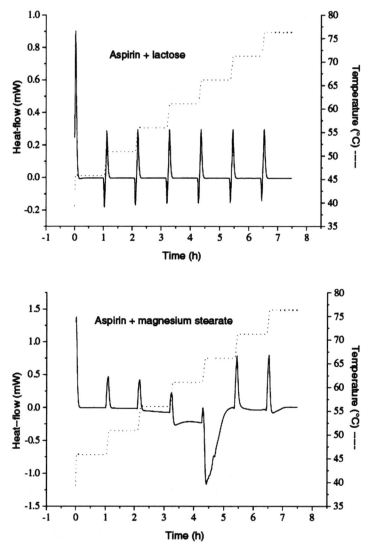

FIGURE 9.10 HSDSC traces for 1:1 mixtures of aspiring with lactose (top) and magnesium stearate (bottom) following a step-isothermal temperature program. Aspirin/lactose shows no apparent incompatibility whereas aspirin/magnesium stearate shows a considerable heat response. (Wissing, 2000, unpublished data.)

9.5 SUMMARY

HSDSC is a relatively recent (from a commercial point of view) development of standard DSC and was developed initially as a tool to enable biochemical scientists to study the thermal behavior of biological molecules in their native state. This was achieved by constructing calorimeters that combined very accurate control of block temperature and concurrent slow-heating rates with relatively large sample sizes. The use of slow-heating rates allows good calorimetric resolution whereas the large sample size, as well as enhancing calorimetric sensitivity, allows the direct study of dilute solutions of macromolecules. As the number of commercially available instruments has increased, so has the number of applications, and the literature contains many examples in which HSDSC data have given new insights into the behavior of many systems. Most of these data concern interactions among, as well as the thermal behavior of, biologically important molecules. With the introduction of batch-cell instruments, it has become possible to study solid-state and heterogeneous samples, leading to the pharmaceutically relevant application of the direct study of formulations. Industrially, two important uses of HSDSC are rapid stability testing and testing excipient compatibility. Stability tests conducted using HSDSC can be performed over a lower temperature range than is possible on standard DSC instruments, leading to fewer errors in data extrapolation. Excipient compatibility screens may be performed in just a few hours. As with all forms of calorimetry, HSDSC does not allow direct molecular interpretation of data but does allow, in most cases, a complete thermodynamic description of a particular system to be obtained. The examples discussed in this chapter highlight potential applications of HSDSC to pharmaceutical and biological fields, although there are many other applications that lie outside these areas. The ability of HSDSC instruments to hold large sample quantities and to allow the direct study of virtually any system that will fit in the cell, in conjunction with more sophisticated data-analysis software, should ensure that the technique becomes widely used in both the research and manufacturing environments.

REFERENCES

1. Ladbury, J.E. and Chowdhry, B.Z., *Biocalorimetry: Applications of Calorimetry in the Biological Sciences*, Wiley, Chichester, 1998.
2. Noble, D., *Anal. Chem.* 67, 323A, 1995.
3. Sturtevant, J.M., *Annu. Rev. Phys. Chem.* 38, 463, 1987.
4. Chowdhry, B.Z. and Cole, S.C., *TIBTECH* 7, 11, 1989.
5. Wiseman, T., Williston, S., Brandts, J., and Lin, L., *Anal. Biochem.* 179, 131, 1989.
6. Coleman, N.J. and Craig, D.Q.M., *Int. J. Pharm.* 135, 13, 1996.
7. O'Neill, M.A.A., Gaisford, S., Beezer, A.E., Skaria, C.V., and Sears, P., *J. Therm. Anal. Cal.* 84, 301, 2006.
8. Craig, D.Q.M. and Newton, J.M., *Int. J. Pharm.* 74, 33, 1991.
9. Finnin, B.A., O'Neill, M.A.A., Gaisford, S., Beezer, A.E., Hadgraft, J., and Sears, P., *J. Therm. Anal. Cal.* 83, 331, 2006.
10. Florence, A.T. and Attwood, D., *Physicochemical Principles of Pharmacy*, 2nd ed., Macmillan Press Ltd., London, 1988, pp. 291–292.
11. Schmolka, I.R., *J. Am. Oil Chem. Soc.* 54, 110, 1977.
12. Beezer, A.E., Loh, W., Mitchell, J.C., Royall, P.G., Smith, D.O., Tute, M.S., Armstrong, J.K., Chowdhry, B.Z., Leharne, S., Eagland, D., and Crowther, N.J., *Langmuir* 10, 4001, 1994.
13. Buckton, G., Chowdhry, B.Z., Armstrong, J.K., Leharne, S., Bouwstra, J.A., and Hofland, H.E.J., *Int. J. Pharm.* 83, 115, 1992.
14. Buckton, G., Armstrong, J.K., Chowdhry, B.Z., Leharne, S., and Beezer, A.E., *Int. J. Pharm.* 110, 179, 1994.

310   Thermal Analysis of Pharmaceuticals

15. Linse, P., *Macromolecules* **26**, 4437, 1993.
16. McDonald, C. and Wong, C.K., *J. Pharm. Pharmacol.* **26**, 566, 1974.
17. Wanka, G., Hoffmann, H., and Ulbricht, W., *Colloid Polym. Sci.* **268**, 101, 1990.
18. Wu, G., Chu, B., and Schneider, D.K., *J. Phys. Chem.* **98**, 12018, 1994.
19. Wu, G., Chu, B., and Schneider, D.K., *J. Phys. Chem.* **99**, 5094, 1995.
20. Alexandridis, P. and Hatton, T.A., *Colloid Surf. A. Phys.Eng. Abst.* **96**, 1, 1995.
21. Gaisford, S., Beezer, A.E., Mitchell, J.C., Loh, W., Finnie, J.K., and Williams, S.J., *J. Chem. Soc. Chem. Commun.* **18**, 1843, 1995.
22. Gaisford, S., Beezer, A.E., and Mitchell, J.C., *Langmuir* **13**, 2606, 1997.
23. Reddy, N.K., Fordham, P.J., Attwood, D., and Booth, C., *J. Chem. Soc. Faraday Trans.* **86**, 1569, 1990.
24. Brown, W., Schillén, K., and Almgren, M., *J. Phys. Chem.* **95**, 1850, 1991.
25. Mitchard, N.M., Beezer, A.E., Rees, N.H., Mitchell, J.C., Leharne, S., Chowdhry, B.Z., and Buckton, G., *J. Chem. Soc. Chem. Commun.* **13**, 900, 1990.
26. Mitchard, N.M., Beezer, A.E., Mitchell, J.C., Armstrong, J.K., Chowdhry, B.Z., Leharne, S., and Buckton, G., *J. Phys. Chem.* **96**, 9507, 1992.
27. Gaisford, S., Beezer, A.E., Mitchell, J.C., Bell, P.C., Fakorede, F., Finnie, J.K., and Williams, S.J., *Int. J. Pharm.* **174**, 39, 1998.
28. Gaisford, S., Thermodynamic and Kinetic Investigations of a Series of Pharmaceutical Excipients, Ph.D. dissertation, University of Kent, Canterbury, Kent, 1997.
29. Privalov, P.L. and Gill, S.J. *Adv. Prot. Chem.* **39**, 191, 1988.
30. Riesen, R. and Widmann, G., *Thermochim. Acta* **226**, 275, 1993.
31. Sturtevant, J.M., Holtzer, M.E., and Holtzer, A., *Biopolymers* **31**, 489, 1991.
32. Holtzer, M.E., Askins, K., and Holtzer, A., *Biochemistry* **25**, 1688, 1986.
33. Connelly, P.R., *Curr. Opin. Biotechnol.* **5**, 381, 1994.
34. Sturtevant, J.M., *Curr. Opin. Struct. Biol.* **4**, 69, 1994.
35. Breslauer, K.J., Freire, E., and Straume, M., *Methods Enzymol.* **211**, 533, 1992.
36. Stearman, R.S., Frankel, A.D., Freire, E., Liu, B., and Pabo, C.O., *Biochemistry* **27**, 7571, 1988.
37. Breslauer, K.J., Frank, R., Blöcker, H., and Marky, L.A., *Proc. Natl. Acad. Sci. U.S.A.* **83**, 3746, 1986.
38. Chaires, J.B. and Sturtevant, J.M., *Proc. Natl. Acad. Sci. U.S.A.* **83**, 5479, 1986.
39. Vesnaver, G., Chang, C.-N., Eisenberg, M., Grollman, A.P., and Breslauer, K.J., *Proc. Natl. Acad. Sci. U.S.A.* **86**, 3614, 1989.
40. Plum, G.E., Park, Y.W., Singleton, S., Dervan, P.B., and Breslauer, K.J., *Proc. Natl. Acad. Sci. U.S.A.* **87**, 9436, 1990.
41. Albon, N. and Sturtevant, J.M., *Proc. Natl. Acad. Sci. U.S.A.* **75**, 2258, 1978.
42. Zhang, F. and Rowe, E.S., *Biochemistry* **31**, 2005, 1992.
43. Taylor, K.M.G. and Morris, R.M., *Thermochim. Acta* **248**, 289, 1995.
44. Siegel, D.P. and Banschbach, J.L., *Biochemistry* **29**, 5975, 1990.
45. Koenigbauer, M.J., Brooks, S.H., Rullo, G., and Couch, R.A., *Pharm. Res.* **9**, 939, 1992.
46. Robson, H.J., Craig, D.Q.M., and Deutsch, D., *Pharm. Res.* **13**, S-332, 1996.
47. Wissing, S., Craig, D.Q.M., Barker, S.A., and Moore, W.D., *Int. J. Pharm.* **199**, 141, 2000.

10 Thermorheological (Dynamic Oscillatory) Characterization of Pharmaceutical and Biomedical Polymers

David S. Jones

CONTENTS

10.1 INTRODUCTION

Rheology, derived from the Greek words *rheo* (to flow) and *logos* (science), may be defined as the science of deformation and flow. Owing to its many applications, rheology is a fundamental component of many scientific disciplines, e.g., polymer engineering and fabrication, pharmaceutical formulation and manufacturing, and the formulation of detergents and paints (1,2). In particular, it is understood that rheological properties are important determinants of product performance. For example, in the pharmaceutical and biomedical device industries, rheological considerations are required to ensure optimal product mixing, extrusion, injection molding, filling into final containers, quality control of starting materials, and final product. Furthermore, the rheological properties of pharmaceutical and biomedical devices will directly affect their clinical performance, e.g., removal of topical formulations from containers and application and retention at the site of administration, drug release properties, rigidity (and hence patient comfort), and resistance to mechanical failure (3–9). Consequently, in the design and formulation of pharmaceutical and biomedical systems, a proper knowledge of their rheological properties is required to ensure optimal performance.

10.2 THE NATURE OF THE RHEOLOGICAL RESPONSE: ELASTICITY, VISCOSITY, AND VISCOELASTICITY

Prior to the description of the rheological properties of pharmaceutical and related systems, it is important to describe and fully understand the terms *elasticity*, *viscosity*, and *viscoelasticity*.

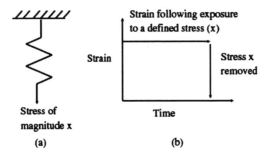

FIGURE 10.1 Rheological properties of an ideal solid: (a) depicts the rheological representation of an ideal solid, whereas (b) displays the relationship between stress, strain, and time for an ideal solid.

10.2.1 SOLIDS AND ELASTICITY

Following application of a stress, an ideal solid will undergo deformation, commonly referred to as *strain*. The relationship between the applied stress and the resultant strain, first described by Robert Hooke, is linear; the proportionality constant relating these two parameters is termed the *elastic modulus* (1,2,10). In rheology, an elastic material is represented by a loaded spring (Figure 10.1) whose extension is directly proportional to the applied weight. Furthermore, the deformation and recovery of ideal solids are observed instantaneously following application and removal of the applied stress, respectively. Therefore, there is no time-dependent deformation behavior associated with ideal solids (11,12), as depicted in Figure 10.1b.

The deformation of elastic solids occurs because of the stretching of intermolecular bonds to a point where internal stresses balance the externally applied stress (11,13). At this point, an equilibrium deformation is established. As there is little motion involved in the stretching of bonds, this occurs rapidly, and the equilibrium deformation is established infinitesimally. Deviations from ideal identity occur whenever the elastic limit of the solid material is exceeded and irreversible sample deformation results, i.e., breakage of chemical bonds (2,14). Irreversible sample deformation leading to fracture forms the basis of tensile testing (11).

There are also elastic materials that do not obey Hooke's law and exhibit a nonlinear relationship between the applied stress and the resulting strain; however, a discussion of these materials is outside the scope of this chapter. For further details, the texts by Ferry (15), Ward (11) and Ward and Hadley (12) may be consulted.

10.2.2 VISCOSITY

The application of stress to ideal liquids results in a process referred to as *flow*. Specifically, Newton proposed that the applied stress is proportional to the rate of strain (the velocity gradient), the proportionality constant in this case being referred to as the *viscosity*. Therefore, the response of ideal liquids to an applied stress is time dependent, i.e., dependent on the rate of strain and not the strain itself (1,2,16). The deformation of the liquid will continue indefinitely until the stress is removed, and, unlike ideal solids, the liquid will not return to its original undeformed state

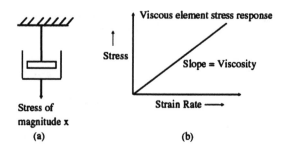

FIGURE 10.2 Rheological properties of an ideal liquid: (a) depicts the rheological representation of an ideal liquid in terms of a fluid-filled piston and dashpot, whereas (b) displays the relationship between stress and the rate of strain for an ideal liquid.

following removal of the applied stress. For a Newtonian liquid, the viscosity is constant and independent of the rate of shear. In rheology, Newtonian liquids may be represented by a dashpot or piston (Figure 10.2). Application of a stress to the system will cause the piston to extend indefinitely at a rate dependent on both the stress and the viscosity of the liquid. When the stress is removed, the piston will come to rest immediately but will not return to its original position (10,12,16,17). The flow behavior associated with Newtonian liquids may be ascribed to the greater mobility of the molecules in a liquid. In this case, an equilibrium rate of flow is established when the external stress is balanced by the frictional forces of the molecules that act to oppose deformation.

Deviations of liquids from Newtonian behavior are frequently observed for pharmaceutical and biomedical systems. In these, the relationship between stress and the rate of strain is nonlinear, examples of which include pseudoplastic (shear thinning), dilatant (shear thickening), plastic, and Bingham and Ostwald systems (1,17). Such systems are commonly referred to as *non-Newtonian systems*.

10.2.3 VISCOELASTICITY

In practice, there are few materials employed in the formulation of pharmaceutical and biomedical systems that conform to either Hooke's law or Newtonian theory (16,18). Most polymeric systems exhibit behavior in which the applied stress is proportional to both the resultant strain and the rate of strain, i.e., exhibit varying degrees of both elastic and viscous behavior simultaneously. Such materials are known as *viscoelastic* (the term *elastoviscous* is sometimes used when referring to materials that exhibit predominantly viscous behavior).

Common characteristics of viscoelastic systems include (2,14,16,18):

- The inability to maintain a constant deformation following application of a constant stress. Such systems display a phenomenon referred to as *creep*, in which sample deformation continues as a function of time. This is illustrated in Figure 10.3 (16,19).
- The ability of a viscoelastic liquid to store a portion of applied energy and use this to partially recover from deformation induced by flow.

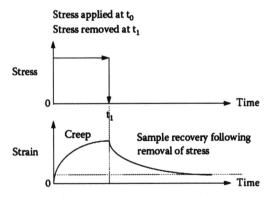

FIGURE 10.3 Diagrammatic representation of creep in polymeric systems.

The type of rheological response observed for a material will depend on the time scale of observation, and how this compares with the natural time of the material. This is an indication of the time for spontaneous molecular diffusion. For example, solid materials have very long natural times, commonly in the order of thousands of years. Therefore, most rheological events observed experimentally (taking place over a period of seconds, hours, minutes, or even days) are comparatively fast, giving rise to an elastic response. However, the movement of geological strata over millions of years confirms that such materials do in fact exhibit viscous behavior. Conversely, liquids may have very short natural times, perhaps as low as 10^{-13} sec. Therefore, normal everyday events and experimental observations are comparatively slow, giving rise to predominantly viscous behavior. However, if it were possible to observe events occurring more rapidly than in natural time, one might see liquids exhibiting elastic behavior (2,12,15). The most interesting groups of materials are those with natural times on the order of seconds, minutes, and hours that will exhibit viscoelastic responses to everyday events. Importantly, many polymeric materials employed in the design and formulation of pharmaceutical and biomedical systems fall into this category.

10.2.4 Rheological Representations of Viscoelasticity

Previously, the conventional rheological representations of ideal solids and liquids were described as an elastic spring and a dashpot composed of a piston compressing into a purely viscous liquid, respectively. Each spring is assumed to possess rigidity, a measure of the resistance to deformation, which is defined as the ratio of applied force to displacement, whereas each dashpot possesses a frictional resistance, i.e., the ratio of force to velocity of displacement. Therefore, the terms *rigidity* and *frictional resistance* are analogous to shear modulus and viscosity, respectively (15,18). As the rheological properties of viscoelastic materials encompass both elastic and viscous characteristics, it is common practice to represent viscoelastic materials as combinations of springs and dashpots in series or parallel. The most basic models often used to describe the behavior of viscoelastic materials are the Maxwell model and the Voigt (Kelvin) model, illustrated in Figure 10.4.

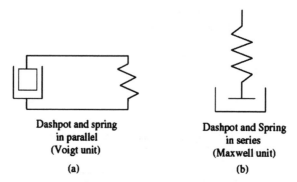

Dashpot and spring
in parallel
(Voigt unit)

(a)

Dashpot and Spring
in series
(Maxwell unit)

(b)

FIGURE 10.4 Rheological representation of viscoelasticity: (a) depicts the Voigt unit, whereas (b) depicts the Maxwell unit.

10.2.4.1 The Maxwell Model

The Maxwell model comprises an elastic spring (of modulus G) in series with a dashpot containing a purely viscous liquid (of viscosity η) (12,15,18). Following application of an external force to this system, the stress in the spring is equal to that in the dashpot and, furthermore, the total strain in the system (γ_{total}) is the sum of the strain in the spring (γ_s) and that in the dashpot (γ_d).

Hence:

$$\gamma_{total} = \gamma_s + \gamma_d \tag{10.1}$$

In accordance with Hooke's law, the spring will undergo instantaneous deformation (strain, γ) following application of a shear stress (σ):

Thus:

$$\gamma_s = \frac{\sigma}{G} \tag{10.2}$$

where G refers to the elastic modulus.

In a similar fashion, the response of the dashpot to the applied stress will follow Newton's law and thus:

$$\frac{d\gamma}{dt} = \frac{\sigma}{\eta} \tag{10.3}$$

Therefore, the time dependency of the total strain in the Maxwellian model may be written as follows (2,18):

$$\frac{d\gamma_{total}}{dt} = \frac{d\gamma_s}{dt} + \frac{d\gamma_d}{dt} = \frac{d\sigma / dt}{G} + \frac{\sigma}{\eta} \tag{10.4}$$

This equation may be conveniently rearranged to yield:

$$\sigma + \tau_M \frac{d\sigma}{dt} = \eta \frac{d\gamma}{dt} \tag{10.5}$$

The term τ_M replaces η/G and is called the *relaxation time*.

The rheological consequences of the Maxwell model are apparent in stress relaxation phenomena. In an ideal solid, the stress required to maintain a constant deformation is constant and does not alter as a function of time. However, in a Maxwellian body, the stress required to maintain a constant deformation decreases (relaxes) as a function of time. The relaxation process is due to the mobility of the dashpot, which in turn releases the stress on the spring. Using dynamic oscillatory methods, the rheological behavior of many pharmaceutical and biological systems may be conveniently described by the Maxwell model (for example, Reference 7, Reference 17, References 20 to 22). In practice, the rheological behavior of materials of pharmaceutical and biomedical significance is more appropriately described by not one, but a finite or infinite number of Maxwell elements. Therefore, associated with these are either discrete or continuous spectra of relaxation times, respectively (15,18).

10.2.4.2 Voigt–Kelvin Model

The Voigt–Kelvin model comprises an elastic spring and a dashpot (containing a purely viscous liquid) in parallel (11,15). In the parallel arrangement of the Voigt–Kelvin model, the deformation of the spring is always equal to that in the dashpot and, accordingly, the total stress (σ) of the system is described by the sum of the stresses in the elastic spring (σ_s) and viscous dashpot (σ_d).

The Voigt–Kelvin model may be mathematically described as follows:

$$\sigma = G\gamma + \eta \frac{d\gamma}{dt} \tag{10.6}$$

It may be shown that following application and maintenance of a stress (σ), the resultant deformation (γ) as a function of time (t) is described by the following equation:

$$\gamma = \left(\frac{\sigma}{G} \right) \left[1 - \exp\left(\frac{-t}{\tau_K} \right) \right] \tag{10.7}$$

The term τ_K replaces the ratio η/G and is referred to as the *retardation time*. In practice, the retardation time determines the rate at which the sample deforms following application of the stress. In a Hookean solid, the retardation time is zero as the deformation following an applied stress is instantaneous. Thus, the retardation time describes the retarding effects of the viscous properties of the dashpot to sample deformation following the application of a stress to this system (2,11,15,18). Furthermore, it may be observed that the relative contributions of the elastic properties of the spring (G) and the viscous properties of the dashpot (η) will directly influence the retardation time. Thus, if the viscous properties of the dashpot are small in comparison to the elastic properties of the spring, the retardation time will be short and there will be rapid recovery of the strain. Conversely, if the viscous properties of the dashpot predominate, then the retardation and, hence, recovery time of the system will be significant.

Once more, the rheological behavior of many pharmaceutical and biomedical materials is more appropriately described by a number of Voigt units connected in series (18). The model illustrated in Figure 10.4 describes the rheological behavior of a viscoelastic solid as, in this case, the elastic contribution is sufficient to ensure that there is no unlimited, nonrecoverable viscous flow. However, if the spring in one of the units possesses zero elasticity (i.e., $G = 0$), then nonrecoverable viscous flow will be observed, and the material is better described as a viscoelastic liquid or, alternatively, an elastoviscous system.

10.3 DYNAMIC MECHANICAL METHODS FOR THE ANALYSIS OF THE VISCOELASTIC PROPERTIES OF PHARMACEUTICAL AND BIOMEDICAL MATERIALS

10.3.1 INTRODUCTION

There are several techniques that may be employed to rheologically or mechanically characterize polymeric systems (1,14,16,23). These include:

1. Flow rheometry, in which a nonoscillatory shearing stress is applied to the sample and the resultant rate of deformation (shear) measured. By plotting the shear stress against the rate of shear, characteristic rheograms are obtained from which information concerning the resistance to flow (viscosity) may be obtained.
2. Thermomechanical analysis, in which a nonoscillatory stress is applied to the sample and the resultant deformation (strain) measured as a function of temperature.
3. Thermodilatometry, in which the sample is exposed to a range of temperatures and changes in the physical dimensions of the sample recorded.
4. Oscillatory rheometry, a nonsample destructive technique in which an oscillatory stress is applied to the sample over a range of frequencies and the resultant strain determined, usually at a defined (controlled) temperature.
5. Dynamic mechanical thermal analysis, a non-sample-destructive technique in which an oscillatory stress is applied to the sample and the resultant strain determined as a function of both frequency and temperature. Examples of this technique include thermal-ramped oscillatory rheometry and conventional dynamic thermal mechanical analysis.

Oscillatory rheometry and dynamic mechanical thermal analysis, both termed *dynamic oscillatory methods*, allow for the convenient, accurate, and rapid quantification of the viscoelastic properties of pharmaceutical and biomedical systems. In light of this, considerations of the theory, practice, and applications of these methods will form the basis of this chapter.

Prior to a more detailed description of these methods, it is appropriate at this point to provide definitions (and units of measurement) of the commonly employed terms in dynamic oscillatory analysis (16):

Stress: the force per unit area (Pa) required to deform the sample.

Strain: the amount by which the sample is deformed. In oscillatory rheometry, the strain is horizontal (torsional) and is measured in radians, whereas in dynamic mechanical thermal analysis, the strain is termed *amplitude* (distance) owing to the vertical nature of the deformation.

Damping: the ability of a material to dissipate applied mechanical energy into heat (dimensionless quantity).

Modulus: the resistance of a material to deformation (Pa).

10.3.1.1 Mathematical Description of Linear Viscoelasticity

It is commonplace for rheologists to investigate the linear viscoelastic properties of materials and, indeed, this is certainly the case in the pharmaceutical and related sciences. Bird et al. (24) and Barnes et al. (2) have suggested several reasons for this, including the ability to derive speculative molecular structures of materials from their rheological response in the linear viscoelastic region and, additionally, the ability to relate the parameters derived from the linear viscoelastic response to quality control procedures and, in some instances, to clinical response (7,9,19,25–27). Furthermore, the mathematical principles associated with the linear viscoelastic response are less complex than those for nonlinear viscoelasticity, thus ensuring relatively simple interpretations of results.

Inherent in the mathematical treatment of linear viscoelasticity is the Boltzmann superposition principle (15), which, in simple terms, states that the deformation resulting at any time is directly proportional to the applied stress. This is illustrated in Figure 10.5.

Consequently, in this superposition, doubling of the applied stress results in a twofold increase in the strain, and so on. Furthermore, the rheological parameters associated with the material under investigation, e.g., viscosity and rigidity modulus, are assumed to be constant, i.e., unaffected by changes in strain or strain rate within the linear viscoelastic region. In practical terms, a range of stresses may be employed for dynamic oscillatory testing as the magnitude of the viscoelastic parameters associated with the material will be independent of the applied stress, provided the stress selected conforms to linear viscoelasticity.

Comprehensive treatments of the mathematical basis of linear viscoelasticity have been previously described by several authors (e.g., 11,12,15,28) and, consequently, only an overview of the theory is described here.

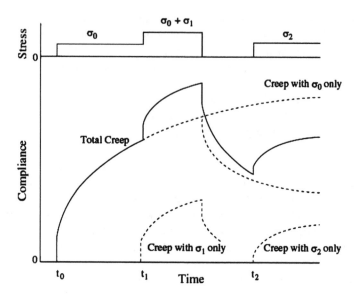

FIGURE 10.5 Application of the Boltzmann superposition principle to a creep experiment. (Modified from Vasquez-Torres, H. and Cruz-Ramos, C.A., *J. Appl. Polym. Sci.* 54, 1141, 1994.)

Typically, in dynamic oscillatory testing, a sinusoidal (oscillatory) small-amplitude stress is applied to the sample and the mechanical response measured as functions of both oscillatory frequency and, in some instances, temperature.

At any given time (t) in an oscillation cycle, the stress (σ) can be defined using the following equation:

$$\sigma = \sigma_o \cos(\omega t) \tag{10.8}$$

where σ_o is the amplitude of the stress and ω is the frequency of oscillation.

In the case of an ideal elastic solid (which obeys Hooke's law), the stress is directly proportional to the strain (γ), i.e.,

$$\sigma = G\gamma \tag{10.9}$$

where G is the elastic modulus for the material.

The strain in such a material at any given time (t) in an oscillation cycle can therefore be described by the following equation:

$$\gamma = \frac{\sigma}{G} = \frac{\sigma_o}{G}\cos(\omega t) = \gamma_o \cos(\omega t) \tag{10.10}$$

Thus, for an ideal solid, the maximum strain is observed at the same instant as the maximum applied stress (Figure 10.6). The applied energy has therefore been employed to ensure instantaneous recovery from deformation.

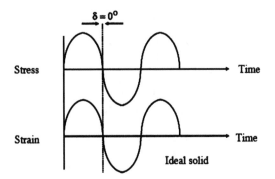

FIGURE 10.6 Relationships between stress and time and between strain and time for an ideal solid in oscillatory analysis.

In the case of a Newtonian liquid, the stress (σ) is directly proportional to the rate of shear ($\dot{\gamma}$), i.e.,

$$\sigma = \eta\dot{\gamma} \quad \text{or} \quad \sigma = \eta\frac{\partial\gamma}{\partial t} \tag{10.11}$$

where η is the viscosity of the liquid.
Therefore

$$\sigma_o\cos(\omega t) = \eta\frac{\partial\gamma}{\partial t} \tag{10.12}$$

This can be rewritten as:

$$\frac{\sigma_o}{\eta}\cos(\omega t) = \frac{\partial\gamma}{\partial t} \tag{10.13}$$

Integrating with respect to time, the following equation may be obtained:

$$\gamma = \frac{\sigma_o}{\eta\omega}\sin(\omega t) = \frac{\sigma_o}{\eta\omega}\cos\left(\omega t - \frac{\pi}{2}\right) \tag{10.14}$$

Therefore:

$$\gamma = \gamma_o\cos\left(\omega t - \frac{\pi}{2}\right) \tag{10.15}$$

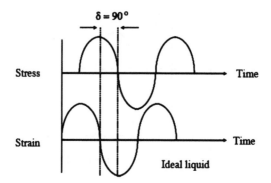

FIGURE 10.7 Relationships between stress and time and between strain and time for an ideal liquid in oscillatory analysis.

Therefore, a Newtonian liquid will exhibit a 90° phase lag to an applied stress; i.e., the strain lags behind the stress by an angle of $\frac{\pi}{2}$ radians (i.e., 90°) as illustrated in Figure 10.7.

The strain in a viscoelastic material that has both elastic and viscous elements will lag behind the applied stress by an angle (δ), where $0 < \delta < \frac{\pi}{2}$. This is illustrated in Figure 10.8.

Oscillatory analysis is used to determine the phase angle (δ) and the amplitude of the strain (γ_o) that results from the application of an oscillatory stress of amplitude (σ_o). From these, it is possible to vectorially resolve the response into elastic and viscous components.

For a perfectly elastic material that obeys Hooke's law, the elastic modulus (G) is defined as the ratio of the stress to the strain. As previously described, the strain

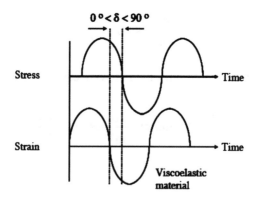

FIGURE 10.8 Relationships between stress and time and between strain and time for a viscoelastic material in oscillatory analysis.

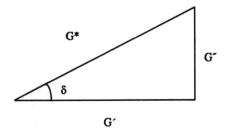

$$\text{Where:} \qquad |G*| = \frac{\text{Stress Amplitude } (\sigma_o)}{\text{Strain Amplitude } (\gamma_o)}$$

FIGURE 10.9 Vector diagram of the relationships between the elastic (storage) modulus (G'), viscous (loss) modulus (G'') and the phase angle, depicting the complex modulus ($G*$) for viscoelastic materials.

produced in a viscoelastic material when subjected to a sinusoidal stress will have both elastic and viscous components. In this case, the ratio of the stress amplitude (σ_o) to the strain amplitude (γ_o) is called the *complex modulus* ($G*$). The strain lags the stress by an angle δ. This can be represented by a vector, as shown in Figure 10.9.

The *elastic modulus* (G) is the portion of the strain that is in phase with the stress; i.e., it results from elastic deformation. Elastic deformation is recovered on removal of the stress and, therefore, the elastic modulus represents the amount of energy stored per cycle. For this reason, it is often referred to as the *storage modulus*.

The *viscous (loss) modulus* (G'') is the portion of the strain that is 90° out of phase from the stress; i.e., it results from viscous deformation. Viscous deformation is not recovered on removal of the stress. Instead, the energy required to produce viscous deformation is dissipated as heat due to friction. Accordingly, the viscous modulus is often referred to as the *loss modulus*. Typically, the Maxwellian model suggests that at high oscillatory frequencies, the behavior of viscoelastic systems approaches perfect elasticity, as the mobility of the dashpots is negligible in comparison to that of the springs. In this scenario, the loss modulus approaches zero and corresponds to the absence of atomic or molecular mobility that would normally be capable of dissipating energy in the form of heat. However, it is worthwhile to note that this behavior is rarely displayed by polymeric systems. Conversely, at low oscillatory frequencies, the viscous modulus of viscoelastic liquids is directly proportional to the applied frequency. The proportionality constant is referred to as the *Newtonian steady-flow viscosity* (η_o) and has a magnitude of 1 on a logarithmic plot. Furthermore, for viscoelastic solids, the relationship between G'' and oscillatory frequency, once more over a range of low frequencies, is also directly proportional.

The *loss tangent* (tan δ) is a dimensionless term that describes the ratio of the loss modulus (G'') to the storage modulus (G'), and this may be usefully employed to convey information concerning the structure of the polymeric system in question. This is frequently referred to as the *mechanical damping parameter*. To substantiate the relevance of this viscoelastic parameter, several examples of the different types of relationships between oscillatory frequency and the loss tangent are described (15).

For dilute polymeric solutions, both the solvent and dissolved polymer contribute to the loss modulus whereas only the dissolved polymer contributes to the storage modulus. Therefore, the predominance of the loss modulus to the viscoelastic properties ensures that, in these systems, the loss tangent is large (>1). Similarly, at low frequencies, tan δ for un-cross-linked polymers is large, and within this frequency region, the loss tangent is inversely proportional to the oscillatory frequency. In both these examples, it is important to note that the viscous properties predominate. For amorphous polymers (cross-linked or non-cross-linked) in the regions of low frequency, the loss tangent ranges from approximately 0.2 to 3, whereas glassy or crystalline and cross-linked polymers exhibit loss tangent values of about 0.1 and 0.01, respectively (15). Thus, the wide range of tan δ values exhibited by these various polymer categories is a direct consequence of their structural differences, and, in particular, the ability of these polymers to utilize or dissipate applied energy. The pharmaceutical importance of the loss tangent has been further illustrated by Davis (20). In this, the author described the logarithmic plots of tan δ against frequency as "consistency spectra" that were individual to each pharmaceutical system. In such fashion, the loss tangent may be used, e.g., to examine the effects of physical and chemical factors on the rheological properties of such systems. However, it should be noted that to correctly observe the maximum in the loss tangent, a wide range of oscillatory frequencies may be required (about 10–20 decades). This is difficult to experimentally perform and generally requires the use of both creep and dynamic methods and the application of mathematical models.

10.3.2 Methods for Determination of Dynamic Mechanical Properties of Polymeric Systems

As previously described, dynamic mechanical analytical techniques provide valuable information concerning the viscoelastic properties of polymeric systems, in particular, over defined ranges of frequencies of oscillatory stress and temperature. Owing to the development of a wide range of sampling handling facilities (geometries), the reported uses of dynamic mechanical methods have dramatically increased, establishing such techniques in the pharmaceutical and related industries. There are a number of different methods by which the dynamic mechanical properties of polymeric systems may be measured (12,14,29), including:

- Torsion pendulum
- Wave propagation methods
- Forced vibration nonresonance methods

10.3.2.1 The Torsion Pendulum Method

This is a relatively simple technique that is primarily employed to quantify the viscoelastic properties of polymeric rods. In this, the polymeric sample is clamped vertically between a fixed upper clamp and a horizontal bar, the latter being fixed with moveable weights. The horizontal bar is twisted by the application of a suitable force, released, and allowed to oscillate naturally. The amplitude of the oscillations will gradually decrease as a function of time and the logarithmic decrement (Λ), i.e., the natural logarithm of the ratio of the amplitude of successive oscillations, is measured. Notably, by altering the distance of separation of the two weights attached to the horizontal bar, both the moment of inertia (I) and the period of oscillation may be altered to suit the consistencies of selected polymer samples.

The equation of motion of an elastic rod may be described as follows:

$$I\ddot{\theta} + \tau\theta = 0 \qquad (10.16)$$

where τ is the torsional rigidity of the rod.

Furthermore, the torsional rigidity of the rod may be related to the torsional modulus (or shear modulus, G) of the rod as follows:

$$\tau = \frac{\pi r^4 G}{2l} \qquad (10.17)$$

where l and r are the length and radius of the rod, respectively.

Under the conditions of the experimental method, the sample will exhibit simple harmonic motion, and thus the angular frequency (ω) may be related to the shear modulus (G):

$$\omega = \sqrt{\frac{\pi r^4 G}{2Il}} \qquad (10.18)$$

For viscoelastic materials, the torsion modulus is more complicated than that for elastic solids and may be described by the following equation:

$$G^* = G' + iG'' \qquad (10.19)$$

Under conditions of small damping, the elastic (storage) modulus (G') equals the torsional modulus of the rod (G) and thus the equation for simple harmonic motion may be rewritten:

$$\omega^2 = \frac{\pi r^4 G'}{2Il} \qquad (10.20)$$

Importantly, the logarithmic decrement (Λ) may be related to the loss tangent:

$$\frac{G''}{G'} = \frac{\Lambda}{\pi} = \tan\delta \qquad (10.21)$$

Generally, the operative frequency range for the torsional pendulum method is 0.01 to 50 Hz, the upper limit of the frequency defined by the dimensions of the oscillatory frequency relative to the dimensions of the sample. At higher investigative temperatures, the polymeric materials may undergo extensional deformation (creep) due to the weight of the inertia bar. Under such circumstances, a modified torsion pendulum apparatus may be used in which an inertia disk is attached directly onto the end of the sample (12). In addition, this method is frequently employed to measure the torsion modulus at low frequencies.

10.3.2.2 Wave Propagation Methods

These methods may be conveniently categorized into two general groups according to the magnitude of applied oscillatory frequency, namely, the kilohertz frequency range and the megahertz (ultrasonic) frequency range.

The frequency range for the former category is 1 to 10 kHz. In this, monofilament samples are deformed longitudinally with one end attached to a diaphragm, typically, a loudspeaker. Changes in the signal amplitude and phase along the length of the sample are measured using a piezoelectric crystal. Typically, the phase is directly proportional to the distance along the sample, the proportionality constant being referred to as the *propagation constant* (k). Importantly, it may be shown that the propagation constant and the attenuation coefficient (α) may be related to the storage and loss modulus.

Thus:

$$G' = \frac{\omega^2}{k^2}\rho$$

$$G'' = \frac{2\alpha c}{\omega} \tag{10.22}$$

$$\tan\delta = \frac{2\alpha c}{\omega}$$

where ρ is the density of the sample and c is the longitudinal wave velocity.

For further information on this technique, the texts by Ward (11) and Ward and Hadley (12) may be consulted.

In the second category, the viscoelastic properties of polymeric materials, particularly oriented polymers and composites, may be determined by the application of oscillatory frequencies in the megahertz frequency range. These methods are more complicated in design, and the interpretation of results is more complex. Further information on these methods is given in the excellent texts by Ward (11) and Ward and Hadley (12).

10.3.2.3 Forced Vibration Methods

Typically, this method represents what is considered to be classical dynamic thermal mechanical analysis (14) and, accordingly, most of the more recently published information on the viscoelastic properties of polymeric materials has been obtained using this method. As a result, a more comprehensive explanation of this method will be provided. Typically, samples are subjected to an oscillatory stress or deformation (strain) over a defined frequency range (usually 0.01–100 Hz). As described previously, following the imposition of, e.g., an oscillatory strain on a viscoelastic material, the strain will lag behind the stress by a phase angle (δ), provided the measurements have been derived from the region of linear viscoelasticity and, additionally, that there is no appreciable variation of stress along the length of the sample (14,16). It is therefore customary to ensure that the length of the sample is short compared to the wavelength of the stress waves (<10%) (12).

Most reports concerning the use of fixed vibration-based systems describe the mechanical properties of polymeric samples of defined dimensions, e.g., rectangular and cylindrical (30–32), reflecting the sample-handling requirements of these machines. However, instrumentation is now commercially available that may be effectively employed to characterize the viscoelastic properties of samples exhibiting a wide range of physicochemical properties and consistencies, primarily owing to the availability of several types of measuring geometries (16). One example of such instrumentation is the *dynamic mechanical analyzer* (DMA) 2980 (33). Because this is the instrument employed in the authors' laboratories, further discussions of the fixed vibration method of analysis will be directed toward this instrument.

In accordance with other forced vibration methods, mechanical analysis of samples involves the application of a fixed stress or oscillation on the material over a defined temperature range. The various specific measuring geometries are employed, first, to hold the samples in place and, second, to interface with the driving motor of the instrument. Hence, the oscillatory stress is provided by a driving motor and is transmitted to the sample via the clamping system. Accurate control of the temperature of the system is performed by enclosing the stainless steel clamps (with attendant sample) within a furnace. In a similar fashion to other thermal analysis methods, subambient temperatures may be achieved within the furnace by the use of a liquid nitrogen cooling facility. A diagrammatic representation of the design of the dynamic mechanical thermal analyzer 2980 is shown in Figure 10.10.

As explained previously, the versatility of the DMA 2980 and other related commercially available instrumentation to the mechanical characterization of pharmaceutical and related polymeric systems is a direct consequence of the availability of a wide range of sample-clamping systems. The sample consistencies that may be analyzed range from viscous liquids to materials of high modulus and low damping properties. Furthermore, a range of sample presentations may be conveniently analyzed using these geometries, e.g., dilute gels, thin films, single and bundled fibers, and cylinders. The use of older instruments was restricted primarily to samples of defined geometry, e.g., rectangular sheets of known thickness and cylinders. There are several clamp geometries available for use with modern instruments, including:

- Single- and dual-cantilever clamps
- Three-point bending clamp
- Shear sandwich clamp

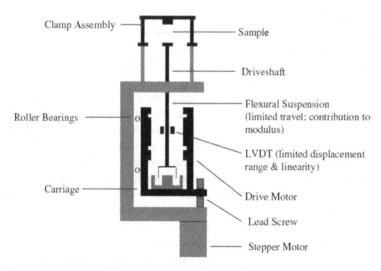

FIGURE 10.10 Diagrammatic representation of the design of the dynamic mechanical analyzer 2980. (Adapted from Dynamic Mechanical Analyser 2980 Operator's Manual, TA Instruments, New Castle, DE, 1996.)

- Compression clamp
- Film tension and fiber tension clamps

10.3.2.3.1 Single- and Dual-Cantilever Clamps

These clamping systems are primarily employed for the characterization of the mechanical properties of relatively weak to moderately stiff materials, e.g., supported thermoseting resins, elastomers, and thermoplastic materials. In the single-cantilever system, the sample is firmly clamped using the outer cantilever and allowed to reside within the central region of the clamp, i.e., the measuring head. Conversely, in the dual-cantilever clamp, the sample is clamped between the two outer cantilevers and, once more, allowed to reside within the central region of the clamp. In both cases, the clamp system is attached to the driver motor via the measuring head of the clamp. Therefore, this portion, and thus the sample, will directly respond to the vertical oscillatory motion of the driving motor. It is customary to define the distance over which the oscillation occurs. A diagram of these clamping systems is presented in Figure 10.11.

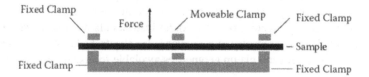

FIGURE 10.11 Diagrammatic representation of the single and dual cantilever geometries. (Adapted from Dynamic Mechanical Analyser 2980 Operator's Manual, TA Instruments, New Castle, DE, 1996.)

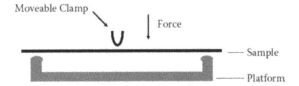

FIGURE 10.12 Diagrammatic representation of the three-point bend geometry. (Adapted from Dynamic Mechanical Analyser 2980 Operator's Manual, TA Instruments, New Castle, DE, 1996.)

The choice of either the single- or dual-cantilever systems depends on the mechanical properties of the samples. It is recommended by many manufacturers that dual-cantilever clamps be used for the analysis of weak elastomers and supported resins, whereas single-cantilever clamps are preferred for the characterization of amorphous polymers and elastomers through the glass transition and for the analysis of materials that possess high thermal expansion. In general, we would further recommend these clamping systems for the characterization of pharmaceutical packaging materials and conventional tubular and planar polymeric medical devices (e.g., wound dressings, transdermal patch systems, urethral and ureteral catheters, and endotracheal tubes) that possess sufficiently high modulus. As a general rule, the shear clamp does not require the sample to provide a restoring force to deformation and, therefore, samples with high damping characteristics may be analyzed using this technique.

10.3.2.3.2 Three-Point Bending Clamp

Three-point bending clamps are used primarily for the analysis of stiff, highly elastic materials, e.g., polymethylmethacrylate, metals, and ceramics. In this, the sample is rested on the two fixed clamps of the clamping system; hence, the use of rigid materials in this mode of analysis, over which is placed the moveable clamp. Once more, the moveable clamp is attached to the drive motor, allowing a direct response to the vertical oscillatory motion of this component. A diagram of this clamping system is presented in Figure 10.12. It is important that the samples analyzed using this method exhibit sufficient elasticity to restore the moveable clamp to its original position.

10.3.2.3.3 Shear Sandwich Clamp

In a similar fashion to the cantilever clamping systems, the shear clamp does not require the sample to provide a restoring force to deformation and, therefore, samples with high damping characteristics may be analyzed using this technique. As a result, this measuring geometry is frequently employed to characterize the viscoelastic properties of samples of low elasticity, ranging from unsupported viscous liquids, e.g., solutions of polymers, pharmaceutical creams, to elastomers above the glass transition temperature, e.g., silicone. In the shear sandwich clamp, two samples of equal size and shape are placed on either side of the central portion of the moveable clamp. As before, the moveable clamp is oscillated vertically at a defined frequency over a defined distance and the modulus of the samples determined. The manufac-

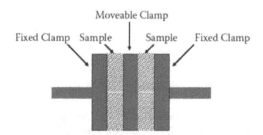

FIGURE 10.13 Diagrammatic representation of the shear sandwich geometry. (Adapted from Dynamic Mechanical Analyser 2980 Operator's Manual, TA Instruments, New Castle, DE, 1996.)

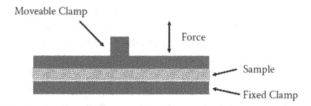

FIGURE 10.14 Diagrammatic representation of the compression clamp geometry. (Adapted from Dynamic Mechanical Analyser 2980 Operator's Manual, TA Instruments, New Castle, DE, 1996.)

turers recommend that the users of this clamping system pay particular attention to the clamping conditions. Typically, stiff samples have been observed to slip and, additionally, the apparent modulus of an elastomeric material may change significantly with clamping force (33). A diagram of this clamping system is presented in Figure 10.13.

10.3.2.3.4 Compression Clamp
The compression clamp may be conveniently employed to measure the viscoelastic properties of materials exhibiting low to medium modulus; i.e., samples that offer a restoring force to compression (low damping properties). Examples of samples that may be analyzed in this fashion include pharmaceutical gels, toothpastes, and adhesives. In this mode of measurement, the sample is introduced between two plates, an upper fixed plate and a lower "movable" plate that is connected to the driver motor. The sample is subsequently deformed (to a defined amplitude) by the vertical oscillatory motion of the lower plate. The manufacturers recommend that the sample have as high a thickness: diameter ratio as possible in light of the restrictions of sample preparation and instrument limits (maximum allowed thickness is about 10 mm) (33). A diagram of this clamping system is presented in Figure 10.14.

10.3.2.3.5 Film Tension and Fiber Tension Clamps
These clamps are employed for the mechanical characterization of both films and fibers in tensile mode. Typically, the top of the film and fiber sample is mounted within the upper (static) portion of the clamp and the bottom section held by the

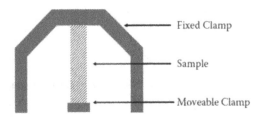

FIGURE 10.15 Diagrammatic representation of film on and fiber tension geometry. (Adapted from Dynamic Mechanical Analyser 2980 Operator's Manual, TA Instruments, New Castle, DE, 1996.)

movable arm that is connected to the driver motor. As before, the sample is deformed to a defined amplitude by the vertical oscillatory motion of the lower movable arm. These clamping systems are employed for the mechanical analysis of polymeric films and fibers of a wide range of consistencies, e.g., elastomers, and rigid films. A diagram of this clamping system is presented in Figure 10.15. Furthermore, examples of the use of the film tension clamping system for the thermorheological characterization of poly(ε-caprolactone) films, designed as coatings of urinary medical devices, have been reported by the authors (34,35).

10.3.2.3.6   Controlled Temperature Oscillatory Rheometry

As previously stated, in dynamic mechanical analysis the sample is subjected to a sinusoidal (vertical) stress and the resulting strain measured as functions of both oscillatory frequency and temperature. The resultant relationship between these two parameters is used to obtain information on the viscoelastic properties of the material (16). In controlled temperature oscillatory rheometry, the same principles apply; however, in this technique the oscillatory stress is applied to the sample in a horizontal (torsional) fashion over a range of oscillatory frequencies. Temperature control is facilitated by the use of a water bath that is connected directly to the measuring geometry that locates the sample. Oscillatory rheometry is primarily employed for the mechanical characterization of samples that possess low to medium modulus. One of the most common systems for the determination of the oscillatory properties of such samples is the Cone and Plate Rheometer. Typically, a range of geometries, ranging from 2 to 6 cm, are available for the rheological characterization of polymeric systems using this method. In general, samples of lower consistency require the use of larger-diameter geometries to ensure adequate sensitivity and, conversely, materials possessing significant modulus may be conveniently analyzed using plates of smaller diameter. The use of larger plates for samples that possess considerable viscous properties ensures the accurate application of small torques to such samples. The choice of the most appropriate geometry is optimally performed following determination of the linear viscoelastic region of the sample under investigation. The linear viscoelastic region is initially identified as the region in which applied stress is directly proportional to strain (and the storage modulus [G'] remains constant) following torque sweep over a defined modulus (Pa) at defined frequencies (generally, the lower and upper limits that will be employed in subsequent analyses).

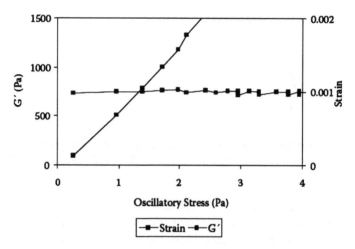

FIGURE 10.16 Linear viscoelastic region of bioadhesive pharmaceutical gels.

An example of this is shown in Figure 10.16. Consequently, the plate associated with the largest linear viscoelastic region should be selected.

10.4 PHARMACEUTICAL AND BIOMEDICAL APPLICATIONS OF DYNAMIC (OSCILLATORY) RHEOLOGICAL ANALYSIS

Dynamic mechanical techniques have been widely employed in the polymer and related industries; however, the applications of such techniques for the characterization of pharmaceutical and certain biomedical systems have not received similar attention. Therefore, in this section, an overview of reported and future applications of these techniques for the characterization of pharmaceutical and biomedical systems is provided.

In general, owing to both the range of both experimental methods, and indeed, the varied choice of measuring geometries, dynamic mechanical methods may be employed to characterize the viscoelastic (and related) behavior of materials of many sample types, e.g., ranging from viscoelastic liquids to rigid viscoelastic polymeric films. Typically, the following information concerning such polymeric systems may be obtained using oscillatory analysis (16):

1. Quantitative modulus, i.e., storage and loss moduli
2. Glass transitions (primary, secondary, etc.)
3. Rate and extent of polymeric curing
4. Quantification of gelation, e.g., sol–gel transitions
5. Damping properties, i.e., characterization of the ratio of loss to storage moduli at defined temperatures
6. Polymer morphology

7. Interactions between polymeric components or between drug molecules
   and polymeric constituents of pharmaceutical or biomedical systems
8. Dynamic viscosity

It is important to realize that this is not an exhaustive list of applications of
oscillatory methods; however, the information described within these categories is
most frequently employed to describe the oscillatory properties of pharmaceutical
and biomedical systems.

10.4.1 DYNAMIC ANALYSIS OF POLYMERIC GEL SYSTEMS

It is commonly accepted that polymeric gel systems are composed of supermolecular
structures that are formed by cross-linking or by the entanglement of polymeric
chains within a liquid solvent (36,37). The network structure of these systems
directly influences their mechanical properties and, indeed, through alterations in
the network structure (cross-linking and entanglement), gels may be formulated to
offer a wide range of moduli. The use of gels in clinical medicine, e.g., as drug
delivery systems (5,7,35), electrically conductive interfaces (38), implantable drug
delivery systems (39), and as artificial body fluids, e.g., salivary replacements (40),
is widely accepted. Furthermore, the clinical efficacy of pharmaceutical and biomedical gel systems may be directly related to their mechanical and rheological
properties. Therefore, dynamic oscillatory methods are important analytical tools
for both the characterization of such systems, and hence, the optimization of their
formulation to provide clinical efficacy.

10.4.1.1 Characterization of Rheological Properties of
Gel Systems

Dynamic oscillatory methods may also be employed to provide information concerning both the structure and modulus of gel systems. This information may then
be usefully employed in the development of formulations that will offer optimal
clinical and nonclinical (e.g., stability) performance. Examples of these aspects are
provided in the following subsection.

Carbopols are a series of cross-linked carboxyvinyl polymers that have found
extensive use in the pharmaceutical and related industries as a result of their marked
(and consistent) thickening properties at low concentrations (41,42). More recently,
an interest has developed in the inclusion of these polymers in pharmaceutical
products owing to their bioadhesive properties, i.e., their ability to strongly interact
with biological substrates. Because of this adhesive interaction, the retention time
of formulations at the site of application may be enhanced, thus increasing either
the absorption of a drug across a biological membrane or, alternatively, maintaining
an effective concentration of drug within a particular cavity, e.g., rectum and oral
cavity (5,6,8,9,27,43). The viscoelastic properties of Carbopol gels have been shown
to influence their clinical performance by altering, e.g., their bioadhesive properties
(27) and the subsequent rate of drug release from such systems (44). One of the
earliest comprehensive studies on the viscoelastic nature of Carbopols was reported

by Barry and Meyer (42). Using oscillatory rheometry over a range of temperatures, these authors described the relatively negligible change of both elastic and loss modulus over the range of oscillatory frequencies (10^{-2}–10^1 Hz), indicative of the plateau region of viscoelasticity (15). From this and the information derived from creep analysis, the authors concluded that the Carbopols were high-molecular-weight amorphous polymers with long chains. Increasing the concentration of Carbopol was shown to increase both the elastic and loss moduli, indicative of increased frequency of polymer chain entanglement.

The effects of formulation excipients, e.g., solvents, presence of therapeutic agents or salts on the rheology of Carbopol gels have been examined by several authors. From these studies, essential information has been derived concerning the suitability of these polymers as topical drug delivery systems and electrical interfaces. For example, Chu et al. (25) examined the effects of mixed solvent-(binary and ternary) systems composed of propylene glycol, glycerol formal, and water on the rheological properties of Carbopol 934 gels. In this study, it was shown that, in the presence of high concentrations of nonaqueous solvents, Carbopol (0.5–1.5%) existed as low-viscosity solutions, indicative of low interaction between the solvent and the polymer. Increasing the concentration of water in these systems significantly increased both the elastic and loss moduli and decreased the loss tangent. The authors attributed these observations to the greater affinity of the solvent system for the polymer, resulting in polymer chain extension. This study has therefore demonstrated the role of solvent composition, and hence solubility parameter, on the viscoelastic nature of Carbopol gels. More recently, Craig et al. (43) described the effects of a cosolvent (propylene glycol), polymer concentration, and the incorporation of a model drug (chlorhexidine gluconate) on the rheological properties of Carbopol 934 gels. Increasing concentrations of either Carbopol or propylene glycol increased both the storage and loss moduli; however, in the case of propylene glycol, no concentration-dependent effects on these viscoelastic properties were apparent. In addition, the presence of chlorhexidine gluconate was also shown to alter the rigidity of the bonds between adjacent polymer chains. These observations have important implications for the design of topical drug delivery systems based on Carbopol gels. In a subsequent study, Tamburic and Craig (27) illustrated the reciprocal relationship between mucoadhesive strength and formulation loss tangent and highlighted the potential importance of viscoelastic properties and clinical performance. More recently, Bonner et al. (45) and Jones et al. (38) reported the effects of both ethanol concentration and added electrolytes on the viscoelastic properties of Carbopol gels as electrically conducting interfaces. Increasing the concentrations of either ethanol or electrolyte concentration (NaCl, $CaCl_2$, and tetracaine hydrochloride) decreased both the storage and loss moduli of these gels. These effects were attributed to the greater affinities of the electrolytes and ethanol for the solvent sheath associated with the polymer, and hence, the extent of the polymer and solvent interaction was decreased. At higher concentrations of ethanol (50% w/w) or electrolytes (e.g., 0.2 M NaCl), the rheological properties of the polymer systems resembled polymer solutions and, as such, were inappropriate for use as electrically conductive interfaces and drug delivery systems. More recently, Jones et al. (46) described the dynamic oscillatory properties of nonaqueous (propylene glycol, ethylene glycol, and glycerol) gels of polyacrylic acid. Interestingly, within these nonaqueous solvents, polyacrylic acid existed as an entangled gel whose rheological properties were dependent on the nature of the solvent. In particular, the capacity of the solvent to hydrogen-bond with the carboxylic acid groups of the polymer was a major determinant of the resultant rheological properties, with glycerol promoting the greatest elastic structure.

Water-soluble cellulose ether derivatives are polymers that are commonly used in pharmaceutical formulation, e.g., to alter the viscosity of formulations or to control drug release (47). A wide range of cellulose derivatives offering different degrees of substitution is available, e.g., hydroxyethylcellulose, methylcellulose, hydroxypropylcellulose, and hydroxypropylmethylcellulose. The viscoelastic properties of these systems are instrumental in product performance and, accordingly, oscillatory methods have been employed to characterize them. The viscoelastic properties of two cellulose derivatives (sodium carboxymethylcellulose, and hydroxyethylcellulose) were described by Jones et al. (48). These polymers formed gels whose viscoelastic properties were dependent on polymer concentration. Over the entire concentration range examined, the polymer gels exhibited the plateau region of viscoelasticity, from which we can infer that in this case gel formation may be attributed to entanglement of the linear polymer chains. In addition, correlations were apparent between elastic and loss moduli and various textural properties (hardness and compressibility) and, hence, patient acceptability and spreadability. Furthermore, the mucoadhesive properties of these gels, as determined using a tensile method, were inversely related to formulation loss tangent, further illustrating the role of formulation viscoelasticity on clinical efficacy. In a related study, the viscoelastic properties of gels composed of either hydroxypropylmethylcellulose or the polysaccharide xanthan gum were examined with respect to their performance as excipients in oral controlled-release formulations (49). At concentrations arbitrarily selected to represent the hydrated "gel-like" surface of the tablet at the interface between the formulation and gastrointestinal fluids, hydroxypropylmethylcellulose exhibited rheological properties consistent with polymeric solutions. Conversely, aqueous systems containing comparative concentrations of xanthan gum exhibited viscoelastic properties consistent with physically entangled gel systems. The authors concluded that the differences in the rheological behavior of these two polymers were consistent with the greater ability of matrices composed of xanthan gum to retard drug release.

The effects of solvent nature on the viscoelastic properties of cellulose-based polymeric gels have been described by Brown et al. (50). Within the concentration range examined, hydroxypropylmethylcellulose, hydroxyethylcellulose, and sodium carboxymethylcellulose exhibited evidence of polymer chain entanglement, i.e., a gel-like structure, in an entirely aqueous solvent. Introduction and increase of the concentration of alcoholic cosolvents (ethanol, propylene glycol, and glycerol) into these systems significantly increased both the elastic and loss moduli. However, at certain solvent compositions (dependent on the type of alcohol), a maximum in these viscoelastic parameters was observed for each mixed-solvent system. Further increases in the concentration of cosolvent resulted in either sequential or dramatic reductions in each of these moduli. Typically, cellulose polymers undergo swelling in aqueous solvents until fully dispersed, at which point, the systems exhibit maximum viscosity. Then, because of progressive dilution of the solubilized polymer, the viscosity of these systems progressively decreases until equilibrium viscosity is attained (24). The ability of alcoholic cosolvents to enhance both the elastic and loss moduli of cellulose derivatives may therefore be attributed to both the lower concentration of available water in these formulations for hydration and to suppression of dissolution of the swollen polymer following swelling. Furthermore, the loss of viscoelastic structure associated with certain concentrations of cosolvents may be explained by the inability of the cellulose polymer to undergo swelling in such solvent systems owing to their unfavorable solubility parameters.

Further considerations of the role of solvent composition on the viscoelastic properties of cellulose derivatives were described by Lin et al. (26). Using oscillatory rheometry, the viscoelastic nature of hydroxypropylmethylcellulose in a range of mixed-solvent systems composed of glycerol formal or ethanol, propylene glycol, and water was quantified. The elastic and loss moduli were greatest when glycerol formal was employed as a solvent and, furthermore, by careful consideration of the composition of the mixed-solvent system, gels offering a range of moduli could be obtained. Once more, this study highlighted the interrelationship between the viscoelastic properties of this polymer and solvent composition. Using the nondepleted frog palate model, the rate of mucociliary transport of aqueous solutions of hydroxypropyl methylcellulose (HPMC) and other polymeric systems (polyoxyethylene oxide, sodium carboxymethylcellulose, sodium alginate, and polyacrylic acid) may be correlated with the viscoelastic properties of such systems, in particular the loss tangent, in a curvilinear fashion. This study once again highlighted the potential relationship between clinical efficacy, in this case persistence within the nasal cavity, and polymer viscoelasticity. In a related study, the same authors (51) described the effects of ternary solvent composition (glycerol formal, propylene glycol, and water), polymer molecular weight, and salt concentration on the viscoelastic properties of polyethylene oxide solutions designed for nasal administration. Typically, in aqueous solution, low-molecular-weight polyethylene oxide possessed primarily liquidlike behavior, whereas the high-molecular-weight polymer exhibited evidence of polymer chain entanglement, i.e., physical gelation. Both the elastic and loss moduli dramatically increased both as the concentration of polyethylene oxide increased as the proportion of water was increased in ternary solvent systems, the latter phenomenon being attributed to the greater facilitation of polymer–solvent interactions. The authors concluded that the use of a mixed-solvent system that ensures low-viscosity (liquid) properties would be appropriate for nebulization. However, following deposition on the mucous environment of the nasal cavity, the formulation will be transformed into a highly viscoelastic gel because of the interaction with the moisture present at the site of application. This transformation was therefore proposed to enhance the persistence within the nasal cavity.

The relationships between structure and gelling characteristics of agarose, polyvinyl alcohol, and high methoxyl pectin-water gels in mixed-solvent systems composed of dimethyl sulfoxide (DMSO) and water were examined by Watase and Nishinari (52–54). The elastic properties of gels composed of agarose or polyvinylalcohol, DMSO, and water increased as a function of DMSO concentration, up to a maximum value of 0.277 mole fraction. The authors suggested that DMSO interacted strongly with water, thus increasing the effective concentration of polymer; hence, the gelling ability increases. This concentration of DMSO was associated with the optimal number and structure of junction zones (sites of chain entanglement) and also optimal conformation of flexible chain molecules connecting junction zones. At higher concentrations, the concentration of free DMSO increases, thus inhibiting the formation of junction zones owing its strong solvation properties. Similar observations were reported by these authors for gels composed of high methoxyl pectin, DMSO, and water (54).

Oscillatory methods may also be employed to characterize the process of gel formation. In one report, Jauregui et al. (55) studied the viscoelastic properties of two hydroxyethylated starch aqueous systems and identified three concentration-dependent viscoelastic regions. At low concentrations, a fluidlike zone, corresponding to a homogeneous solution in which there were no interactions between polymeric chains, was observed. Second, at intermediate concentrations, the fluid–gel transition zone was apparent in which specific intermolecular interactions between adjacent polymer chains were operative. Finally, at high concentrations, a structured network was present, referred to as the *gel-like zone*, facilitated by hydrogen bond formation between adjacent polymer chains. This study illustrated the gradual structural organization of gel systems and the impact of such structural properties on thermomechanical properties.

10.4.1.2 Characterization of Rheological Properties of Mixed-Polymer Gel Systems

It is commonplace in pharmaceutical formulation to combine two or more polymers within a single dosage form, e.g., gel, semisolid, and solid oral dosage form as, in many instances, this ensures enhanced control of dosage form performance, e.g., stability, drug release rate, spreadability, etc. Once more, both the clinical and nonclinical performance of these systems may be related to their (complex) viscoelastic behavior, and, hence, these systems have been characterized using dynamic oscillatory methods.

The rheological properties and compatibility of mixed gels composed of the polysaccharide κ-carrageenan and various galactomannans (locust bean gum, tara gum, and guar gum) have been described by Muruyama et al. (56). From viscoelastic measurements, it was confirmed that the basic elastic structure of the gel was provided by κ-carrageenan. However, this elastic structure was significantly improved by the addition of locust bean gum (primarily) or tara gum. These effects were accredited to the binding of the "smooth zone" of the galactose portion of these gums to the double helix of κ-carrageenan. Conversely, the presence of guar gum did not enhance the elastic properties of gels of κ-carrageenan. Similarly, the use of oscillatory methods to characterize the rheological properties of mixed-gel systems was provided by Jones et al. (7,19). In this work, the viscoelastic properties of biphasic semisolid systems containing hydroxyethylcellulose (HEC), polyvinylpyrrolidone (PVP), polycarbophil (PC), and chlorhexidine or tetracycline designed for topical application to the oral cavity were described. Typically, increasing the concentrations of each component significantly increased formulation elastic and loss modulus and was accredited to enhanced entanglement of the dissolved (HEC, PVP) and swollen (PC) polymeric chains. At higher concentrations of the polymers, PVP and PC existed primarily as dispersed solids, thus rendering such formulations extremely elastic, i.e., semisolid-like properties were evident (low-loss tangent). Consequently, it was shown that by altering the relative states of hydration of each polymeric component, formulations offering a wide range of viscoelastic properties may be attained. This confirmed the role of polymer state on formulation viscoelasticity. In addition, it was illustrated that inverse correlations between elasticity and bioadhesion and between elasticity and drug release existed. Accordingly, by manipulation of the concentrations and states of the polymeric components, formulations were prepared that exhibited an enhanced range of bioadhesive and drug (tetracycline) release properties (9,19).

10.4.1.3 Characterization of Temperature-Dependent Sol–Gel Transitions in Gel Systems

An interesting property of many polymers is their ability to undergo a temperature dependent transition from a primarily liquidlike state (sol) to a gel possessing elasticity. The temperature at which this transition occurs is termed the *sol–gel* transition. Given the relevance of this transition to both the clinical and nonclinical performance of pharmaceutical systems, various authors have examined this phenomenon using thermoanalytical methods, including thermal oscillatory analysis. Examples of these are now presented.

Poloxamers are commercially available block copolymers of polyoxyethylene and polyoxypropylene that have been employed for the topical administration of drugs to, e.g., the skin (57), eye (58), and the periodontal pocket (59). These materials undergo a temperature-dependent sol–gel transition that dramatically affects their rheological properties and, hence, their clinical efficacy (21). Jones et al. (22) employed dynamic thermal methods to characterize the effects of formulation components, e.g., molecular weight of the poloxamer system, nature of the solvent system, and the presence of therapeutic drug substances, on their thermorheological properties. The characteristic sol–gel transition of Poloxamer 407 (25% w/w) in water is graphically displayed in Figure 10.17.

From these studies, several interesting observations were reported. For example, decreasing the average molecular weight of poloxamer increased the sol–gel temperature whereas increasing the concentration of poloxamer or the concentration of cosolvent (ethanol, propylene glycol, or glycerol) decreased this parameter. In addition, the viscoelastic properties of these poloxamer systems at temperatures exceeding the sol–gel transition were dependent on these properties. Furthermore, this study highlighted the disparate effects of selected therapeutic agents (chlorhexidine diacetate and tetracaine hydrochloride) on the sol–gel transition and viscoelastic properties of aqueous poloxamer systems. These phenomena may be attributed to the effects of the aforementioned formulation variables on the ability of adjacent poloxamer chains to desolvate and form cross-linked aggregates. Subsequently, these formulations would be expected to exhibit different clinical efficacies and,

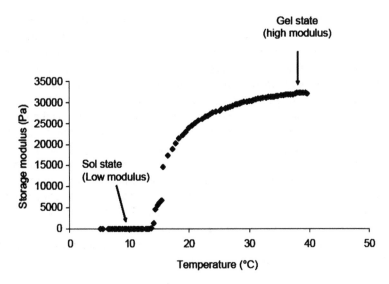

FIGURE 10.17 The relationship between storage modulus and temperature for an aqueous poloxamer 407 formulation (25% w/w).

indeed, patient acceptability. Similarly, the sol–gel transitions of other pharmaceutical and related systems have been reported using dynamic mechanical methods, e.g., cellulose in an ammonia/ammonia thiocyanate solvent system (60), hydroxyethylated starch aqueous systems (55), and agarose in a dimethyl sulfoxide/water solvent system (52).

10.4.2 DYNAMIC ANALYSIS OF EMULSIFIED SYSTEMS

Emulsified systems are composed of two insoluble liquid phases, in which one of the phases (disperse phase) is homogeneously dispersed throughout the continuous phase (17). Such systems are unstable, and the dispersed droplets will spontaneously reform into a single layer to satisfy the thermodynamic requirements of the system. Maintenance of the homogeneity of the dispersed phase may be achieved by using surfactants. Although there are several cooperative mechanisms that explain the stabilizing effects of surfactants, it is apparent that their effect on the viscoelastic structure of emulsified systems is a major contributor to this phenomenon. In a series of publications, Barry and coworkers described the effects of several mixed-emulsifier systems on the viscoelastic properties of disperse systems using dynamic thermorheological methods. For example, emulsions containing sodium dodecyl sulfate and cetyl alcohol were observed to increase in consistency from primarily liquidlike systems to semisolid creams as the concentrations of emulsifying agents were increased. This was attributed to the formation of viscoelastic networks at higher temperatures (61). Similarly, Barry and Saunders (21) described the formation of viscoelastic networks at low temperatures in emulsions stabilized by cetomacrogol 1000-cetostearyl alcohol. This highlighted the differences in the mechanism of formation of surfactant networks in ionic and nonionic systems. Furthermore, the same authors (62,63) described the influence of quaternary alkyl chain length on the viscoelastic nature of emulsions composed of alkylmethylammonium bromide and cetostearyl alcohol mixed emulsifying system. Interestingly, emulsions prepared with medium-chain-length derivatives (C_{14} and C_{16}) produced viscoelastic networks, whereas those prepared with short (C_{12}) chain lengths possessed tenuous networks; those with long chain lengths (C_{18}) were not as elastic as predicted. The differences in viscoelastic properties were related to the effects of chain length on the formation of smectic structures. In a related study (64), alterations in the alcohol chain length (cetostearyl, cetyl, stearyl, and myristyl) were observed to significantly affect the rheology of cetomacrogol creams, indicating the complex physical interactions between formulation components in emulsified systems. Further insights into the emulsion microstructure of disperse systems were obtained by the complementary use of low-frequency dielectric spectroscopy and oscillatory techniques (65). In combination, these techniques ensured a complete characterization of the viscoelastic networks of two model creams, one ionic and the other nonionic, in which the former system possessed a greater network structure.

10.4.3 DYNAMIC ANALYSIS OF SOLID POLYMERIC SYSTEMS OF PHARMACEUTICAL AND BIOMEDICAL IMPORTANCE

Solid polymeric systems are extensively employed in the design of pharmaceutical and biomedical systems. In the pharmaceutical industry, solid polymeric systems are employed as components of both packaging systems and pharmaceutical dosage forms. In the latter category, polymeric films are included in dosage forms to anchor transdermal systems to the skin, to act as drug reservoirs in transdermal dosage forms and implants, whereas in the case of solid dosage forms, polymeric films are present on the surface of tablets to offer controlled drug release, taste masking, and protective functions. Furthermore, polymeric systems form the basis of the vast majority of medical device implants, e.g., dental prostheses, catheters, ureteral stents, and endotracheal tubes. Such systems are designed to possess appropriate mechanical properties that ensure maximum performance, e.g., to ensure efficient mastication and conductance of body fluids (urine, peritoneal dialysate, and cerebrospinal fluid) and to offer effective protection to compromised sites (wound dressings). Alterations in the viscoelastic properties of solid polymeric systems that are components of either pharmaceutical or biomedical systems may therefore compromise their clinical performance and, in certain cases, may result in more serious consequences, e.g., failure to adequately control drug delivery and implant fracture *in vivo* (e.g., Reference 3). Therefore, characterization of the viscoelastic properties of these systems using dynamic oscillatory methods is an important consideration in their design. These methods may be usefully employed to provide information concerning their rheological behavior, e.g., glass transition temperatures and formulation effects on this parameter, compatibility of polymer blends, quantification of moduli, damping properties, and rate of curing of polymeric systems.

There have been several studies of the thermorheological properties of polymeric systems of pharmaceutical and biomedical importance. However, rather than describing all studies in detail, some selected examples have been described in the accompanying section to provide an overview of the effective use of dynamic mechanical techniques for the evaluation of the thermorheological properties of solid polymeric systems.

10.4.3.1   Characterization of Glass Transition Temperatures of Solid Polymers

The glass transition is an important temperature-dependent physicochemical property of amorphous polymers and in amorphous regions of partially crystalline polymers (23,66). Thus, at this defined temperature the rheological behavior of the polymer changes from a glasslike state, in which there is restricted motion of the polymeric chains, to a rubberlike state in which there is a loss in rigidity (i.e. increased relaxation) as a result of enhanced polymeric chain mobility (12,15). Many polymers exhibit more than one temperature-dependent relaxation behavior, which may be attributed to their structural nature, e.g., temperature-dependent mobility of pendent side groups. The ability of dynamic thermal methods to accurately identify and quantify the range of relaxation phenomena in polymeric systems has ensured that dynamic thermal mechanical analysis is one of the techniques of choice for this purpose (15). In dynamic mechanical thermal analysis, the glass transition is accepted as the temperature at which either a maximum in the mechanical damping parameter (tan δ) or loss modulus (G'') occurs. An example of the use of dynamic mechanical analysis to identify the glass transition temperatures of a congeneric series of polymethacrylate esters is presented in Figure 10.18 (67).

Three glass transitions may be identified in this example, termed α, β, and γ. The transitions are categorized according to their associated temperature; hence, the primary (α) transition is that which occurs at the highest temperature, with subsequent transitions described in order of decreasing temperature. It may be recalled that in this study the author has described the glass transition as the maximum value in the plot of loss modulus and temperature. Therefore, whenever dynamic thermal methods are employed to quantify glass transitions, it is important to describe the method of measurement, i.e., either the loss modulus or tan δ, of this parameter. It is worthwhile to reflect on the sources of the various glass transitions in polymeric systems. In the examples of polymethacrylate esters, the primary transition (α) may be attributed to the temperature-dependent increased mobility of the main linear polymeric chains, whereas the secondary (β) and tertiary (γ) transitions are a result of either side-group mobility and end-group motions of the main polymeric chains. Similarly, multiple transitions may be observed for the amorphous polymer polystyrene and may be related to main-chain and localized mobility within the polymer (68).

In addition, crystalline polymers may also possess thermal transitions that may be attributed to relaxations in the amorphous regions between lamellae or, alternatively, to melting and premelting phenomena. To illustrate these phenomena, the viscoelastic properties of polystyrene and high relative molecular mass polyethylene oxide fast-cooled from the molten state are displayed in Figure 10.19 and Figure 10.20 (68). Furthermore, the relative degree of crystallinity in an amorphous polymer may be determined using dynamic mechanical methods based on the comparative

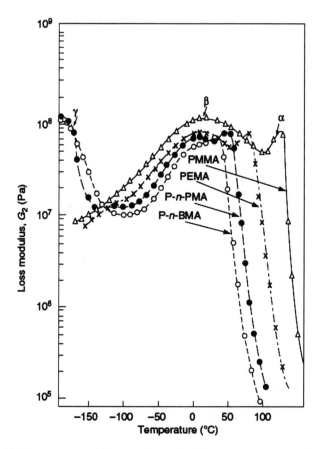

FIGURE 10.18 Thermorheological properties of polymethacrylate ester. (Reprinted from Wetton, R.E. et al., *Thermochim. Acta* 175, 1, 1991. With permission.)

behavior of the storage modulus at the glass transition temperature. In this, amorphous polymers exhibit a characteristic sharp decline in storage modulus at the glass transition temperature, whereas systems exhibiting a gradual drop in storage modulus associated with a number of transitions would be expected to possess semicrystalline behavior (14).

An example of a semicrystalline polymer that exhibits a glass transition and undergoes melting is poly(ε-caprolactone). This polymer has received considerable attention both for drug delivery (69–72) and biomedical applications (34,35). In our laboratories, we have examined the thermorheological properties of poly(ε-caprolactone), blends of poly(ε-caprolactone), and sequential interpenetrating networks of poly(ε-caprolactone) and poly(hydroxyethylmethacrylate) as biomedical materials. An example of the thermorheological properties of poly(ε-caprolactone) is shown in Figure 10.21. Several key features may be observed: First, at the glass transition, there is a decrease in the magnitude of the storage modulus. Typically, the reduction in the storage modulus over this region is smaller than for amorphous polymers (12). Second, a peak in both the loss modulus and the loss tangent may be observed at

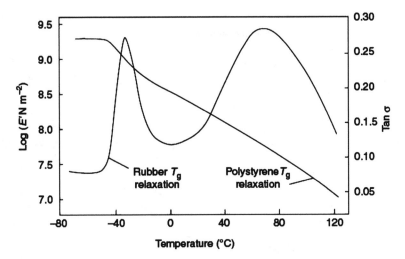

FIGURE 10.19 Dynamic mechanical thermal analysis of a block copolymer of SBR rubber and polystyrene. This illustrates two phases; however, owing to the broad loss modulus peak, it may be assumed that some mixing has occurred in the polystyrene phase. (Reprinted from Wetton, R.E. et al., *Thermochim. Acta* 175, 1, 1991. With permission.)

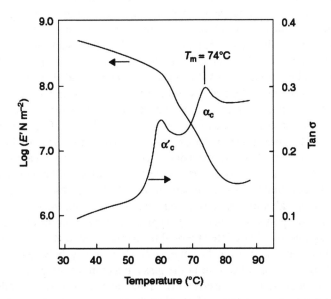

FIGURE 10.20 Dynamic mechanical thermal analysis of polystyrene that has been fast-cooled from the molten state, illustrating two relaxation events that may be attributed to crystal disordering. (Reprinted from Wetton, R.E. et al., *Thermochim. Acta* 175, 1, 1991. With permission.)

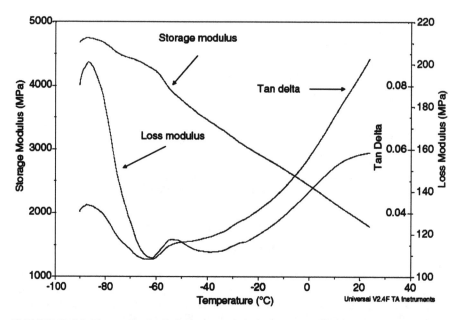

FIGURE 10.21 Thermorheological properties of poly(ε-caprolactone) as determined using dynamic mechanical thermal analysis (film geometry, 50-μm amplitude, and 1 Hz).

about −55°C. This temperature is referred to as the *glass transition* and is attributed to increased molecular mobility within the amorphous regions of the polymer. Interestingly, the shape of the peak in the loss modulus and the loss tangent is rather broad, which is indicative of a broad molecular relaxation spectrum.

In associated studies, Jones et al. (34,35) examined the effects of storage, average polymer molecular weight, and the presence of a dispersed antimicrobial agent (polyvinylpyrrolidone-iodine, [PVP-I]) on the mechanical and viscoelastic properties of poly(ε-caprolactone). The authors showed that alteration in the ratio of high- to low-molecular-weight polymer, storage at 30°C and 75% relative humidity, and the incorporation of the polymeric antimicrobial complex did not affect the viscoelastic properties and the glass transition temperature of poly(ε-caprolactone). Interestingly, the authors illustrated that the mechanical properties of poly(ε-caprolactone), as determined using tensile analysis, were adversely affected by the molecular weight blend and storage time but not the incorporation of PVP-I. Based on these observations, it was concluded that there was no interaction between the polymeric antimicrobial agent and poly(ε-caprolactone). Typically, in dynamic mechanical thermal analysis (and indeed, other related thermal techniques), interaction between a small organic or inorganic molecule and a polymer or, alternatively, between two polymers may be visualized by an alteration in the glass transition of the polymers.

Based on this, it is no surprise that dynamic mechanical methods may be employed to quantify the effects of plasticizers on the glass transition temperature and the compatibility of such plasticizers with the polymeric components. Such agents are common components of polymeric systems and are included to reduce the glass transition temperature of the polymer, frequently to temperatures below which the polymer would be expected to perform. Thus, plasticizers alter the rheological properties of the polymer from the glassy to rubbery state, the latter being associated with greater flexibility. Dynamic mechanical methods may therefore be employed to quantify the reduction in glass transition temperature associated with the presence of the plasticizer. However, the resolution of the loss tangent peak (associated with the glass transition) is an important measure of compatibility between plasticizer and polymer. High resolution occurs whenever there is good compatibility between polymer and plasticizer, whereas poor resolution (peak broadening) is observed whenever the plasticizer has a limited solubility in the polymer system, i.e., poor compatibility (12,15).

10.4.3.2 Effects of Cross-Linking on Polymer Rheology

Cross-linking agents are included in polymeric films frequently to enhance their mechanical properties, e.g., enhance elasticity and increase durability. However, the incorporation of cross-linking agents into polymeric systems may also adversely influence their performance by altering the rate of drug diffusion through such systems or, alternatively, by substantially modifying the glass transition temperature. Consequently, quantification of the effects of such agents on the rheological properties of polymers is frequently assessed during formulation, development, and fabrication. For this purpose, dynamic mechanical methods are frequently employed to evaluate the effects of cross-linking agents on the glass transition and viscoelastic properties of polymeric systems. Cross-linking agents directly reduce the main-chain mobility of polymer systems and, additionally, reduce the distance between polymer chains (i.e., reduced free volume). Therefore, the glass transition of cross-linked systems is generally greater than their non-cross-linked counterparts. For example, Rials and Glasser (73) reported an increase in the glass transition of hydroxypropylcellulose films following cross-linkage with toluene diisocyanate. Oysted (74) described the effects of a range of chemically related dimethacrylate cross-linking agents, namely, ethyleneglycoldimethacrylate, 1,3-propanedioldimethacylate, 1, 4-butanedioldimethacrylate, diethyleneglycoldimethacrylate, or triethyleneglycoldimethcrylate, on the dynamic mechanical properties of multiphase acrylate systems formed by polymerization of a mixture of liquid methacrylate monomers (methylmethacrylate) and polymethylmethacrylate (PMMA). Such polymeric systems are frequently employed as bone cements and oral prostheses. The author reported increased glass transition values and increased values of storage modulus as the concentrations of cross-linking agents increased in systems that had been autopolymerized. Conversely, the effects of the various cross-linking agents on the storage modulus and glass transition temperatures of samples that had been heat polymerized were not as marked. This study illustrated the effects of the polymerization method on the rheological properties of polymeric systems designed for biomedical application. Similarly, Cascone (75) reported increased glass transition and storage modulus values for cross-linked gelatin films compared to their un-cross-linked counterparts.

Dynamic mechanical methods may be also used to characterize time-dependent changes in the elastic modulus during polymer curing and cross-linking, from which

information concerning both the rate and extent of the process may be determined
(76,77). The successful use of dynamic mechanical thermal analysis to evaluate both
the curing rate and optimal curing temperature of selected polymeric coatings and,
additionally, the effects of catalysts on these processes were described by Skrovanek
(77). This approach has been adopted by other authors (e.g., Reference 78 to Reference 80). More recently, the use of dynamic oscillatory methods for the analysis
of both the rate and extent of curing of tin-catalyzed silicone and also for the
quantification of the viscoelastic properties of the cured elastomer has been examined
(81). Curing may be identified as a plateau in the relationship between storage
modulus and time. The study provided information concerning the optimum conditions for curing of the elastomer, e.g., concentration of catalyst (stannous octanoate)
and temperature of curing. In addition, the effects of two model antimicrobial agents,
namely, hexetidine and triclosan, on both the curing process and the viscoelastic
properties of the cured polymer were investigated. Interestingly, both agents significantly (adversely) affected both the rate and extent of curing and the viscoelastic
properties of the cured product in a concentration-dependent fashion. The effects of
hexetidine on this process were of particular note. The effects of hexetidine on the
curing process of silicone observations are presented in Figure 10.22. Furthermore,
in this figure, the lack of effect of another nonantibiotic antimicrobial agent, chlorhexidine, on this process are presented, thereby highlighting the deleterious effect
of hexetidine.

10.4.3.3 Thermorheological Properties of
Biomedical Polymers

As a result of the wide range of mechanical properties exhibited by synthetic
polymers, these materials have found extensive application within the medical device
(primarily) and pharmaceutical industries. The success of a medical device biomaterial depends on the ability of that system to exhibit biocompatibility, e.g., resistance
to infection, encrustation, the host immune response and, also, to retain the appropriate mechanical properties throughout the period of implantation. Loss of the
mechanical properties of these systems may result in impaired function, e.g.,
decreased drainage of body fluids and, in the worst scenario, fracture resulting in
(surgical) removal of the fragments of the failed device. As a result, considerations
of the thermorheological properties of biomedical devices, e.g., using dynamic
mechanical methods, are commonplace in the development of such systems. Indeed,
following reports of the fracture of polyurethane ureteral stents *in vivo*, the mechanical properties of polyurethane ureteral stents that had been removed from patients
following a series of dwell times were determined using tensile testing and dynamic
mechanical analysis (3). Through the effective use of these techniques, the authors
reported that there were no noticeable changes in the viscoelastic and tensile properties of ureteral stents following use *in vivo*.

   The use of dynamic mechanical analysis for the development of polymeric
systems as medical devices has been described by several authors. In one report,
Silver et al. (82) examined the physical properties and hemocompatibility of poly-
urethanes containing polyethylene oxide macrogols. Using this technique, a series

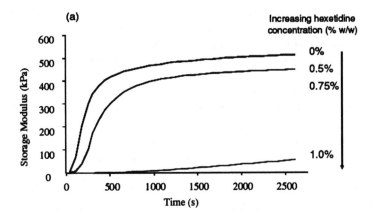

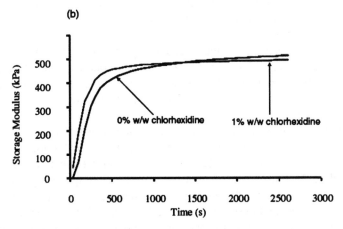

FIGURE 10.22 The effect of the incorporation of hexetidine (a) or chlorhexidine (b) on the curing properties of medical grade (tin-catalyzed) silicone. (From Jones, D.S., unpublished data. With permission.)

of transitions were observed in polyurethane polymers containing polyethylene glycol 600, 1450, and 8000, at approximately −120, −70, −30, 80, and 150°C. The transitions occurring at 30, 80, and 150°C were attributed to the soft-phase glass transition, hard-phase glass transition, and hard-phase mixing. On the other hand, the transitions at −120 and −70°C were attributed to local chain motions in the tetramethylene portion of the chain extender and in the soft-phase areas that had been infiltrated with water. Therefore, dynamic mechanical analysis provided the authors with an indication of the morphological properties of the series of biopolymers. Similarly, the successful use of dynamic mechanical analysis to characterize the thermorheological properties of poly(urethane-urea) biopolymers was reported by Ikeda et al. (83). These authors described the effects of solvent evaporation rate on the microphase-separated structure of this polymer. Thus, a range of transitions were observed and were accredited to the micro-Brownian motion of the soft segment of the polymer, the melting of the soft segment phase, the interaction between the polyether oxygen and the NH moiety of the hard segment, and the melting of the hard-segment domain. The authors were therefore able to conclude that a microphase-separated structure was formed in each sample following evaporation of the solvent.

Sun et al. (84) examined the phase transitions of poly(ethylene terephthalate-co-parahydroxybenzoic acid) liquid crystals using dynamic mechanical analysis and polarizing microscopy. The authors reported notable changes in viscoelastic behavior at defined temperatures that were associated with certain phase transitions. The coanalysis of results from both polarizing microscopy and dynamic mechanical analysis allowed identification of the crystal-nematic disordering and nematic-isotropic transition in the parahydroxybenzoate-rich region and, additionally, the crystal melting and the metastable isotopic-nematic transition in the poly(ethylene terephthalate)-rich regions. As it is likely that in the future more medical device biomaterials will be composed of liquid crystals, this study has highlighted the application of dynamic mechanical analysis for the characterization of the thermorheological properties of such systems.

In light of the importance of the mechanical properties of polymers employed as dental prostheses, e.g., dentures, temporary crown materials, bridge materials, and denture soft lining materials, there have been several reports that have examined their dynamic mechanical properties. For example, Clarke (85–87) described the dynamic mechanical properties of a range of dental biomaterials, namely, heat-cured poly(methylmethacrylate)-based materials, bis-phenol A-related resins, and heterocyclic methacrylates. In particular, the author described their thermal viscoelastic spectra, the effects of formulation additives on these properties, and the various glass transition values for these materials. Furthermore, by plotting the logarithm of the oscillatory frequency against the reciprocal of either the primary (α) or secondary (β) glass transition temperatures, the author calculated activation energies, i.e., the energy required to induce relaxation phenomena, within the designated samples and examined the effects of formulation additives, e.g., fillers and cross-linking agents, on the energy requirements for such transitions.

More recently, Vaidyanathan and Vaidyanathan (88) reported on the dynamic mechanical properties of commercially available denture base resins (Triad, Lucitone 199, and Acron MC) that had been cured using heat, microwave, or visible light. Thus, this study allowed an evaluation of the effects of different energy sources, an important processing condition, on the mechanical properties of the final product. Interestingly, the dynamic mechanical properties of visible-light-cured denture bases were significantly different from those cured using either microwave or heat and, importantly, were considered to be potentially inappropriate for the desired clinical applications. The observed differences were attributed to the effects of filler loading and cross-linking agents in the visible light cure resin. There were no differences between the mechanical properties of microwave-cured or heat-cured systems. Dynamic mechanical analysis has also been employed to characterize the viscoelastic properties of commercially available denture-lining materials, first, to arrive at a greater understanding of structure–property relationships of these systems and, second, to examine the mechanical properties under experimental conditions that simulate the rate and type of deformation that these materials would encounter *in vivo*. For example, the mechanical properties of commercially available soft liners have been described by Kalachandra et al. (89) and Waters et al. (90). In the study by Kalachandra et al. (89), the effects of water sorption on the viscoelastic properties of four commercial materials, namely, Moloplast B™ (a silicone-based system), Novus™ (a phosphazine), Kurepeet™ (a fluropolymer), and Super Soft™ (an acrylate) were examined. The authors described insignificant differences in the storage modulus between the nonhydrated and hydrated states of Molloplast B™, Kurepeet™, and Super Soft™. In addition, despite their obvious differences in chemical composition, the storage moduli of these biomaterials were numerically similar at 37°C. However, hydrated Novus™ was observed to possess a significantly lower G' in comparison to the nonhydrated state, which was attributed to the high water content of the hydrated material. Conversely, Waters et al. (90) described differences in some of the viscoelastic properties of alternative dental materials and, importantly, highlighted the potential clinical relevance of these differences. Finally, Sadiku and Biotidara (91) described the elastic properties of blends of polymethylmethacrylate and polyvinylchloride and compared these to polymethylmethacrylate alone. Introduction of polyvinylchloride into polymethylmethacrylate was observed to increase the elastic modulus and hence improve the mechanical properties of this biomaterial for dental applications.

Polyhydroxyethylmethacrylate (PHEMA) and its copolymers may be conveniently referred to as *hydrogels*, i.e., highly swollen solids that are water-swollen, cross-linked hydrophilic polymers (28). Due to their biocompatibility, these materials have received much attention over the past few decades as components of biomedical devices, including soft contact lenses (92), implantable drug delivery systems, (38) and urinary catheters or ureteral stents (93). Once more, an important stage in the design of such systems involves the mechanical analysis of such materials; however, owing to their high water content, water loss may occur during the temperature cycle in dynamic mechanical analysis. Therefore, it is important when testing these materials that water loss is minimized by coating the samples with a hydrophobic gel, e.g., petroleum gel or silicone vacuum grease for temperatures up to 45 and 85°C, respectively. Furthermore, it is important that the dimensions of the sample under analysis are rationally selected according to the modulus of the material. An excellent example of the dynamic mechanical analysis of hydrogels was reported by Lusig et al. (94) who described the dynamic mechanical properties of hydrogels composed of poly(2-hydroxyethylmethacrylate) and poly(2-hydroxyethylmethacrylate-co-methylmethacrylate). In this study, the sample dimensions (6 × 1 × 0.3 cm) were selected according to their modulus and all samples were hydrated in water, then coated with highly viscous petroleum gel and stored in separate containers of paraffin oil at room temperature for at least 4 d prior to analysis. The authors described the viscoelastic moduli of PHEMA, polymethylmethacrylate (PMMA), and a copolymer of these two materials as functions of water content, temperature, and frequency of oscillation. In particular, water was observed to strongly affect the mechanical properties of PHEMA and P(HEMA-co-MMA), reducing the α, β, and γ transitions and also the storage moduli as the water content was increased. This study also provided an insight into the nature of transitions in these systems. As the role of hydrogels in modern medicine increases in popularity, it is expected that many more studies concerning the dynamic mechanical properties will be reported in the scientific literature.

Dynamic mechanical analysis has also been employed for the characterization of suture materials (fibers) using the tension mode. For example, von Fraunhofer and Sichina (95) characterized the viscoelastic properties of four commercially available suture materials, namely, Prolene™, Maxon™, Vicryl™, and silk using the aforementioned technique. In particular, the authors described the effects of time on the modulus of such materials, an important determinant of clinical performance. For example, excessive loss of modulus may result in hemorrhage around a wound, whereas the relaxation of sutures will allow improved recovery of the wound, i.e., without scarring. In this study, the monofilament Prolene™ (polypropylene) was observed to exhibit a β transition that was associated with the amorphous region of the polymer, whereas the monofilament Maxon™ possessed an α transition close to room temperature. Thus, it was suggested that these materials would exhibit alterations in modulus as a function of time. Conversely, silk and Vicryl™ possessed α transitions at higher temperatures that were stable with respect to time at room temperature. The authors concluded that, owing to their greater stability, Vicryl and silk would be preferred for ligation of blood vessels as these materials would minimize the risks of delayed hemorrhaging. Conversely, the viscoelastic properties of Prolene and Maxon would be more suitable for wound apposition owing to their ability to relax during the healing process and, hence, would promote better wound healing. This study has clearly demonstrated the importance of accurate characterization of the viscoelastic properties of biomedical systems and their clinical relevance.

Cellulose polymers (and indeed, other polysaccharides) are common components of pharmaceutical dosage forms and biomedical devices and, as a result, their thermorheological properties have been the subject of several studies. For example, using differential scanning calorimetry and dynamic mechanical analysis, three distinct relaxation transitions were identified for dioxane or acetone-cast hydroxypropylcellulose films (73). The authors attributed these transitions to three phases, namely, amorphous, crystalline, and intermediate, the last named phase arising from the liquid crystal mesophase formed in solution. Dynamic mechanical analysis has also been successfully employed to determine glass transitions within other cellulose polymers, e.g., hydroxyethylcellulose, hydroxypropylmethylcellulose, and hydroxypropylcellulose (96). The thermorheological properties of soluble collagen, a natural polysaccharide, were described by Cascone (75). Chitin is a polysaccharide that is obtained from the cuticle of the marine crustacean that, along with its deacetylated derivative, chitosan, have attracted attention over the past decade as components of pharmaceutical dosage forms and biomedical devices as a result of their biocompatibility (97). In a recent study, the thermal properties of chitin and its hydroxypropyl derivative (hydroxypropyl chitin) were examined using differential scanning calorimetry and dynamic mechanical analysis (98). The introduction of the hydroxypropyl moiety was observed to enhance the solubility of chitin in a number of organic solvents and was therefore thought to offer possibilities for the formulation of pharmaceutical and biomedical systems. Typically, the primary (α) transitions for chitin and hydroxypropylchitin, both cast from formic acid as 5% w/w solutions, were 236 and 252°C, respectively, and were due to relaxations associated with the polymer main chain. The greater primary transition temperature associated with hydroxypropylchitin was suggested to be due to the restriction of the relaxation process by the bulky hydroxypropyl group. Secondary (β) transitions were observed at 133 and 143°C for chitin and the hydroxypropyl derivative and were attributed by the authors to the presence of acetamide groups on each polymer. A further transition was observed at 105°C for hydroxypropylchitin that was only observed using dynamic mechanical analysis and that was attributed to the relaxation of the hydroxypropyl moiety. Such information is therefore important in the design of pharmaceutical and biomedical systems composed of these polysaccharides.

10.4.3.4   Thermorheological Properties of Pharmaceutical Polymers

Whereas there have been several applications of dynamic mechanical analysis for the characterization of the viscoelastic and mechanical properties of biomedical polymers, the characterization of pharmaceutical polymers has not received similar attention. One possible reason for this disparity has been the relative unavailability of appropriate sample geometries for pharmaceutical systems. However, the availability of newer geometries should address this problem. In spite of these difficulties, several studies have successfully employed dynamic mechanical analysis for the characterization of pharmaceutical systems and some of these are described in the following paragraph.

The Eudragit™ polymers are a series of acrylate-based polymers that have found extensive use in pharmaceutical formulations, e.g., as components (coatings) of oral drug delivery systems, transdermal drug delivery systems, and as adhesives for wound dressings and transdermal systems (99). As a result of these applications, the mechanical properties of various members of the Eudragit series have received attention. For example, the dynamic mechanical properties of Eudragit films containing the antimicrobial agent metronidazole and designed for wound dressings were reported by Jones (100). Interestingly, the incorporation of metronidazole (e.g., 2.4% w/w) into films composed of Eudragit™ RL100 or Eudragit™ RS100 reduced the main α transitions from 52 to 36°C and from 42 to 32°C, respectively. The authors concluded that the presence of metronidazole was sufficient to plasticize the polymeric film. More recently, Hare et al. (101) characterized the viscoelastic properties of films of Eudragit E100, a commonly used adhesive and, additionally, the effects of incorporated drug (ibuprofen) on these properties. At 25°C, Eudragit films devoid of drug exhibited characteristic Maxwellian viscoelastic properties as a function of oscillatory frequency. Incorporation of ibuprofen (up to 50 μg cm^{-2}) significantly (and sequentially) reduced the storage modulus and increased the loss modulus of the Eudragit films. In addition, as the storage modulus was decreased (and loss modulus increased) by the addition of increasing amounts of ibuprofen, the adhesive properties were reduced, suggesting reduced clinical performance. The effects of ibuprofen on the rheological properties of Eudragit E100 films are presented in Figure 10.23.

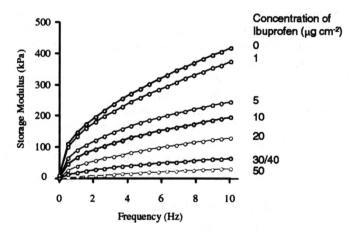

FIGURE 10.23 Effect of incorporated ibuprofen on the rheological properties of Eudragit™ E100 Films. (From Hare, L. et al., *J. Pharm. Pharmacol.* 50, S165, 1998. With permission.)

10.4.3.5 Thermorheological Properties of Mixed Polymeric Systems

Frequently, polymers are physically combined or blended to fulfill a number of purposes, including improvement of the mechanical properties, enhancement of biocompatibility, or to modify drug release properties. There are several methods by which this may be achieved, e.g., the physical blending of two or more polymers, the use of composites, and the formation of interpenetrating networks (IPNs), i.e., the combination of two or more different polymer networks that consist of purely physical entanglement of the polymer chains (5,16,102). Following their preparation, dynamic mechanical methods may be employed to characterize the physicochemical properties in terms of compatibility between constituent polymers, their Tg values, and rheological properties. In a more recent study (5), a series of IPNs composed of polymethylmethacrylate (PMMA) and polyurethane (PU) were prepared for use as ureteral stent biomaterials. These materials were designed to offer acceptable resistance to extrinsic compression (kinking) following implantation *in vivo*. Characterization of these systems using dynamic mechanical analysis allowed quantification of their storage and loss moduli, from which specific ratios of PMMA to PU were identified that possessed optimal properties to resist compression and yet offered comfort to the patient following implantation. Interestingly, characterization of the relaxation processes (glass transitions) using dynamic mechanical analysis provided an insight into the compatibility of the two polymers. Thus, the observed increase in the glass transition temperatures of the IPN as a function of PMMA composition was indicative of good compatibility between the two polymers.

Lafferty (103) examined the dynamic mechanical properties of cast films of Eudragit NE30D and Eudragit NE30D or Aquacoat™ systems. Typically, films composed of Eudragit exhibited a primary glass transition at about 30°C. Introduction of Aquacoat into the Eudragit film resulted in the identification of two glass transitions, one at about 30°C (ascribed to the Eudragit component) and the other at about 115°C because of the presence of Aquacoat. Alteration of the relative amounts of either polymeric component was observed to alter the intensity of the glass transition peaks. The existence of two glass transition temperatures in the mixed films may be interpreted as an example of limited miscibility (compatibility) of the two components. As these polymer blends are frequently employed as film coats on tablets, the compatibility of such systems is an important consideration for the selection of final coating compositions. Similarly, the dynamic mechanical properties of blends of poly(DL-lactide) and poly(ε-caprolactone) aliphatic polyesters that have received interest as components of drug delivery systems and biomedical devices owing to their biodegradability were described by Tsuji and Ikada (104). The moduli of the various blends and pure components alone were determined over a wide temperature range (–100 to about 60°C), from which it was concluded that phase separation of the two polymeric components occurred following evaporation of the casting solvent (dichloromethane). Furthermore, Vazquez-Torres and Cruz-Ramos (105) employed dynamic mechanical analysis to describe the viscoelastic properties and glass transition temperatures of films of poly(ε-caprolactone) with either cellulose triacetate, cellulose acetate butyrate, or cellulose diacetate that had been prepared by casting and evaporation of the appropriate solvent (dichloromethane for blends containing cellulose triacetate and cellulose acetate butyrate or a mixture of dichloromethane and acetone [70:30] for blends with cellulose diacetate). In light of the characteristic relationships between both the viscoelastic properties and glass transition temperatures and polymeric composition of the various blends, the authors concluded that blends of poly(ε-caprolactone) and either cellulose triacetate or cellulose diacetate were immiscible, whereas blends of poly(ε-caprolactone) and cellulose acetate butyrate were partially immiscible.

10.5 CONCLUSIONS

In this chapter, the theory, measurement, and applications of dynamic oscillatory methods for the useful characterization of the thermorheological properties of pharmaceutical and biomedical polymeric systems have been presented. The particular advantages and disadvantages of each method of measurement have been described, and, significantly, examples of the quality of information derived from dynamic oscillatory methods have been provided to highlight the versatility of these techniques. In particular, the use of oscillatory techniques to quantify polymer viscoelastic properties as functions of temperature, glass transition temperature, sol–gel transitions, rate and extent of polymer curing, polymer morphology, and polymer–polymer compatibility have been described. It is accepted that these techniques are employed in the biomedical device industry and, indeed, many examples described in this chapter relate to the characterization of such devices. The potential roles of dynamic oscillatory methods for the characterization of pharmaceutical systems have also been described. In particular, the availability of newer sample geometries that allow for convenient dynamic analysis and can accommodate a range of product geometries should overcome this problem and thus improve the acceptability of this technique to the pharmaceutical industry. As a result of this, the number of publications that describe the use of dynamic oscillatory methods for the characterization of novel pharmaceutical systems, e.g., matrix tablets, polymeric coating on the surface of tablets, drug interactions with plastic packaging and intravenous fluid bags, and, indeed, systems of biological interest, e.g., mucin and elastin, should rapidly increase in the coming years. Therefore, both the quality and range of information that may be obtained from dynamic oscillatory methods should further establish their roles within the pharmaceutical and biomedical industries.

REFERENCES

1. Lenk, R.S., *Polymer Rheology*, Applied Science Publishers Limited, London, 1978.
2. Barnes, H.A., Hutton, J.F., and Walters, K., *An Introduction to Rheology*, Elsevier, Amsterdam, 1996.
3. Gorman, S.P., Jones, D.S., Bonner, M.C., Akay, M., and Keane, P.F., *Biomaterials* 8, 631, 1997.
4. Jones, D.S., Woolfson, A.D., and Brown, A.F., *Pharm. Res.* 14, 450, 1997.
5. Jones, D.S., Bonner, M.C., Akay, M., Keane, P.F., and Gorman, S.P., *J. Mater. Sci.: Mater. Med.* 8, 713, 1997.
6. Jones, D.S., Woolfson, A.D., Brown, A.F., and O'Neill, M.J., *J. Controlled Release*, 49, 71, 1997.
7. Jones, D.S., Woolfson, A.D., and Brown, A.F., *Pharm. Res.* 15, 1131, 1998.
8. Jones, D.S., Irwin, C.R., Woolfson, A.D., Djokic, J., and Adams, V., *J. Pharm. Sci.* 88, 592, 1999.
9. Jones, D.S., Irwin, C.R., Brown, A.F., Woolfson, A.D., Coulter, W.A., and McClelland, C., *J. Controlled Release* 67, 357, 2002.
10. Goodwin, J.W. and Hughes, R.W., *Rheology for Chemists: An Introduction*, The Royal Society of Chemistry, London, 2000.
11. Ward, I.M., *Mechanical Properties of Solid Polymers*, John Wiley & Sons, Chichester, 1983.
12. Ward, I.M. and Hadley, D.W, *An Introduction to the Mechanical Properties of Solid Polymers*, John Wiley & Sons, Chichester, 2000.
13. Ashby, M.F. and Jones, D.R.H., *Engineering Materials 1: An introduction to Their Properties and Applications*, 2nd ed., Butterworth-Heinemann, Oxford, 2000.
14. Craig, D.Q.M. and Johnson, F.A., *Thermochim. Acta* 248, 97, 1995.
15. Ferry, J.D., *Viscoelastic Properties of Polymers*, 3rd ed., John Wiley & Sons, New York, 1980.
16. Jones, D.S., *Int. J. Pharm.* 179, 167, 1999.
17. Martin, A.A., *Physical Pharmacy*, 4th ed., Lea and Febiger, Philadelphia, 1993.
18. Barry, B.W., Rheology of pharmaceutical and cosmetic semisolids, in *Advances in Pharmaceutical Sciences*, Vol. 4, Bean, H.S., Beckett, A.H., and Carless, J.E., Eds., Academic Press, London, 1974, pp. 1–72.
19. Jones, D.S., Brown, A.F., and Woolfson, A.D., *J. Pharm. Sci.* 90, 1978, 2001.
20. Davis, S.S., *J. Pharm. Sci.* 60, 1351, 1971.
21. Barry, B.W. and Saunders, G.M., Rheology of systems containing cetomacrogol 1000-cetostearyl alcohol. 1. Self-bodying action, *J. Colloid Interface Sci.* 38(3), 616–625, 1972.
22. Jones, D.S., Brown, A.F., and Woolfson, A.D., *J. Appl. Polym. Sci.* 87, 1016, 2002.
23. Ford, J.L. and Timmins, P., *Pharmaceutical Thermal Analysis*, Ellis Horwood, Chichester, 1989.

24. Bird, R.B., Curtiss, C.F., Armstrong, R.C., and Hassager, O., *Dynamics of Polymeric Liquids: Fluid Mechanics*, Vol. 1, 2nd ed., John Wiley & Sons, New York, 1987, pp. 55–168.

25. Chu, J.S., Yu, D.M., Amidon, G.L., Weiner, N.D., and Goldberg, A.H., *Pharm. Res.* 9, 1659, 1992.

26. Lin, S.Y., Amidon, G.L., Weiner, N.D., and Goldberg, A.H., *Int. J. Pharm.* 95, 57, 1993.

27. Tamburic, S. and Craig, D.Q.M., *J. Controlled Release*, 37, 59, 1995.

28. Anseth, K.S., Bowman, C.N., and Brannon-Peppas, L., *Biomaterials*, 17, 1647, 1996.

29. Read, B.E., Dean, G.D., and Duncan, J.D., Determination of dynamic moduli and loss factors, in *Physical Methods of Chemistry: Determination of Elastic and Mechanical Properties*, Vol. VII, Rossiter, B.W. and Baetzold, R.C., Eds., John Wiley & Sons, London, 1991.

30. Lofthouse, M.G. and Burroughs, P.J., *Therm. Anal.* 13, 439, 1978.

31. Connop, A., Huson, M.G., and McGill, W.J., *J. Them. Anal.* 24, 223, 1982.

32. Akay, M., Rollins, S.N., and Riordan, E., *Polymer* 29, 37, 1988.

33. Dynamic Mechanical Analyser 2980 Operator's Manual, TA Instruments, New Castle, DE, 1996.

34. Jones, D.S., Djokic, J., McCoy, C.P., and Gorman, S.P., *Plast. Rubbers Comp.* 29, 371, 2000.

35. Jones, D.S., Djokic, J., McCoy, C.P., and Gorman, S.P., *Biomaterials* 23, 4449, 2002.

36. Clarke, A.H. and Ross-Murphy, S.B., *Adv. Polym. Sci.*, 83, 57, 1987.

37. Ross-Murphy, S., *J. Text. Stud.* 26, 391, 1995.

38. Jones, D.S., Bonner, M.C., and Woolfson, A.D., *Proc. 16th Pharm. Tech. Conf.* 16, 379, 1997.

39. Rao, J.K., Ramesh, D.V., and Rao, K.P., *Biomaterials* 15, 383, 1994.

40. Jones, D.S., Brown, A.F., Woolfson, A.D., and McKeever, P.A., *J. Pharm. Pharmacol.* 51, S310, 1999.

41. Barry, B.W. and Meyer, M.C., *Int. J. Pharm.* 2, 1, 1979.

42. Barry, B.W. and Meyer, M.C., *Int. J. Pharm.* 2, 27, 1979.

43. Craig, D.Q.M., Tamburic, S., Buckton, G., and Newton, J.M., *J. Controlled Release* 30, 213, 1994.

44. Buckton, G. and Tamburic, S., *J. Controlled Release*, 20, 29, 1992.

45. Bonner, M.C., Jones, D.S., and Woolfson, A.D., *J. Pharm. Pharmacol.* 49, S132, 1997.

46. Jones, D.S., Muldoon, B.C.O., Woolfson, A.D., and Sanderson, F.D., *J. Pharm. Sci.*, in press.

47. Peppas, N.A., *Hydrogels in Medicine and Pharmacy: Polymers*, Vol. II, CRC Press, Boca Raton, FL, 1987, pp. 115–160.

48. Jones, D.S., Woolfson, A.D., and Brown, A.F., *Int. J. Pharm.* 151, 223, 1997.

49. Talukdar, M.M., Vinckier, I., Moldenears, P., and Kinget, R., *J. Pharm. Sci.*, 85, 537, 1996.

50. Brown, A.F., Jones, D.S., and Woolfson, A.D., *J. Pharm. Pharmacol.* 50, S157, 1998.

51. Yu, D.M., Amidon, G.L., Weiner, N.D., and Goldberg, A.H., *J. Pharm. Sci.* 83, 1443, 1994.

52. Watase, M. and Nishinari, K., *Polym. J.* 20, 1125, 1988.

53. Watase, M. and Nishinari, K., *Polym. J.* 21, 567, 1989.

54. Watase, M. and Nishinari, K., *Carbohyd. Polym.* 20, 175, 1993.

55. Jauregui, B., Munoz, M.E., and Santamaria, A., *Int. J. Biol. Macromol.* 17, 49, 1995.

56. Muruyama, A., Ichikawa, Y., and Kawabata, A., *Biosci. Biotechnol. Biochem.* 59, 5, 1995.
57. Suh, H. and Jun, H.W., *Int. J. Pharm.* 129, 13, 1996.
58. Miller, S.C. and Donovan, M.D., *Int. J. Pharm.* 12, 147, 1992.
59. Esposito, E., Carotta, V., Scabbia, A., Trombelli, L., D'Antona, P., Menegati, E., and Nastruzzi, C., *Int. J. Pharm.* 142, 9, 1996.
60. Frey, M.W., Cuculo, J.A., and Khan, S.A., *J. Polym. Sci. Part B: Polym. Phys.* 34, 2375, 1996.
61. Barry, B.W., *J. Colloid Interface Sci.* 28, 82, 1968.
62. Barry, B.W. and Saunders, G.M., *J. Colloid Interface Sci.*, 35, 689, 1971.
63. Barry, B.W. and Eccleston, G.M., *J. Pharm. Pharmacol.* 25, 394, 1973.
64. Eccleston, G.M., *J. Pharm. Pharmacol.* 29, 157, 1977.
65. Tamburic, S., Craig, D.Q.M., Vuleta, G., and Milic, J., *Int. J. Pharm.* 137, 243, 1996.
66. Callister, W. D., *Materials Science and Engineering: An Introduction*, 4th ed., John Wiley & Sons, New York, 1997, pp. 465–510.
67. Heijober, J., *Physics of Non-Crystalline Solids*, North-Holland, Amsterdam, 1965.
68. Wetton, R.E., *Anal. Proc.* 416, 1981.
69. Pitt, C. G., Marks, T. A., and Schindler, A., Biodegradable drug delivery systems based on aliphatic polyesters: application to contraceptives and narcotic antagonists, in *Controlled Release of Bioactive Materials*, Baker, R., Ed., Academic Press, New York, 1980, pp. 19–43.
70. Pitt, C., Polycaprolactone and its copolymers, in *Biodegradable Polymers as Drug Delivery Systems*, Chasin, M. and Langer, R., Eds., Marcel Dekker, New York, 1990, pp. 71–120.
71. Medlicott, N.J., Tucker, I.G., Rathbone, M.J., Holborow, D. W., and Jones, D.S., *Int. J. Pharm.* 143, 25, 1996.
72. Medlicott, N.J., Rathbone, M.J., Tucker, I.G., Holborow, D.W., and Jones, D.S., *J. Controlled Release* 61, 337, 1996.
73. Rials, T.G. and Glasser, W.G., *J. Appl. Polym. Sci.* 36, 749, 1988.
74. Oysted, H., *J. Biomed. Mater. Res.* 24, 1037, 1990.
75. Cascone, M.G., *Polym. Int.* 43, 55, 1997.
76. McCrum, N.G., Read, B.E., and Williams, G., *Anelastic and Dielectric Effects in Polymeric Solids*, John Wiley & Sons, London, 1967.
77. Skrovanek, D.J., *Prog. Org. Coat.* 18, 89, 1990.
78. Mangion, M.B.M. and Johari, G.P., *J. Polym. Sci. Part B, Polym. Phys.*, 28, 1621, 1990.
79. Mangion, M.B.M. and Johari, G.P., *Macromolecules* 23, 3687, 1990.
80. Mangion, M.B.M. and Johari, G.P., *J. Polym. Sci. Part B, Polym. Phys.*, 29, 437, 1991.
81. Jones, D.S., unpublished data.
82. Silver, J.H., Myers, C.W., Lim, F., and Cooper, S.L., *Biomaterials* 15, 695, 1994.
83. Ikeda, Y., Tabuchi, M., Sekiguchi, Y., and Miyake, Y., *Macromol. Chem. Phys.* 195, 3615, 1994.
84. Sun, Y., Lin, Y.G., Winter, H.H., and Porteus, R.S., *Polymer* 30, 1257, 1989.
85. Clarke, R.L., *Biomaterials* 10, 494, 1989.
86. Clarke, R.L., *Biomaterials* 10, 549, 1989.
87. Clarke, R.L., *Biomaterials* 10, 630, 1989.
88. Vaidyanathan, J. and Vaidyanathan, T.K., *J. Mater. Sci.: Mater. Med.* 6, 670, 1995.
89. Kalachandra, S., Minton, R.J., Takamata, T., and Taylor, D.F., *J. Mater. Sci.: Mater. Med.* 6, 218, 1995.
90. Waters, M., Jagger, R., Williams, K., and Jerolimov, V., *Biomaterials* 17, 1627, 1996.

91. Sadiku, E.R. and Biotidara, F.O., *J. Biomater. Appl.* 10, 250, 1996.
92. Wichterle, O., in *Encyclopaedia of Polymer Science and Technology*, Vol. 15, Bikales, N.M., Ed., Interscience Publishers, New York, 1971.
93. Tunney, M.M., Jones, D.S., and Gorman, S.P., *Int. J. Pharm.* 151, 121, 1997.
94. Lusig, S.R., Caruthers, J.M., and Peppas, N.A., *Polymer* 32, 3340, 1991.
95. von Fraunhofer, J.A. and Sichina, W.J., *Biomaterials* 13, 715, 1992.
96. Kararli, T.T., Hurlbut, J.B., and Needham, T.E., *J. Pharm. Sci.* 79, 845, 1990.
97. Garvin, C.P., Jones, D.S., and Gorman, S.P., *Proc. 17th Pharm. Tech. Conf.* 1, 356, 1998.
98. Kim, S.S., Kim, S.J., Moon, Y.D., and Lee, Y.M., *Polymer* 35, 3212, 1994.
99. Lehman, K.O.R., Chemistry and application properties of polymethacrylate coating systems, in *Aqueous Polymeric Coatings for Pharmaceutical Dosage Forms*, McGinity, J.W., Ed., Marcel Dekker, New York, 1997, pp. 1–77.
100. Jones, C.E., Preparation and Characterisation of Polymer Films for the Release of Metronidazole, Ph.D. thesis, School of Pharmacy, University of London, 1990.
101. Hare, L., Jones, D.S., and Woolfson, A.D., *J. Pharm. Pharmacol.* 50, S165, 1998.
102. Crawford, R.J., *Plastics Engineering*, Butterworth-Heinemann, Oxford 1998.
103. Lafferty, S.V., Evaluation of Properties of Polymers Used as Controlled Release Membranes, Ph.D. thesis, School of Pharmacy, University of London, 1992.
104. Tsuji, H. and Ikada, Y., *J. Appl. Polym. Sci.* 60, 2367, 1996.
105. Vasquez-Torres, H. and Cruz-Ramos, C.A., *J. Appl. Polym. Sci.* 54, 1141, 1994.
106. Wetton, R.E., Marsh, R.D.L., and van de Velde, J.G., *Thermochim. Acta* 175, 1, 1991.

11 The Use of Thermally Stimulated Current Spectroscopy in the Pharmaceutical Sciences

N. Boutonnet-Fagegaltier, A. Lamure,
J. Menegotto, C. Lacabanne, A. Caron,
H. Duplaa, and M. Bauer

CONTENTS

11.1 INTRODUCTION TO THERMALLY STIMULATED CURRENT SPECTROSCOPY

It is well known that the majority of active drug substances are delivered to the human body as solid dosage forms. Consequently, a considerable amount of effort has been devoted to the physical characterization of the drugs themselves and, to a lesser extent, the excipients, as it is recognized that the physical state of drugs and excipients can have a major influence on the processing viability, the physical and chemical stability, and the bioavailability of solid dosage forms such as tablets, capsules, and freeze-dried products (1). The characterization of the physical state of a solid in the pharmaceutical development process is mandatory to obtain a robust formulation and, in this respect, the regulatory authorities require these characteristics to be well documented in any application submitted for approval. Particle size distribution, particle shape (habitus), porosity, surface state and, of course, crystalline polymorphism, are now systematically taken into account in the design of new formulations.

It has also been recognized, however, that other aspects of the physical structure deserve to be investigated to better understand the physicochemical behavior of the pharmaceutical solid forms. In particular, the behavior of amorphous drug state and the relevance of crystalline defects have been the subject of an intense investigation (see, for example, Reference 2 through Reference 4). If the material is produced in a largely or completely crystalline state, different processing following the crystallization process (e.g., drying, milling, micronization, etc.) may disturb the physical integrity of the drug or excipient by creating crystalline defects and amorphous material, which in turn may dramatically impact the previously mentioned pharmaceutical properties.

Up to now, the degree of crystallinity has been typically investigated using x-ray powder diffraction (XRPD), differential scanning calorimetry (DSC), microcalorimetry, and even density measurements (5). Although such approaches are interesting and useful from a qualitative and quantitative point of view, these techniques give relatively little information on the molecular motions pertaining to the different phases. There is therefore a need for spectroscopic techniques that allow in-depth structural insights into the different phases constituting the solid product. In this respect, the dielectric properties of the molecules could provide, through adequate methodologies, a sound description of these molecular motions. Classical isothermal dielectric spectroscopy has been recently applied to a variety of pharmaceutical systems (6,7), but this method does not provide a very detailed analysis of the internal molecular dynamics.

We describe here a relatively new form of dielectric spectroscopy known as thermally stimulated current (TSC) spectroscopy that has to date been applied to the study of inorganic materials (insulators, semiconductors) and organic macromolecules (collagen or elastin) (8) but does not appear to have been applied extensively to small organic molecules (9).

11.1.1 THEORY OF THERMALLY STIMULATED CURRENT (TSC) SPECTROSCOPY

Techniques such as differential thermal analysis, DSC, thermostimulated luminescence, electronic emission, conductivity, or polarization are based on the principle of measuring processes stimulated by temperature; a number of texts are available that outline such thermostimulated processes and spectroscopies (see for example Reference 10 to Reference 12). Similarly, information regarding the applications of TSC spectroscopy has been published in a bibliographic review by Lavergne and Lacabanne (13) and a chapter of the book *Dielectric Spectroscopy of Polymers* on polymer analysis by TSC (14).

11.1.1.1 Development of the Technique

The TSC method was first developed by Bucci and Fieschi in 1964 (15). The technique was initially used to characterize point defects in simple crystals. Later, it was applied to a wide variety of samples, including inorganic materials (insulators as halide crystals, polycrystals, or amorphous materials; semiconductors in crystalline or amorphous state), or organic materials (small molecules in noncrystalline or crystalline states, amorphous or semicrystalline synthetic macromolecules, and natural macromolecules) (12–14).

11.1.1.2 Principles of the Method

11.1.1.2.1 The Complex TSC Spectrum

As shown in Figure 11.1, to obtain a complex TSC spectrum, a static electric field E is applied to the sample at a polarization temperature labeled T_p for a time t_p, which is necessary to obtain polarization saturation, i.e., the equilibrium polarization. Afterward, the sample is cooled down to a temperature T_0 in such a way that the dielectric relaxation proceeds extremely slowly, so that after removal of the field the sample retains a frozen-in polarization. The depolarization current, I_d, caused by the return to equilibrium of dipolar units, is then recorded by increasing the temperature at a constant rate from T_0 up to the final temperature T_f, where $T_f > T_p$. The plot of I_d as a function of temperature is a complex TSC spectrum.

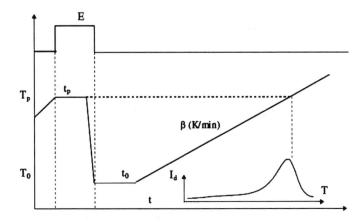

FIGURE 11.1 TSC principle for complex spectrum.

For a given depolarization current I_d, the current density j is given by,

$$j(T) = \frac{I_d(T)}{a} \tag{11.1}$$

where a is the sample area. For facilitating comparison of results, a "dipolar" conductivity σ can be used

$$\sigma(T) = \frac{j(T)}{E} = \frac{I_d(T)}{s \cdot E} = \frac{I_d(T) \cdot t}{s \cdot U} \tag{11.2}$$

where s is the sample thickness, E is the field, and U is the potential.

11.1.1.2.2 The Elementary TSC Spectrum

The whole TSC spectrum cannot generally be described by only one Debye-like relaxation process (i.e., with a single relaxation time). However, a complex TSC spectrum can be resolved experimentally by using the fractional polarization method (16,17). With this procedure, each elementary spectrum is described by a single relaxation time, allowing Bucci–Fieschi analysis (15,18). The experimental principle is illustrated in Figure 11.2 and involves applying the electric field E over a limited temperature window ΔT. E is applied to the sample at the temperature T_p for a time t_p to allow the dipoles with relaxation times $\tau(T_p)$ lower than t_p to reorientate. However, the temperature is then only lowered by ΔT to T_d. At T_d, the field is cut off and the sample is held at T_d for t_d so that dipoles with relaxation times $\tau(T_d)$ lower than t_d can relax. Finally, the sample is quenched to $T_0 \ll T_d$ so that only dipoles with relaxation times $\tau [(T_p, T_d)/2] \sim t_d$ remain oriented at T_0. Linear heating is then performed up to T_f above T_p during which the rate of decay of the frozen-in

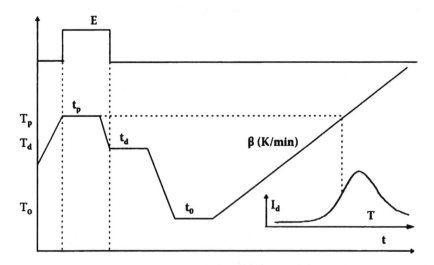

FIGURE 11.2 TSC principle for elementary spectrum.

polarization (i.e., the depolarization current) is recorded. Such a procedure results in orientating only a narrow distribution of relaxation times over a ΔT temperature window around T_p. By increasing the T_p value by constant steps along the temperature axis, the complete series of elementary processes is obtained.

11.1.1.3   Data Analysis

11.1.1.3.1   Relaxation Time
When a dielectric material is submitted to a direct current (DC) field, the constituent permanent dipoles orientate to the field direction, and a macroscopic polarization is established. After an infinite polarization time, the polarization reaches its saturation value, P_0, given, at low field, by

$$P_0 \cong \frac{N\mu^2 E}{3kT_p} \tag{11.3}$$

where μ is the dipole moment, N the dipole concentration (per unit volume), k the Boltzmann constant, and T_p the polarization temperature. In the case of depolarization under zero-field conditions, $P(t)$ is obtained from

$$\frac{dP}{dt} + \frac{P}{\tau(T)} = 0 \tag{11.4}$$

where $\tau(T)$ is the dielectric relaxation time τ at the temperature T. When the sample is polarized at T_p and cooled down to $T_0 \ll T_p$, such that $\tau (T_0)$ is high enough to prevent depolarization in the usual time scales, and hence P_0 is frozen in. If the sample is heated up at a constant rate β such that

$$\beta = \frac{dT}{dt} \tag{11.5}$$

a thermostimulated depolarization occurs. The temperature dependence of the polarization is given by

$$P(T) = P_0 \exp\left[-\int_{T_0}^{T} \frac{1}{\beta} \frac{dT'}{\tau(T')}\right] \tag{11.6}$$

where T is the running temperature, and $P(T)$ the remaining polarization. $P(T)$ is obtained experimentally by integrating the current density $j(T)$ between T and T_f, the final temperature at which the current vanishes, i.e.,

$$P(T) = \frac{1}{\beta} \int_{T}^{T_f} j(T')dT' \tag{11.7}$$

The relaxation time τ can be calculated experimentally without any assumptions regarding its temperature dependence. Indeed, from Equation 11.4 and Equation 11.6

$$\tau(T) = \left| \frac{P}{dP/dt} \right|_T \tag{11.8}$$

Experimentally, all the relaxation times are included between 100 sec and 5000 sec at the maximum of the elementary peaks (T_m). So, by analogy with dynamic electrical experiments, an equivalent frequency v_{eq} can be attributed to the TSC technique, i.e., $2\pi v_{eq} \cdot \tau(T_m) = 1$ and $v_{eq} \approx 10^{-3}$ Hz. This low-equivalent frequency explains the high resolution of TSC and constitutes one of the most attractive features of the technique. Moreover, TSC permits the operator to perform a rapid characterization of materials and is an alternative approach to conventional dielectric techniques over this frequency range.

11.1.1.3.2 Activation Parameters

Analysis of elementary peaks gives a temperature-dependent relaxation time τ (T), which generally follows an Arrhenius-like dependence

$$\tau(T) = \tau_0 \exp \frac{\Delta Ea}{kT} \tag{11.9}$$

where ΔEa is the activation energy and τ_0 is the preexponential factor. By considering the relaxation time as the inverse of the frequency of jump between two activated states, the empirical Equation 11.8 can be approximated from that derived by Eyring in the theory of activated states (19)

$$\tau(T) = \frac{h}{kT} \exp \frac{-\Delta S^*}{k} \exp \frac{\Delta H^*}{kT} \tag{11.10}$$

where ΔH^* and ΔS^* are, respectively, the activation enthalpy and entropy, and h is the Planck constant. The ΔH^* variable is related to the barrier height between two activated states, and ΔS^* is related to the ratio of the number of available sites between activated and inactivated states. The Eyring equation gives a theoretical interpretation of the Arrhenius parameters. Moreover, the use of the Eyring formalism allows transformation of the preexponential factor τ_0 of the Arrhenius law into an entropic factor. Because no difference exists between the Arrhenius and Eyring formalism in the limit of common accuracy (20), and because experimental results cannot distinguish which model is more suitable, equivalence between Equation 11.9 and Equation 11.10 is obtained from

$$\tau_0 = \frac{h}{kT} \exp \left(\frac{-\Delta S^*}{k} \right) \tag{11.11}$$

11.1.1.3.3 Compensation Law

When using the previous analysis, each elementary process is fitted according to an Arrhenius equation (Equation 11.8), where $(\tau_0, \Delta H)$ are obtained from a least-squares fitting of $\log \tau$ vs. $(1/T)$. In some cases (21), extrapolation of the $\log \tau$ vs. $(1/T)$ plots shows that several Arrhenius lines converge toward a single point, called the *compensation point* (T_c, τ_c). At the compensation temperature T_c, generally associated with a transition temperature, all relaxation times τ take the same value τ_c. This behavior indicates that the corresponding relaxation times obey a compensation law defined as

$$\tau(T) \approx \tau_c \exp\left[\frac{\Delta H}{k}\left(\frac{1}{T} - \frac{1}{T_c}\right)\right] \tag{11.11}$$

Therefore the preexponential factor τ_0 is linked to the activation energy ΔH by

$$\tau_0 = \tau_c \exp-\frac{\Delta H}{kT_c} \tag{11.12}$$

Consequently, the plot of $\log \tau_0$ vs. ΔH constitutes a usual representation of the compensation rule. The origin intercept and the slope of the lines deduced from a linear regression of $\log \tau_0$ vs. ΔH are used to obtain the compensation parameters, T_c and τ_c.

A widely used interpretation of the compensation law is based on a two-site model proposed by Hoffman et al. (22) to describe crystalline relaxations in *n*-paraffins. Molecular movements are assumed to involve an entire short-chain molecule, the length of the molecules corresponding to the thickness of crystallites. Under these assumptions, the relaxation time is expressed by an Eyring equation (Equation 11.9), with

$$\Delta H^* = \Delta H_{eg}^* + n\Delta H_{cs}^* \tag{11.13}$$

and

$$\Delta S^* = \Delta S_{eg}^* + n\Delta S_{cs}^* \tag{11.14}$$

The subscripts *eg* and *cs* refer to end-group contribution and to elementary contribution of constitutive segments, respectively, and *n* is the number of segments per molecular chain. This model was applied satisfactorily to *n*-paraffins, but also to *n*-esters and *n*-ether. A linear variation of both the activation enthalpy and entropy as a function of *n* has been observed experimentally (22). One may designate as T_c the parameter that joins the intrinsic components describing the elementary contributions of the crystalline relaxation, i.e.,

$$\Delta H_{cs}{}^* = T_c \cdot \Delta S_{cs}{}^* \qquad\qquad (11.15)$$

Equation 11.9 may then be rewritten as the compensation law of Equation 11.11. This concept of hierarchical correlation involves serial processes. The various microscopic degrees of freedom constitute a hierarchy: The fastest units (simple movements) are the first concerned and govern the slowest entities (complex movements), and so on, for more and more complex levels. Consequently, a molecular complex movement is only possible when the neighboring structural units, being more mobile, are packaged in a given configuration (23). To confirm the existence of a compensation phenomenon and to estimate the compensation temperature uncertainty, Krug et al. (24) have proposed a test, which was applied by Teyssèdre and Lacabanne (25) to the dielectric relaxation associated with the glass transition of copolymers P(VDF-TrFE).

11.1.1.3.4 Molecular Origin of the Polarization Process

The polarization peaks due to the TSC current may have different origins:

- In materials that contain molecular or ionic dipoles, orientational polarization can occur. This polarization depends on the part of interaction of the dipoles with their surroundings that can be represented by an effective viscosity. It is attributed, in molecular compounds, to the restricted rotation of the whole molecule or of a part of it and, in ionic crystals, to ions jumping between neighboring sites (ion-vacancy pair). The criteria adopted to identify the large number of dipolar relaxation phenomena are:
  - The proportionality of the TSC current with the polarizing field
  - The independence of the TSC spectrum with the electrode–sample interface
- Space charge polarization is due to the presence of excess charge carriers (electronic or ionic). A macroscopic charge transfer that may be intrinsic (heterocharges) or extrinsic (homocharges) is observed between the electrodes. This polarization is more complex than the dipolar one because it depends on a great number of parameters.
- The interfacial or Maxwell–Wagner–Sillars (MWS) polarization is characteristic of heterogeneous systems. It is due to the piling up of space charges near the interfaces between zones of different conductivities.

11.1.2 Practical Measurements

In the work described here, TSC spectroscopy experiments have been performed using a TSC and Relaxation Map Analysis (RMA) dielectric spectrometer developed in our laboratory and available from Settaram. The TSC cell is shown in Figure 11.3. Powdered compounds (some tens of milligrams) were compressed into flat disks of 0.8-cm diameter and approximately 1 mm thick. The sample was placed between stainless steel disks, which ensures electrical contact, thus allowing polarization and current measurement. The maximum applied voltage was 500 V, with an accuracy of 0.1 V. The depolarization current was recorded using a very

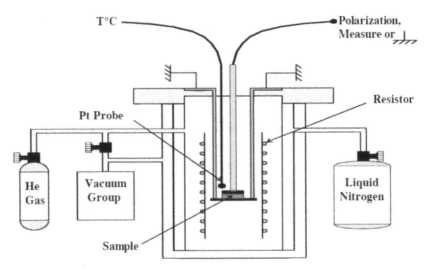

FIGURE 11.3 Diagrammatic illustration of a TSC cell.

sensitive ammeter that can detect intensities between 10^{-17} and 10^{-8} A, with an accuracy of 2×10^{-16} A.

The assembly was placed into a controlled temperature cell. Before experiments, the measuring cell and the isolation chamber were pumped out to a vacuum pressure of 10^{-6} Torr. The measurement cell was filled in with dry He exchange gas at room temperature and the temperature range of the sample varied between -160 to 250°C as indicated.

11.2 EXAMPLES OF PHARMACEUTICAL USES

11.2.1 Excipient Analysis

Various polymers used as excipients may be analyzed by TSC. In this section, we will present, as an example, results obtained for polyethylene glycol (PEG) with a molecular weight $M_w = 10,000$. Using differential thermal analysis (DTA), the thermal profile shows, at lower temperatures, two small steps at around -70 and -50°C (Figure 11.4) that are associated with a double glass transition. The weak magnitude of these events is due to the high crystallinity of the PEG 10,000 polymer. Moreover, the thermal profile also exhibits two endothermic peaks around 0 and 65°C, labeled T_α and T_m, respectively. The T_α peak depends on the thermal history; hence, this transition is generally associated either with crystalline rotation and oscillation as is found for polyethylene (26) or with the melting associated with crystals defects. The T_m peak can be unambiguously assigned to the melting process.

Using TSC, it is important to first note that the dielectric response of PEG (Figure 11.5) is dependent on the polarization temperature T_p. It is necessary to restrict the temperature range to that below the PEG melting point to avoid irreversible changes of the sample and, consequently, a severe increase of the conduc-

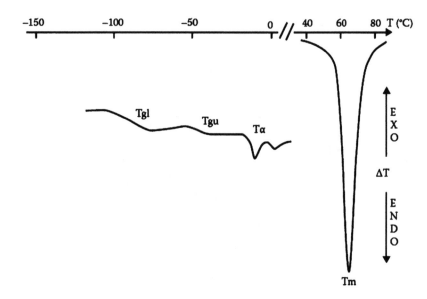

FIGURE 11.4 DTA transitions of polyethylene glycol.

tivity near the melting. By analogy with the DTA results, the two events observed below 0°C can be attributed to the dielectric manifestation of the double glass transition of the polymeric excipient. When the polarization temperature is optimized, these modes are well defined and intense (Figure 11.5). Moreover, a third event labeled T_α is observed by TSC around 0°C. By analogy with results on semicrystalline polymers (27), this event might be due to a process that is effectively a precursor to the melting process, occurring in the crystalline phase. Complementary experiments by the fractional-polarization methods have confirmed the preceding hypothesis.

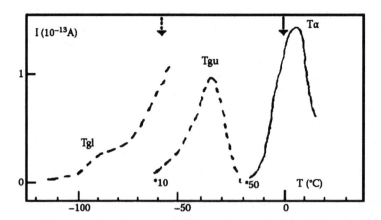

FIGURE 11.5 TSC relaxations of polyethylene glycol.

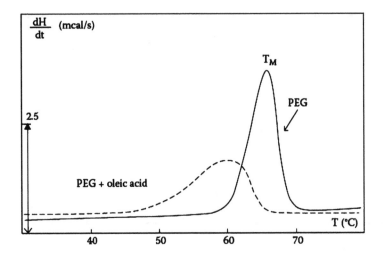

FIGURE 11.6 Melting of polyethylene glycol and polyethylene glycol/oleic acid dispersions.

TABLE 11.1
Thermal Characteristics of PEG–Oleic Acid Mixtures

	Glass Transition		Melting	
	T_g (°C)	ΔC_p (J/g · deg)	T_m (°C)	ΔH (J · g)
PEG	51	40	65.7	194
Oleic acid	—	—	7.1	111
Mixture	53	38	2.4/60.6	112/172

11.2.2 DRUG–EXCIPIENT INTERACTIONS

To mimic a drug–excipient interaction for illustrative purposes, we will present results obtained using oleic acid dispersions in PEG (14% w/w). First, DSC studies yielded information on the evolution of crystalline entities. More specifically, by adding the drug, the crystalline organization of PEG is disrupted; the melting temperature is shifted toward a lower temperature (from 66 to 61°C) and the endothermic peak becomes larger (Figure 11.6). On the other hand, the oleic acid melting is also shifted toward lower temperatures (from 7 to 2°C) when the PEG is added (Table 11.1). This melting evolution indicates a poor crystallization of the active substance, the presence of smaller crystallites, or a eutectic formation. However, DSC profiles show no significant differences in the glass transition temperatures (Table 11.1).

These results are confirmed by the TSC study. The solid dispersion relaxations associated with the dielectric manifestation of the glass transition appear to be similar to those of PEG (Figure 11.5 and Figure 11.7). However, differences in peak intensity can be noted; in particular, we observe an increase of both T_{gl} and T_{gu} intensities. These evolutions denote a slight modification of the amorphous phase; the oleic acid probably penetrates in the PEG and swells the polymer chains. As the DSC and TSC

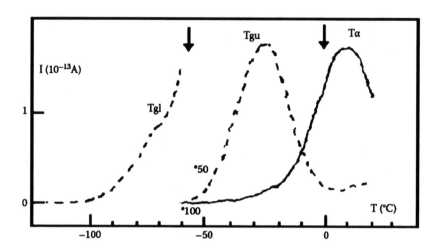

FIGURE 11.7 TSC relaxations of polyethylene glycol/oleic acid dispersion.

responses are modified for the binary systems in comparison to the individual components for both crystalline and amorphous phases, it is logical to suggest that the excipient is interacting with the drug.

11.2.3 DRUG–WATER INTERACTIONS

A further important interaction, namely, that between an active substance and water, can be studied by DSC and TSC (28). To illustrate this case, we will present results obtained with a drug manufactured in three states labeled hydrated hemihydrate (HEMI), monohydrate (MONO), and freeze-dried (FD) forms. The last is mostly amorphous and absorbs water in nonstoichiometric quantities. The other two forms are highly crystalline and contain specific ratios of water per active molecule. Thermogravimetric experiments have shown that water is linked between two active molecules for the hemihydrate, whereas the distribution is homogeneous and the ratio unimolecular for the monohydrate. Moreover, two different hemihydrate batches obtained by two different industrial processes have been analyzed. Complementary studies have shown that the batch labeled PR_1 is more crystalline and more stable than the PR_2 batch in terms of the transition to the monohydrate form.

The hydration state of these drugs has been controlled by pumping the sample for 1 min with a roughing pump to remove extraneous water and choosing room temperature as the maximum temperature used to prevent dehydration of bound water. Figure 11.8 presents the superposition of their corresponding TSC spectra. The temperature position of the relaxation maxima clearly shows differences in structure and interaction between water and the drug. The relaxation differences between the two PR_1 and PR_2 hemihydrate batches reflect an evolution of the physical structure with the manufacturing process, probably linked to the crystallinity.

To identify the origin of these differences, complementary studies have been performed on dehydrated stable compounds. This dehydrated state has been obtained by heating the sample to 80°C and applying a secondary vacuum for 2 h. Comple-

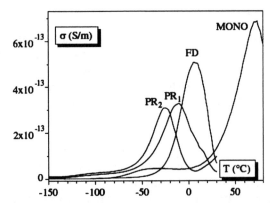

FIGURE 11.8 Superposition of TSC spectra of hemihydrates (PR$_1$, PR$_2$ batches), monohydrates, and freeze-dried (FD) forms of a drug.

mentary experiments by TGA have confirmed the absence of residual water in these samples. For this sample, the relaxations (Figure 11.9) remain stable in position and intensity irrespective of any additional dehydration procedures, hence, a single dehydration step permits one to obtain a stable state. Moreover, with this procedure the relaxation magnitude became linear with the electric field, indicating that these single dipolar relaxations are intrinsic to the drug. It is important to note that the TSC relaxation peak is shifted toward higher temperatures with dehydration for the four forms of this drug (Figure 11.9). This phenomenon is often observed on polymers with the glass transition (11,29,30), and on associated relaxations [mechanical (29,31–33) or dielectric (11,34–37)], when water or small organic molecules (plasticizers) are introduced. Consequently, water may play an analogous plasticizer role in the active substances; intercalation of water between drug molecules probably facilitates molecular mobility and relaxations by breaking of cohesive physical bonds, and by lowering potential barriers, thereby facilitating the relaxation processes.

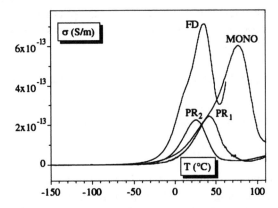

FIGURE 11.9 Superposition of TSC spectra of hemihydrates (PR$_1$, PR$_2$ batches), monohydrates, and freeze-dried (FD) forms of a dehydrated drug.

11.2.4   BATCH-TO-BATCH CONTROL

To illustrate the ability of TSC to discriminate between different physical structures, the example of the two preceding hemihydrate batches has been chosen (28). As shown in Figure 11.8 and Figure 11.9, each batch presents a single intrinsic relaxation process by TSC, plasticized by water with a maximum temperature characteristic of the batch. As dehydration has the same consequence for the two hemihydrate batches, in terms of translation direction and magnitude, these modes might have the same origin.

    To identify the origin of this process, the fractional polarization method has been applied. The polarization temperature has been shifted by constant steps of 5°C from –25 to 75 and 65°C, respectively, for the PR_1 and PR_2 batches, with a polarization window of 10°C. Figure 11.10 illustrates an example of the complex spectrum and the set of elementary peaks obtained for the PR_1 batch. The elementary peak envelope reproduces the complex spectrum and the single relaxation time $\tau(T)$ of each elementary peak has been determined according to the Bucci–Fieschi analysis described earlier. The semilogarithmic plot of τ vs. $1/T$ (Figure 11.11) shows that all relaxation times obey an Arrhenius law, irrespective of the batch. Moreover, the extrapolation of the linear relationships shows a convergence toward a compensation point, indicating the cooperativity of the dipolar movements. This phenomenon is also illustrated by the existence of a straight line on the semilogarithmic plot of the preexponential factor τ_0 vs. the activation enthalpy ΔH (Figure 11.12). In a given batch, the existence of a single compensation indicates the homogeneity of the batches. The values of the compensation parameters (T_c, τ_c) are, respectively, for the PR_1 and PR_2 batches, (139°C, 3×10^{-2} sec) and (133°C, 8×10^{-3} sec). The similitude of compensation temperatures confirms that the isolated molecular movements have the same origin. Moreover, the comparison of the two fine structures underlines an important difference between the two batches, namely, that the compensation line is shifted toward lower preexponential factor values for the PR_2 batch. This evolution means that the corresponding surrounding dipolar entities have a higher entropy (Equation 11.6 and Equation 11.7), so there is an increase in local disorder. This result means that the physical structure of the drug at a microscopic level can be responsible for the macroscopic behavior (optical microscopic aspect, stability).

11.2.5   RELAXATIONS ASSOCIATED WITH FIRST-ORDER TRANSITIONS

Two examples of molecular mobility liberated near two different first-order transitions, namely a solid–solid polymorphic transition and melting, are described:

11.2.5.1   Solid–Solid Transition

Studies were conducted on a model proprietary drug system (28). First, the DSC response of the model drug shows an endothermic peak at 155°C. Complementary studies (x-ray powder diffraction, optical microscopy, and TGA) allow us to attribute this event to a solid–solid transition. By TSC, this drug exhibits a relaxation located at 108°C after polarization at $T_p = 110$°C. This is followed at about 130°C

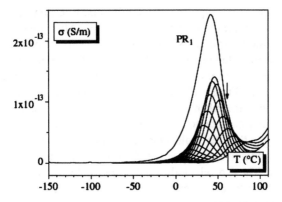

FIGURE 11.10 Superposition of the complex and elementary TSC spectra for the PR$_1$ batch of the hemihydrate.

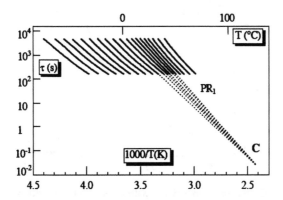

FIGURE 11.11 Arrhenius diagram for the PR$_1$ batch of the hemihydrate after dehydration.

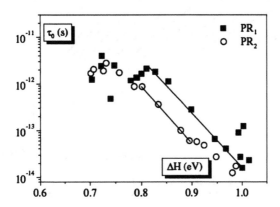

FIGURE 11.12 Compensation diagram for the PR$_1$ and PR$_2$ batches of the hemihydrate after dehydration.

by a sharp increase in the dipolar conductivity due to a flowing of charges when
the phase transition comes into play (Figure 11.13). After going through the transition, the relaxation magnitude is decreased by a factor of two and the maximum
is shifted toward lower temperatures. The fractional polarizations method and the
analysis of the fine structure have been used to evaluate the relaxation time distribution. As shown in Figure 11.14, on a semilogarithmic plot of τ vs. $1/T$ all
relaxation times obey an Arrhenius law. Moreover, the extrapolation of these straight
lines indicates a convergence to a compensation point. The compensation time τ_c
= 10^{-3} sec indicates that the molecular movements involved are localized and at
the compensation temperature (T_c = 209°C) those movements are correlated with
a phase transition. These results indicate that this dielectric relaxation occurs some
50° below the phase transition and consists of precursor cooperative movements
of the solid–solid transition.

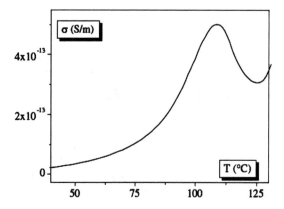

FIGURE 11.13 TSC relaxations associated with a solid–solid transition.

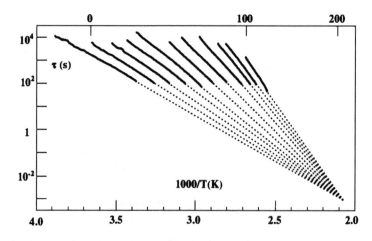

FIGURE 11.14 Compensation diagram of a polymorphic drug.

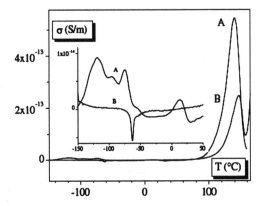

FIGURE 11.15 TSC relaxation associated with the melting.

11.2.5.2 Melting

The TSC study of another drug will illustrate the link between high-temperature relaxation and melting. This drug is characterized by two tautomeric forms, labeled A and B, which exhibit a melting at 182 and 186°C, respectively. In this case, the complex TSC spectrum (Figure 11.15) at high temperature only presents a single relaxation mode situated at 140 and 145°C, respectively, for A and B. The shift between the relaxations, identical with the transitions difference, could be associated with the crystalline network difference. By using the fractional polarization method, the behavior of relaxation times has confirmed that these relaxations are linked to melting. In both forms, the relaxation times are correlated by a compensation law and the parameters (T_c, τ_c) of the two forms A and B, $(202°C, 2 \cdot 10^{-3} \text{ sec})$ and $(189°C, 2 \cdot 10^{-2} \text{ sec})$, respectively, indicate that these relaxations are due to cooperative movements that are precursors of the melting process.

11.2.6 PHYSICAL STRUCTURE OF A HOMOGENEOUS AMORPHOUS DRUG

The amorphous state of the previous drug has been obtained by quenching the system from the molten state. In this case, the DSC profile exhibits a single thermal event, namely, a heat capacity step at 69°C associated with a glass transition. The absence of an endothermic peak in the melting zone (180°C) proves that the amorphous drug does not crystallize in the time scale of the experiment. Complementary studies have shown that partial crystallization occurs after 20 h at room temperature.

After polarization at 85°C, the TSC complex spectrum (Figure 11.16) presents an intense dipolar relaxation at 72°C. According to its temperature position, this mode can be attributed to the dielectric manifestation of the glass transition. This hypothesis has been confirmed by analyzing the fine structure of this relaxation. As shown on Figure 11.17, all experimental relaxation times obey an Arrhenius law and converge toward a single point. This small organic molecule presents, like amorphous synthetic (38–41) or natural (42) macromolecules, a relaxation located

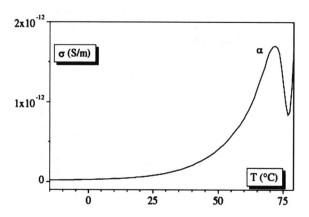

FIGURE 11.16 TSC relaxation of a homogeneous amorphous drug.

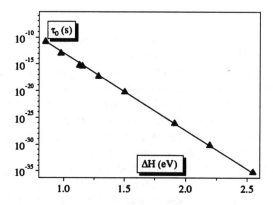

FIGURE 11.17 Fine structure of TSC spectrum of a homogeneous amorphous drug.

in the same temperature range as the glass transition, and consists of cooperative movements. The compensation parameters $T_c = 80°C$, $\tau_c = 11$ sec indicate that this relaxation has the same origin as the glass transition, and the cooperativity of all elementary processes in a single movement proves that the amorphous phase is homogeneous. By studying the dielectric manifestation of the glass transition, TSC gives information on the nature and the homogeneity of the amorphous phase.

11.2.7 PHYSICAL STRUCTURE OF A HETEROGENEOUS AMORPHOUS PHASE

In this case, the DSC profile of the chosen drug presents two events, namely, a weak step in the signal around −17°C and an endothermic peak at 150°C, associated respectively with a glass transition and a solid–solid transition. After polarization at $T_p = 80°C$ (Figure 11.18), TSC studies indicated two modes labeled β_1 and β_u, located at −20 and 30°C, respectively. It is important to note that the magnitude of the β_1 mode is 5000 times less than the β_u mode, and that their position is independent of

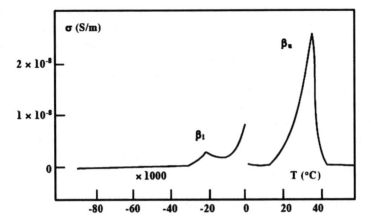

FIGURE 11.18 TSC spectrum of a semicrystalline drug.

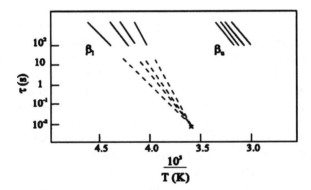

FIGURE 11.19 Arrhenius diagram of a heterogeneous amorphous phase.

the polarization temperature. These modes are intrinsic to the drug. To identify their origin, the fractional polarizations method has been applied and, as shown in Figure 11.19, all relaxation times obey an Arrhenius law. Moreover, the relaxation times isolated in the β_1 mode obey a compensation law with compensation parameters T_c = 7°C, $\tau_c = 7 \ 10^{-3}$ sec. By analogy with results on semicrystalline polymers (21,43), this β_1 relaxation can be attributed to the dielectric manifestation of the "true" glass transition (9).

As for the β_u relaxation, the activation enthalpies deduced from the analysis of the fine structure remain roughly constant. This relaxation phenomenon is also observed in semicrystalline polymers, slightly above the glass transition temperature, and it is generally attributed to a limited molecular mobility in amorphous regions surrounding the crystalline phase. The high crystallinity of the drug implies a small free amorphous phase fraction and an important constrained amorphous phase fraction. This is shown by the large difference in magnitude between the β_1 and β_u modes.

Thus, we may conclude that TSC allows the detection of disorder in highly crystalline compounds. The true free amorphous phase, without short-range order, is generally characterized by a step of the heat capacity at T_g and by the β_1 dielectric relaxation, located around T_g and consisting of cooperative movements with $T_c \sim T_g$. As for the constrained amorphous phase, located at the interphase between true amorphous and crystal, it is only detected by TSC and characterized by the β_u relaxation located some 40° above β_1 (44), and is characterized by isoenergetic processes. Such a biphasic amorphous structure has been observed in natural (42,45) or synthetic [semicrystalline (43,46–51) or amorphous (52)] polymers. According to Wunderlich (53) and Boyer (44,54,55), these amorphous phases are named mobile or rigid in amorphous materials and free or constrained in semicrystalline materials. The position and magnitude of T_{gl} and T_{gu} are dependent on morphology, crystallinity, and crystallization process.

11.2.8 AMORPHOUS CONTENT DETERMINATION

To evaluate the TSC capabilities in terms of the detection of amorphous content in drugs, we have studied different physical blends of crystalline and amorphous phases (Figure 11.20). The TSC spectra of these two extreme states show that only the crystalline state presents dielectric relaxations at low temperature (28). The crystallinity quantification has been achieved by using the complex mode magnitude of the crystalline form, comprising three submodes named LT_1, LT_2, and LT_3. The polarization temperature (T_p) has been chosen to equal 50°C and the final temperature (T_f), 55°C, with $T_f < T_g$ to avoid a flow of the amorphous form above the glass transition, especially in poorly crystalline samples, and to prevent recrystallization. For the sake of clarity, Figure 11.20 represents the superposition of some spectra and indicates that the magnitude of the three low-temperature modes increases with the crystallinity ratio. With this particular drug, the threshold of detection can be estimated around 2.5%; but the higher the mode, the lower this limit. Moreover, this limit could also be lowered by increasing the static electric field or by studying more intense relaxations.

By plotting the magnitude of each submode, corrected by the amorphous baseline vs. the crystallinity, a calibration line was obtained (Figure 11.21). This calibration

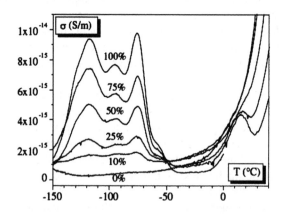

FIGURE 11.20 Evolution of the TSC relaxation with the crystal/amorphous ratio (w/w).

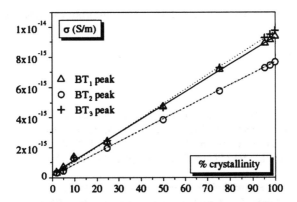

FIGURE 11.21 Correlation between peak magnitude and crystallinity.

line constitutes a powerful control tool to quantify the crystallinity ratio in the drug, and can be used for batch-to-batch control or to follow the recrystallization of the amorphous phase.

11.3 CONCLUSIONS

As drugs and excipients are almost invariably polar, dielectric spectroscopy is well suited for their characterization. With its very low equivalent frequency, TSC spectrometry also has high resolving power. Moreover, TSC spectra avoid the contribution of free charges that might hide small relaxation phenomena. Complex TSC spectra are particularly well adapted to the identification and the control of drugs and excipients. They give interesting information on drug–excipient and drug–water interactions. Nevertheless, the major advantage of the TSC technique is to allow us to resolve complex TSC spectra experimentally into elementary components. In this manner, the dielectric relaxation mode is defined by discrete spectra of relaxation times. For crystalline phases, molecular mobility liberated near first-order transitions such as solid–solid transitions and melting has a distinct behavior profile. In both cases, the dielectric relaxation times follow a compensation law. For amorphous phases, molecular mobility liberated near the glass transition temperature has an analogous behavior but, in this case, the compensation time remains of the order of seconds, i.e., the characteristic time of the glass transition. In semicrystalline materials, the dielectric manifestation of the glass transition consists of two modes:

- The lower component has the previously described behavior and is due to the "true" amorphous phase.
- The upper temperature component is controlled by isoenthalpic mechanisms, and has been associated with amorphous regions under constrains from crystalline zones.

The evolution of the magnitude of both submodes is also important for characterizing the microstructure of semicrystalline materials. Finally, it is worth noting here that TSC allows us to detect 2.5% of an amorphous phase in an otherwise crystalline sample.

REFERENCES

1. Chulia, D., Deleuil, M., and Pourcelot, Y., Ed., Powder technology and pharmaceutical processes, in *Handbook of Powder Technology*, Elsevier, New York, 1994, p. 19.
2. Hancock, B.C. and Zografi, G., *J Pharm Sci* 86, 1, 1997.
3. Hüttenrauch, R., *Acta Pharm Technol* Suppl. 6, 55, 1978, Deutscher Apotheker Verlag, Stuttgart.
4. Hüttenrauch, R. and Keiner, I., *Int J Pharm* 2, 59 1979.
5. Saleki-Gerhardt, Z., Ahlneck, C., and Zografi, G., *Int J Pharm* 101, 237, 1979.
6. Craig, D.Q.M., *STP Pharma Sci* 6, 421, 1995.
7. Duddu, S.P. and Sokoloski, T.D., *J Pharm Sci* 84, 773, 1995.
8. Mezghani, S., Fauran, M.J., Lamure, A., Maurel, E., and Lacabanne, C., *Acta Pharm Biol Clin* 8, 402, 1995.
9. Fagegaltier, N., Lamure, A., Lacabanne, C., Caron, A., Mifsud, H., and Bauer, M., *J Therm Anal* 48, 459, 1997.
10. Van Turnhout, J., *Thermally Stimulated Discharge of Polymer Electrets*, Elsevier, Amsterdam, 1975.
11. Hedvig, P., *Dielectric Spectroscopy of Polymers*, Adam Hilger Ltd., Bristol, 1977.
12. Chen, R. and Kirsh, Y., *Analysis of Thermally Stimulated Processes*, Pergamon Press, Paris, 1981.
13. Lavergne, C. and Lacabanne, C., *IEEE Electrical Insulation Magazine* 9, 5, 1993.
14. Teyssèdre, G., Mezghani, S., Bernès, A., and Lacabanne, C., Thermally stimulated currents of polymers, in *Dielectric Spectroscopy of Polymers*, Hedvig, P., Ed., 1997, pp. 227–258.
15. Bucci, C. and Fieschi, R., *Phys Rev Lett* 12, 16, 1964.
16. Chatain, D., Gautier, P., and Lacabanne C., *J Polym Sci Phys Ed* 11, 1631, 1973.
17. Hino, T., *Jpn J Appl Phys* 12, 611, 1973.
18. Bucci, C., Fieschi, R., and Guidi, G., *Phys Rev* 148, 816, 1966.
19. Eyring, H., *J Chem Phys* 4, 283, 1994.
20. Mc Crum, N.G., Read, B.E., and Williams, G., *Anelastic and Dielectric Effects in Polymeric Solids*, John Wiley & Sons, London, 1967.
21. Lacabanne, C., Lamure, A., Teyssèdre, G., Bernès, A., and Mourgues, M., *J Non-Cryst Solids* 172–174, 884, 1994.
22. Hoffman, J.D., Williams, G., and Passaglia, E., *J Polym Sci C* 14, 173, 1996.
23. Pérez, J., Physique et mécanique des polymères amorphes, in *Techniques et documentations*, Lavoisier, Paris, 1992.
24. Krug, R.R., Hunter, W.G., and Grieger, R.A., *J Phys Chem* 80, 2335, 1976.
25. Teyssèdre, G. and Lacabanne, C., *J Phys D: Appl Phys* 28, 1478, 1995.
26. Lang, M.C., and Noël, C., *J Polym Sci, Polym Phys Ed* 15, 1319, 1997.
27. Teyssèdre, G., Grimau, M., Bernès, A., Martinez, J., and Lacabanne, C., *Polymer* 35, 4397, 1994.
28. Fagegaltier-Boutonnet, N., Analyse de la structure physique de molécules à visée thérapeutique par Courants Thermo-Stimulés, Ph.D. thesis, Toulouse University, France, 1998.

29. Johnson, G.E., Bair, H.E., Matsuoka, S., Anderson, E.W., and Scotti, J.E., Water sorption and its effects on a polymer's dielectric behavior, in *Water in Polymer*, ACS Symposium Series, 127, Rowland, S.P., Ed., Washington D.C., Issue No. 27, 1980, pp. 451–468.

30. Moy, P. and Karasz, F.E., *Polym Eng* 20, 315, 1980.

31. Dufresne, A. and Lacabanne, C., *Polymer* 36, 4417, 1995.

32. Chateauminois, A., Comportement viscoélastique et tenue en fatigue statique de composites verre/epoxy; Influence du vieillissement hygrothermique, Ph.D. thesis, Lyon University, France, 1991.

33. Park, Y., Ko, J., Ahn, T.K., and Choe, S., *J Polym Sci B* 35, 807, 1997.

34. Hitmi, N., Etude des transitions dans les composantes minérale et organique des tissus calcifiés par spectroscopie diélectrique basse fréquence, Ph.D. thesis, Toulouse University, France, 1983.

35. Mezghani, S., Etude par spectroscopie diélectrique de la mobilité moléculaire du collagène au cours du vieillssement et de l'irradiation, Ph.D. thesis, Toulouse University, France, 1994.

36. Vanderschuren, J. and Linkens, A., *Macromolecules* 11, 1228, 1978.

37. Hasted, J.B., *Aqueous Dielectrics*, John Wiley & Sons, New York, 1973.

38. Bernès, A., Etude de la métastabilité d'un polymère amorphe (polystyrène atactique) en fonction de son histoire thermodynamique par spectroscopie enthalpique diélectrique, Ph.D. thesis, Toulouse University, France, 1985.

39. Colmenero, J., Alegria, A., Alberdi, J.M., Del Val, J.J., and Ucar, G., *Phys Rev B* 35, 3995, 1987.

40. Boye, J., Etude de réseaux polyépoxy DGEBA-DDM par Fluage ThermoStimulé; corrélation structure-propriétés anélastiques, Ph.D. thesis, Toulouse University, France, 1990.

41. Diffalah, M., Etude des propriétés anélastiques de mélanges PMMA/LATEX et Polyamide66/EPR par Fluage ThermoStimulé, Ph.D. thesis, Toulouse University, France, 1993.

42. Mégret, C., Etude de la structure physique de l'élastine par spectroscopies thermo-stimulées AED/CTS, Ph.D. thesis, Toulouse University, France, 1988.

43. Bernès, A., Chatain, D., and Lacabanne, C., *Thermochim Acta* 204, 69, 1992.

44. Boyer, R.F., Transitions and relaxations in amorphous and semicrystalline polymers and copolymers, in *Encyclopedia of Polymer Science and Technology*, Suppl. Vol. II, John Wiley & Sons, 1977, pp. 745–839.

45. Mégret, C., Guantieri, V., Lamure, A., Pieraggi, M.T., Lacabanne, C., and Tamburro, A.M., *Int J Biol Macromol* 14, 45, 1992.

46. Teyssèdre, G., Bernès, A., and Lacabanne, C., *Thermochim Acta* 226, 65, 1993.

47. Teyssèdre, G., Bernès, A., and Lacabanne, C., *J Polym Sci B: Polym Phys* 31, 2027, 1993.

48. Mourgues-Martin, M., Bernès, A., and Lacabanne, C., *J Therm Anal* 40, 697, 1993.

49. Demont, P., Fourmaud, L., Chatain, D., and Lacabanne, C., Thermally Stimulated Creep for the study of copolymers and blends, in *Polymer Characterization by Interdisciplinary Methods*, Craver, C.D. and Provder, T., Eds., Advances in Chemistry Series 227, Washington, D.C., 1990.

50. Demont, P., Chatain, D., Lacabanne, C., Ronarc'h, D., and Moura, J.C., *Polym Eng Sci.* 24, 127, 1984.

51. Demont, P., Etude des propriétés mécaniques de mélanges et copolymères éthylène-propylène par fluage stimulé par la température, Ph.D. thesis, Toulouse University, France, 1982.

52. Bernès, A., Chatain, D., and Lacabanne, C., *Polymer* **33**, 4682, 1992.
53. Wunderlich, B., *Thermal Analysis*, Academic Press, New York, 1990.
54. Boyer, R.F., *J Macromol Sci Phys B* **8**, 503, 1973.
55. Boyer, R.F., *J Polym Sci: Symp* **50**, 189, 1975.

Index

A

Ablett studies, 94
Activation parameters, 364
Acyclovir, 73
Adiabatic calorimeters, 268
Adsorption, 176–178
Advances, thermal microscopy, 223–224
Advantages and disadvantages
    differential scanning calorimetry, 275
    microcalorimetry, 275
    modulated temperature DSC, 111–112
Agbada and York studies, 212
Alves studies, 157
Ambrogi studies, 216
Amorphous content
    crystals, 274
    determination, 378, *378–379*
Amorphous-crystalline transition, 275–276
Amorphous drugs, 79, *80–86*, 81–85
Analytical balance, equipment, 25–26
Angberg studies, 269, 272
Anisotropic particles, 226
Annealing effects, 19
Antoine equation, 199
Apparatus and measurements, thermogravimetric
    atmosphere control, 158–160
    balance mechanism, 144–148, *145*
    beam balances, 144, 146–148
    concurrent measurements, 154–155
    container, sample, 153–154
    data capture, 163
    decompositions, solid-state, 152–153
    evolved gas analysis, 160–162
    fundamentals, 142–144
    furnace temperature control regimes, 156–157
    intermediates, 151–152
    mass measurement, 148
    measurements, 142–163
    microscopy, 152–153
    progress reaction, 162
    reactant sample, 148–149
    reaction zone, 155–160
    sample examination, 149–153
    solid-state decompositions, 152–153
    solid-state reactions, 151–152

temperature, 155–158
temperature sensors, sample container, 154
Applications, dynamic rheological analysis
    biomedical polymers, 346–351
    cross-linking effects, 345–346, *347*
    emulsified systems, 339–340
    glass transition temperatures, 341–345,
        342–344
    mixed-polymeric systems, 337–338, 352–353
    pharmaceutical polymers, 351, *352*
    polymer gel systems, 333–339
    solid polymeric systems, 340–353
    temperature-dependent sol-gel transitions,
        338–339, *339*
Applications, high-sensitivity DSC, 296–307
    biological molecules, 299–306
    dilute polymer solutions studies, 296–298,
        297–299
    fundamentals, 296
    industrial applications, 306–307, *307–308*
    lipids, *304–305*, 304–306
    nucleic acids, 302–304
    proteins, 299–302, *300, 302*
Applications, pharmaceutical, *134*, 134–135
    amorphous drugs, 79, *80–86*, 81–85
    compression effects, 66, *67*, 68
    drug polymorphism examples, 62–64, *64*
    drugs as hydrates, 70–72, *71–72*
    enantiotropic polymorphs, 55–57, *56, 58–60*,
        59
    excipients, 76–77, *77–78*
    fractional stoichiometries, *73*, 73–74
    fundamentals, 53–54, 96
    glasses, 79–96
    grinding effects, 66, *67*, 68
    hydrate polymorphism, 74
    hydrates, 68–78
    mixtures of polymorphs, 66
    modulated temperature DSC, 129–135,
        133–134
    monotropic polymorphs, 55–57, *56, 58–60*, 59
    nonaqueous solvates, 74–75, *75*
    polymorphism, 55–68
    practical considerations, 59–62, *61–63*
    relaxation behavior, 132–133, *133*
    solvates, 68–78
    spray-dried systems, *87–88*, 87–96, *90–94*